New Polymeric Composite Materials

Environmental, Biomedical, Actuator and Fuel Cell Applications

by
Inamuddin
Ali Mohammad
Abdullah M. Asiri

Polymeric and composite materials are in high demand and their continuing development is making our life style more comfortable.

This book, 'New Polymeric Composite Materials - Environmental, Biomedical, Actuator and Fuel Cell Applications' brings together and broadly explores the latest developments and applications of polymeric and composite materials in the areas such as shape memory polymers, green composites for artificial organs, geomembranes for the safe disposal of waste, removal of heavy metals and dyes, adhesives, sensors and actuators, fuel cells, membrane and environmental sustainability, etc.

New Polymeric Composite Materials

Environmental, Biomedical, Actuator and Fuel Cell Applications

By

Inamuddin[1], Ali Mohammad[1], Abdullah M. Asiri[2]

[1]Department of Applied Chemistry, Aligarh Muslim University, India.

[2]Chemistry Department, Faculty of Science, King Abdulaziz University, Jeddah 21589, Saudi Arabia.

Published by **Materials Research Forum LLC**
Millersville, PA 17551, USA

Published as part of the book series
Materials Research Foundations
Volume 5 (2016)
ISSN 2471-8890 (Print)
ISSN 2471-8904 (Online)

Print ISBN 978-1-945291-08-1
ePDF ISBN 978-1-945291-09-8

Distributed worldwide by

Materials Research Forum LLC
105 Springdale Lane
Millersville, PA 17551
USA
http://www.mrforum.com

Manufactured in the United State of America
10 9 8 7 6 5 4 3 2 1

Table of Contents

Foreword

Polymeric and composite materials are in high demand and their use and development continually ads to make our life style more comfortable. Some examples for the uses of polymeric and composite materials include their use in biomedical and environmental applications, sensors and actuators and fuel cells technologies.

This book, 'New Polymeric Composite Materials: Environmental, Biomedical, Actuator and Fuel Cell Applications' brings together and broadly explores the latest developments and applications of polymeric and composite materials in the areas such as shape memory polymers, green composites for artificial organs, geomembranes for the safe disposal of waste, removal of heavy metals and dyes, adhesives, sensors and actuators, fuel cells, membrane and environmental sustainability, etc.

The book is the result of significant contributions from experts from various interdisciplinary fields of science. It broadly explores the most ample, in-depth, and up-to-date research and reviews. We are thankful to all the contributing authors/coauthors for their valued contributions. We would also like to thank all publishers, authors, and others who granted permission to use their figures, tables, and schemes.

Chapter 1: Shape-memory polymers (SMPs) provide an attractive insight into material science, opening unexplored horizons and giving access to unconventional functions mainly in polymer materials, and thus SMPs play a vital role in the field of biomedical engineering and clinical applications. This chapter includes different functionality polymers and their structural approaches, types of different stimuli responsive shape-memory polymers and their applications in the biomedical field. At the end, future prospects and challenges related to clinical research are discussed in detail.

Chapter 2: Presently, the concept of 'Green Chemistry for Sustainable Development' is a guiding source for development of next generation materials, known as composite green polymers. The present chapter represents various applications of green composite polymers in medical technology with focus on the development of artificial organs.

Chapter 3: The increasing consumerism has resulted in massive waste generation. The proper disposal of this waste is a real challenge and lots of research is going on around the world to develop new techniques or to improvise the existing to face this challenge. Landfill liners or geomembranes are one of the most important components of landfilling which ensure the safe disposal of wastes. This article summarizes the various physical, chemical, thermal and mechanical factors that affect the performance of a geo-membrane. It provides supportive information for the selection of proper geo-membrane material with respect to exposure conditions, expected life and available resources.

Chapter 4: Nowadays, the amount of implantable devices with lots of applications has increased the need for supplying the required energy to sense, monitor, vibrate, pump and report the collected data of such devices. Such a power source should have a long life time and be of ecofriendly nature. It should be capable of producing energy directly from the living organisms' fluids or organic matter. These required properties, light up the idea of using enzymatic biofuel cells (EBFCs). Unfortunately, EBFCs suffer from short life time and poor electron transfer from the enzyme active site to the electrode surface. This issue can be overcome by using certain biocompatible mediators and some enzyme immobilization techniques such as: physical adsorption, electrostatic attraction, covalent coupling and entrapment.

Chapter 5: In this chapter the synthesis of surfactant based sodium dodecyl sulphate zirconium phosphate (SDS-ZrP) intercalated cation exchange material is discussed.

Chapter 6: The two most popular artificial intelligence (AI) techniques namely, artificial neural networks (ANN) and support vector machines (SVM) are applied for predicting the removal efficiency of heavy metals like copper (II), arsenic (III), lead (II), etc. in an adsorption process using low cost biosorbents.

Chapter 7: In this chapter, the usefulness of aqueous microwave assisted methodology has been discussed in respect of synthesis of alginic acid grafted polyacrylamide.

Chapter 8: The synergy between organic and inorganic materials offers the possibility for the development of new generation innovative organic–inorganic composites with diversified applications such as catalysis, photocatalysis, separation technology, energy storage and drug delivery. The present chapter discusses the scientific and technical advancements in the development of organic polymer based TiO_2 nanocomposite membranes for water treatment.

Chapter 9: Electrochemical biosensors have benefited extensively from advances in nanomaterials (e.g., metal nanoparticles, carbon nanotubes and graphene). Among the various biomolecules, cholesterol has aroused considerable interest in recent years on account of its being an important marker in clinical diagnosis. This chapter highlights the recent advances in the field of enzymatic and non enzymatic cholesterol sensors with the use of nanotechnology.

Chapter 10: Retention behaviour of twenty four synthetic dyes on plain and tri-n-butylamine (TBA) impregnated silica gel-G and starch layers in aqueous alcohol media has been studied. Compact and sharp spots led to some useful separations of dyes, having similar chromophore groups on the basis of their differential solubilities and preferential adsorption on these layers.

Chapter 11: The chapter deals with the basic information regarding to the materials and their types used for manufacturing ionic polymer metal composites (IMPCs) actuators using various fabrication processes under optimum experimental parameters and possible applications. Among the variety of electro active polymers recently developed, IPMCs are good candidates for use in bio-related, robotics and aerospace applications because of their biocompatibility. Several fabrication processes, their performance and mechanical characteristics, a number of recent applications of IPMCs have been reported in this chapter.

Inamuddin

Department of Applied Chemistry, Aligarh Muslim University, India

CHAPTER 1

Shape Memory Polymers in Biomedical Applications

Mahadevappa Y. Kariduraganavar, Radha V. Doddamani and Divya D. Achari
Department of Chemistry, Karnatak University, Dharwad 580 003, INDIA
Correspondence Author's Email: mahadevappayk@gmail.com

Abstract

Shape-memory polymers (SMPs) provide an attractive insight into material science, opening unexplored horizons and giving access to unconventional functions mainly in polymer materials, and thus SMPs play a vital role in the field of biomedical engineering and clinical applications. The shape-memory polymers are able to recover fully under load and generate high levels of stress during recovery. When activated under a programmed thermo-chemical cycle, the materials demonstrate excellent shape-fixity and reach complete recovery on re-heating. In view of different structural approaches, a large number of polymers of medical applications are anticipated from resulting shape-memory polymers. Some of the devices are already being utilized in the medical field and others are under examination. One of these potential applications is the removal of a clot in a blood vessel using a laser-activated shape-memory polymer. This chapter includes different functionality polymers and their structural approaches, types of different stimuli responsive shape-memory polymers and their applications in the biomedical field. At the end, future prospects and challenges related to clinical research are discussed in detail. To compile this chapter and to provide adequate information to the readers, we have explored all possible ways including research articles, books, reviews and search engines such as google.com.

Keywords

Shape Memory, Polymers, Biomedical Applications, Polyrotaxanes

Contents

1. Introduction

Shape-memory polymers belonging to a class of stimuli responsive materials can be deformed and fixed into temporary shape under specific stress conditions [1]. These materials subsequently recover their original shape upon external stimuli, such as temperature, pH, light, magnetic or electric field and analytes of interest (e.g., ions and bioactive molecules) [1,2]. As shown in Fig. 1, a closed flower (temporary shape) made of shape memory polymer (SMP) is fixed at a lower temperature and recovers from the closed state to an open flower (original shape) when the temperature is increased above its switchable temperature. For instance, a bold imagination is also shown in Fig. 2, where a lady who treated her face using shape-memory polymers when she was young so the young face can be remembered. Thus, as her age passes, she could recover her face using stimulus [3].

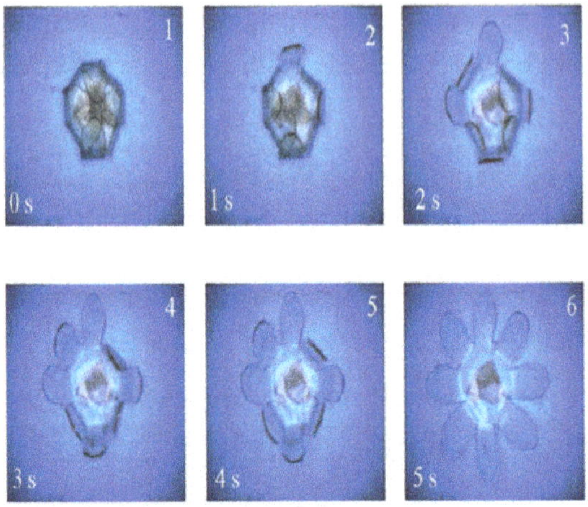

Fig. 1. Shape recovery of SMP (flower made of SMP opens its petals at high temperature). (Reproduced with consent from Ref. [4], Elsevier Publication 2010)

Fig. 2. Appearance recovery in magic mirror (an old woman recovers her young face in a magic mirror). (Adopted with consent from Ref. [3], Elsevier Publication 2010).

Understanding the mechanism of shape-memory polymers, a significant interest has been created during the past two decades in both academia and industry [3]. This has resulted in promising applications in the biomedical field for delivery systems, tissue engineering scaffolds, cell culture supports, bio-separation devices, sensor or actuator systems [5]. To explore these applications, a variety of polymers have been examined as SMP materials, such as polyurethanes, polyacrylates, polynorbornenes, polycaproloctones, polylactides and polyrotaxanes [6-12]. These polymers are capable of exhibiting required properties like biodegradability, biocompatibility, mechanical strength etc. for the use of switchable biomedical devices.

This chapter focuses on shape-memory polymers with special reference to biomedical applications, their classification and future prospects including challenges.

1.1 Historical Background of SMPs

The term "Shape-Memory" was first proposed by Ver-noon in the year 1941 [3]. The term "Shape-Memory Polymer" became better known as shape-memory poly(norbornene), developed by the French Company CDF-Chimie, was commercialized in Japan by Nippon Zeon Company in 1980s under the brand name of Norsorex [6]. However, the industrial applications of the shape-memory polymers were started in 1950s when a chemist named Paul Cook, Founder of Raychem Corporation, invented heat shrink tubing using radiation-crosslinked polyethylene [2]. Although, further efforts to develop SMPs began in the late 1980s, significant progress was made only in the last two decades [3].

Today, a wide range of potential biomedical applications based on SMPs have already emerged and are currently being pursued. The first biomedical application based SMP was demonstrated by Hayashi and colleagues [13], who developed SMP-based catheters, which would soften at body temperature, potentially reducing the risks for soft tissue/organ injuries during their surgical delivery. Maitland et al. [14] designed thermoset polyurethane based micro actuator for treating strokes. Similarly, biodegradable SMPs pertaining to self-expanding and drug eluting stents, self-deployable intragastric implants for treating obesity, self-fitting vascular and coronary grafts, patient-specific customized orthopedic devices, tissue engineering scaffolds and dynamic cell culture substrates have also been developed. This is how, biomedical applications based SMPs have been developed and explored widely in the past few decades [6].

2. Classification of Shape-Memory Polymers

Different types of SMPs have been developed for a wide range of applications [15] and these have been classified based on the type of stimuli and functionality of polymers.

2.1 Shape-Memory Polymers Based on Stimuli Responses

2.1.1 Temperature Sensitive Shape-Memory polymers

Thermo responsive polymers belonging to a class of shape-memory polymers have been the most investigated SMPs due to their ability to respond to a change in temperature [16,17]. In response to the variation in temperature the microstructural features of these polymers are changed [18]. To facilitate the subsequent discussions on the performance of SMPs, some of the important key terms and their meanings used to describe the properties of polymers are summarized in Table 1 [6].

Table 1. Some of the important key terms and their meanings of shape-memory polymers.

Glass transition temperature (T_g)	It is the temperature beyond which a polymer turns from a hard and glass-like state to a rubber-like state. In SMPs, during the transition process, T_g is often defined as the median point of the glass transition range in the heating ramp [19].
Melting temperature (T_m)	It is the temperature at which a material changes from solid to liquid state. In polymers, T_m is the peak temperature at which a semi-crystalline phase melts into an amorphous state.
Transition temperature (T_{trans})	It is the temperature around which a polymer changes from one state to another. T_{trans} could be either melting temperature or glass transition temperature. Above this temperature, the polymers are in their rubbery state. Below the transition temperature, the polymeric materials become brittle and hard. At this point, the rubbery state is replaced by a glassy behavior [20].
Shape deforming temperature (T_d)	It is the working temperature at which SMP is strained to a temporary shape. The relationship of T_d to T_{trans} (below or above) has significant impact on the performance of shape-memory polymers.
Temporary shape fixing temperature (T_f)	It is the working temperature at which the temporary shape of a deformed SMP is fixed. T_f is usually lower than T_{trans}.

Shape recovery temperature (T_r)	It is the working temperature at which an SMP is triggered to recover from its fixed temporary shape. T_r is usually higher than T_{trans} and is often chosen to be the same as the T_d [21].
Shape fixing components	It is defined as the domain or net-point, either covalently crosslinked or physically formed (e.g., via physical entanglement or H-bonding) in an SMP network that maintains the dimensional stability during the deformation and subsequent recovery.
Shape switching components	The polymeric chains in an SMP network consist of switchable units that can switch from one state to another in response to temperature change, which are responsible for the temperature dependent deformation and recovery.

Thermally actuated shape-memory polymers exhibit a characteristic transition temperature, (T_{trans}) above which they can be deformed to a secondary shape from a primary shape by application of external stress. The deformed shape can then be fixed by cooling the polymer below T_{trans} and then removing the deformation stress. The primary shape can subsequently be recovered by heating the polymer above T_{trans} [21].

Shape-memory properties are related to their intrinsic non-crystalline molecular structure. As the glass transition temperature characterizes the polymers. The polymers display different states depending on the temperature range within a few degrees of T_g. In fact, at T_g, significant changes in the mechanical and thermodynamical properties may be observed [20]. The shape-memory effect with respect to temperature is shown in Fig. 3. From the Fig. 3, it is observed that polymers undergo deformation below T_g and then return to their original shape by heating the polymers to higher temperature than T_g. If the polymers are deformed above their glass transition temperature under external force, the deformation will be fixed and maintained after removal of the external force. However, a subsequent heating of the material above its T_g will allow it to recover its original shape [20].

To understand the shape-memory behavior, the following experiment can be performed: the sample was first immersed in a water bath at 80° C, quickly deformed to an "M" shape and quenched to 0° C by immersing in a second bath containing an ice/water mixture. After immersing into a third water bath at 40° C, the sample showed some minor recovery due to the incomplete fixing of the polymer, but a major part of the deformation was retained (since the fixity ratio (R_f) of shape b is still greater than 70%) and a temporary shape b similar to Fig. 4B was fixed. Maintaining the sample temperature at 40° C, the sample was further manipulated by winding it around a glass rod and immersed into the ice/water bath again. This led to fixing of the second temporary shape **a** as shown in Fig. 4A. After immersing the sample with shape a back into the water bath at 40° C, it quickly recovered to shape **b** as shown in Fig. 4B, which further recovered to its permanent shape **c** after immersing into the water bath at 80° C as shown in Fig. 4C [22].

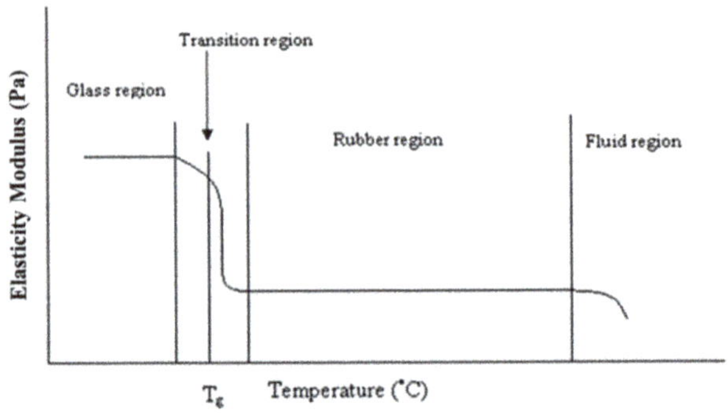

Fig. 3. Elasticity modulus versus temperature of amorphous polymer. (Adopted with consent from Ref. [20], Wiley Online Library Publication 2010).

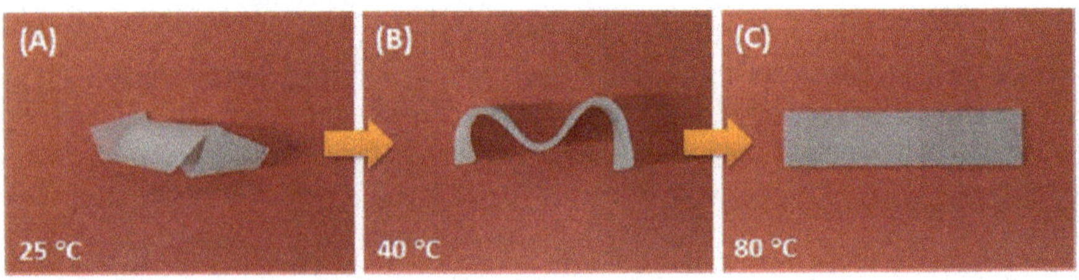

*Fig. 4. Photographs showing the sequential recovery of epoxy/polycaprolactone from temporary shape **a** (A), to temporary shape **b** (B), and to permanent shape **c** (C). (Adoptedwith consent from Ref. [22], Wiley Online Library Publication 2010).*

A unique property of temperature-responsive polymers is the presence of a critical solution temperature, which is the temperature at which the phase of polymer and solution is discontinuously changed according to their composition [23]. It is important not only from the fundamental research point of view, but also from the standpoint of practical applications [24]. If the polymer solution has one phase below a specific temperature, in other words, it becomes insoluble upon heating, due to a lower critical solution temperature (LCST). Otherwise, it is called a higher critical solution temperature (HCST) or an upper critical solution temperature (UCST) [3,25]. Generally, the solubility of most of the polymers increases with an increase in temperature. However, in the case of polymers with a LCST, the solubility decreases as temperature increases and therefore polymers shrink as the temperature increases above the LCST [23]. Thermo-responsive polymer's structures which possess very sensitive balance between the hydrophobic and the hydrophilic groups and a small change in the temperature can create new adjustments [18]. Therefore, the change in LCST can be obtained by incorporating the hydrophilic or hydrophobic groups in a polymeric system.

The polymers with a LCST near to body temperature are the most relevant in the biomedical field. As an example of this kind of polymers, one can highlight the poly(N-substituted acrylamide) family such as poly(N-isopropylacrylamide), poly(N,N'-diethylacrylamide), poly(dimethylamino ethylmethacrylate) and poly[N-L-(1-hydroxymethyl) propyl methacrylamide]. This class of polymers reduce side effects and are therefore preferred in the biomedical field [26,27].

2.1.2 pH Sensitive Shape-Memory Polymers

pH-sensitive polymers are generally polyelectrolytes that have structures with weak acidic or basic groups that either accept or release protons in response to changes in the environmental pH [5]. The acidic or basic groups of polyelectrolytes undergo ionization just like acidic or basic groups of monoacids or monobases [16] in suitable solvents. This group of smart polymers changes their solubility by changing the electrical charges of the polymer molecules [28]. Thus, the transition from a soluble state to an insoluble state is caused due to the decrease of the electrical charge in the polymeric molecules. The electrical charge of the polymer can be decreased by decreasing its pH, neutralizing the electrical charge and reducing the hydrophilicity of the polymer [18]. Polyacidic polymers remain unswollen at low pH, due to the presence of protonated and unionized acidic groups. When the pH of the solution increases, the polymer swells due to the electrostatic repulsion of the negatively charged groups. The opposite behavior is found in case of polybasic polymers, since the ionization of the basic groups increases by decreasing the pH [5]. We have significant variation of pH in different parts of our body that can make the pH sensitive polymers ideal for biomedical applications [18]. Generally, the pH range where a reversible phase transition occurs can be modulated by employing two strategies:

(i) Selection of ionizable moiety with a matching pK_a is important. This can be achieved through proper selection of polyacid or polybase for the desired application.

(ii) Incorporating hydrophobic moieties into the polymer backbone and controlling their nature, amount and distribution. When ionizable groups become neutral, electrostatic repulsive forces disappear within the polymer network through dominated hydrophobic interactions. Thus, the introduction of higher hydrophobic moiety can offer a more compact conformation in the uncharged state and a more accused phase transition. The hydrophobicity of these polymers can be controlled by the copolymerization of hydrophilic ionisable monomers with more hydrophobic monomers with or without pH-sensitive moieties. Typical examples of pH sensitive polymers with anionic groups are poly(carboxylic acids) as poly(acrylic acid) (PAA) or poly(methacrylic acid). Another kind of polyacidic polymers are the polysulfonamides (derivatives of *p*-aminobenzenesulfonamide). A few examples of cationic polyelectrolytes are poly(N,N-diakylaminoethyl methacrylates), poly(lysine) (PL), poly(ethylenimine) (PEI) and chitosan.

2.1.3 Dual Stimuli-Responsive Shape-Memory Polymers

Polymers simultaneously sensitive to dual stimuli are dual stimuli-responsive shape-memory polymers. It is possible to obtain polymers sensitive to both temperature and pH by the simple combination of ionisable and hydrophobic (inverse thermosensitive) functional groups. This has been achieved mainly by the copolymerization of monomers bearing these functional groups, combining thermo-sensitive polymers with polyelectrolytes [semi-interpenetrating network (SIPN) and interpenetrating network (IPN)] or by the development of new monomers that could respond simultaneously to both stimuli [29]. Particularly, polymers responsive to both temperature and pH are gaining a considerable attention recently for their advantages in biotechnological and biomedical applications. In this respect, Leung and coworkers [30] have prepared smart core-shell polymeric microgels based on poly(N-isopropylacrylamide) (PNIPAAm), N, N'-methylenebisacrylamide (MBAAm) and chitosan. These materials exhibited a well-defined core shell structure consisting of temperature-sensitive cores and pH-sensitive shells, made of cationic water-soluble polymers. Kuckling et al. [31] prepared copolymers of N-isopropylacrylamide with acrylamide derivatives bearing carboxylic groups attached to spacers with different chain length, and studied the influence of both temperature and pH on their properties. Sun et al. [32] reported chitosan-based hydrogel films having both temperature and pH sensitivity, by blending chitosan with temperature sensitive poly(N-isopropylacrylamide) and poly(ethylene glycol) (PEG). Hiruta et al. [29] synthesized the dual pH- and temperature-responsive fluorescent terpolymer system ($I_{42}D_{50}S_8$-FL) consisting of PNIPAAm, N,N-dimethylacrylamide (DMAAm) and sulfamethazineacrylamide (SMZ), followed by activation of terminal carboxyl groups with N-hydroxysuccinimide and reaction with 5-aminofluorescein which were applied for the selective imaging of acidic tumor micro-environments.

2.1.4 Other Stimuli-Responsive Shape-Memory Polymers

In addition to, temperature variation or change in pH, there are many other ways to drive the shape recovery in polymers. Other stimuli [33] that can release the stored elastic energy in the elastic component of the polymer to drive recovery of shape are described below:

2.1.4.1 Light Induced Shape-Memory Polymers

The shape memory polymers which have photo sensitive groups are activated by light. These groups act as molecular switches and such polymers are stretched and illuminated by a light of wavelength greater than a fixed wavelength. The photo sensitive groups form crosslinks by which, the polymer is locked in the new shape and retains the temporary shape even after releasing the stress. When this is illuminated by light of lower frequency, the crosslinking cleaves and allows the material to go back to its original state [34].

The incorporation of light sensitive groups as molecular switches in the polymer network enabled the development of light induced SMPs. In this type of SMPs, instead of increasing the sample's temperature, light of different wavelength ranges can be used for the fixation of the temporary and recovery of the permanent shape. Photo-isomerization can also be used to draw out a shape-memory (SM) response. For instance, azobenzene in particular has been extensively studied to convert light into mechanical motion [35]. Here, azobenzene isomerizes from the thermodynamically stable *trans* state to the meta-stable *cis* configuration when irradiated with UV light, and it returns to the *trans* state when irradiated with visible light or heated [36]. In 2005, Lendlein and co-workers [34] demonstrated the use of light as a trigger for temporary crosslinking for SM using the photo-reversible [2+2] cycloaddition of cinnamic groups. Cinnamates, along with similar groups like coumarin, form dimers when irradiated with $\lambda > 260$ nm light and dissociate when irradiated with $\lambda < 260$ nm light as shown in Fig. 5 [35,37]. The authors programmed a temporary shape by mechanical deformation coupled with irradiation to form cinnamic dimer crosslinks; and the restoration of the original shape was achieved using a lower wavelength light to reversibly break the crosslinks. The kinetics of shape fixing was relatively slow, taking an hour of irradiation to achieve 30-50% shape fixity. However, the work represented a dramatic move forward and an important proof-of-concept, showing the potential of light to be used as a stimulus for SMPs. More recently, several researchers have developed new light-active SMPs. Wang et al. [38] developed hyperbranched, highly-entangled polycoumarates with the ability to hold multiple shapes that respond to both light and temperature stimuli.

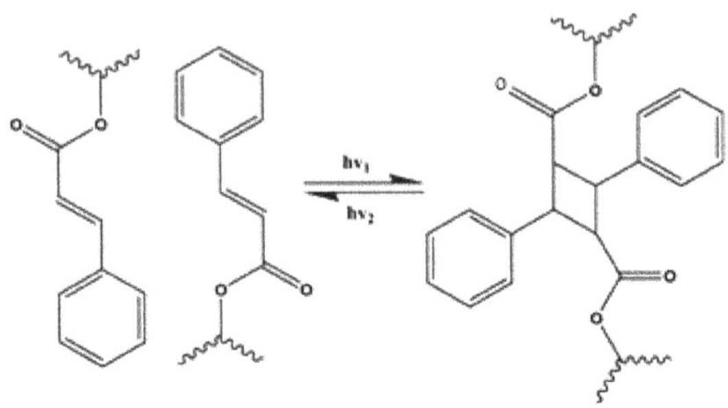

Fig. 5. Photo-reversible crosslinks [2+2] cycloaddition of cinnamic groups.(Reproduced with consent from Ref. [37], The Royal Society of Chemistry Publication 2014).

Lendlein et al. [34] reported two photo-responsive shape-memory polymers, one was the grafted polymer obtained by copolymerization of n-butylacrylate (BA), hydroxyethyl methacrylate (HEMA) and ethyleneglycol-1-acrylate-2-cinnamic acid (HEA-CA) with poly(propylene glycol)-dimethacrylate as crosslinker. In the second polymer, an interpenetrating network was created from BA with poly(propylene glycol)-dimethacrylate as crosslinker. The permanent network was star-poly(ethylene glycol) containing cinnamylidene acetic acid (CAA) terminal groups (SCAA), under the influence of light, SCAA molecules can be reversibly crosslinked, fixing the temporary shape by interpenetrating the permanent network.

2.1.4.2 Magnetically Induced Shape-Memory Polymers

A magnetic field can be used to trigger the shape recovery of SMPs filled with magnetic particles. In comparison with direct or electrical heating, inductive heating by alternating the magnetic field has more advantages especially in medical applications [39]. For thermally-adaptive SMPs, energy is used from the direct heating to trigger the shape change, where as in inductive heating approach, recovery of SMPs is by indirect heating, through the selection of suitable ferromagnetic particle as filler, using this method one can eliminate the problem of overheating. Generally, magnetic responsive SMPs can be produced by employing ferromagnetic fillers which include metal particles, ferromagnetic particles, like iron oxide particles, NdFeB particles, Ni-Mn-Ga single crystals and nickel powders [3, 39-40]. Figure 6 represents the shape-memory polymer (black strip) changing to its permanent shape in the presence of an alternating magnetic field [41]. The magnetic field adaptive shape-memory effect is more advantageous compared to pH and temperature. Thermo-magnetically and electromagnetically-induced shape-memory effects are of great interest for medical applications, because the embedded SMP devices can be actuated by an externally applied magnetic field through a non-contact remote mode. Other promising applications include smart implants, on demand drug delivery, etc. [39]. As these materials are originally proposed for biomedical applications, the SMP matrices are mostly biodegradable and biocompatible polymers, such as poly(D,L-lactide), crosslinking poly(ε-caprolactone), a copolymer, poly[p-dioxanone-co-(ε-caprolactone)], crosslinking oligo(ε-caprolactone)dimethacrylate/butylacrylate, and grafting polymerpoly(ε-caprolactone) diisocyanatoethyl methacrylate (PCLDIMA) and poly(ethylene glycol) mono-methylether-mono-methacrylate (PEGMA) [39].

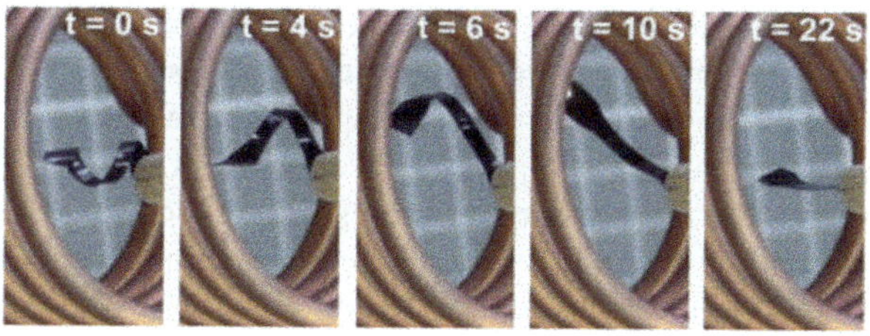

Fig. 6. Sequence of photographs showing the shape-memory polymer (black strip) changing to its permanent shape in the presence of an alternating magnetic field. (Adopted with consent from Ref. [41], Copyright (2006) National Academy of Sciences, U.S.A.).

2.1.4.3 Phase Sensitive Smart Polymers

Other types of smart polymers, suitable for the biomedical field are phase sensitive smart polymers. These can be used to develop biocompatible formulations for controlled delivery of proteins in a conformationally stable and biologically active form. These smart polymeric systems have many advantages over other systems which require less stressful manufacturing conditions for sensitive drug molecules and have a high loading capacity [42].

This category of polymers employs water insoluble biodegradable polymers. Polymers such as poly(D, L-lactide), poly(D,L-lactide-co-glycolide) and poly(D,L-lactideco-ε-caprolactone), when dissolved in pharmaceutically acceptable solvent to which a drug is added, forms a solution or suspension. After injection of the formulation in the body, the water-miscible organic solvent dissipates and water penetrates into the organic phase. This causes phase separation and precipitation of the polymer forming a depot at the site of injection [42].

2.1.4.4 Water/Moisture Activated Shape-Memory Polymers

Although various external stimuli can be used as a recovery trigger, the most widely studied are thermally induced SMPs, which are not always suitable for biomedical applications, because their switching temperature does not fall completely within the range of room temperature to body temperature. Thus, water-induced SMPs may be used to resolve this issue due to the lower transition temperature exhibited by the polymers with water in the polymer matrix. Water-induced SMPs have been found more suitable for biomedical applications, because their shape recovery occur upon exposure to body fluid and consequently, these polymers have received great attention in recent years [42]. In 2005, Huang et al. [43] first observed water or solvent-driven shape recovery effects in many SMPs having glass transition as the switching transition. Water or solvent molecules may penetrate into the amorphous areas of polymers [39], and thereby increase the flexibility of polymer chains due to a plasticizing effect. As a result, the glass transition temperature of the SMPs may drop to a temperature below the room temperature, which leads to the shape recovery of a deformed SMP [39]. One of the methods to obtain an outstanding water-active shape-memory effect is filling a hydrophilic or water soluble ingredient in SMPs [39]. Chen et al. [44] reported a water sensitive biodegradable stent made of chitosan films crosslinked with an epoxy compound. Based on this idea, Luo et al. [45] prepared cellulose nanowhisker/shape-memory polyurethane (SMPU) nanocomposites with outstanding water-active shape-memory effect (SME). The cellulose nanowhisker in SMPUs did not affect the intrinsic thermal-active SME of the SMPU, but enabled the SMPU water-active SME. The water-active SME is tailorable by changing the content of the cellulose nanowhiskers. The water-active SMPUs require hours to lower the glass transition temperature for recovery; while the water-responsive cellulose nanowhisker/elastomer composite needs only 10 min for recovery which is much faster. The water-active cellulose nanowhisker/elastomer composites have several advantages over water-active SMPU based its dropping glass transition temperature [46].

2.1.4.5 Electricity Responsive Shape-Memory Polymers

In the past decade, several researches have conducted the experiments on the mechanism of shape recovery and electrical properties of SMP filled with conductive filler, and their corresponding applications. One of the approaches to make shape recovery triggered, more conveniently by passing an electrical current through SMP composite by employing typical conductive fillers like carbon black (CB) nanoparticle and short carbon fiber (SCF) [47]. Electricity as a stimulus enables resistive actuation of shape-memory polymer filled with fillers. In this way, external heating, which may be unfavorable for specific applications, is used to stimulate conventional shape-memory polymers, can be avoided. The electric triggering of shape-memory polymer composites enlarges their technological potential. Smela [48] reported electrically actuated polymeric materials known as electro-active polymers (EAPs). These are mostly conjugated polymers, useful in regeneration of tissues, biosensors, etc. The devices made of these polymers find applications in the biomedical field as artificial muscle, a technology for biomimetic actuation [48,49].

2.2 Shape-Memory Polymers Based on Functionality

The approaches for designing different shape–memory functions have become more abundant as scientists better understand the SME mechanism of SMPs and their increasing diversification. Thus, the development of shape–memory functions of SMPs is growing rapidly. The shape-memory functionality of SMPs is illustrated in Fig. 7 [3].

Moreover, two and even three different types of shape–memory functions can be achieved simultaneously in the same SMP material [50,52]. An example of this is an SMP network mainly composed of star-shaped hydroxy-telechelic poly(ω-pentadecalactone) (PPD) and poly(ε-caprolactone) (PCL), herein named the PDCL network, that shows one-way, two-way and triple SMEs under different programming [3].

SMPs may adopt one (dual-shape), two (triple-shape) or several (multi-shape) stable temporary shapes and recover their permanent shape (or other temporary shapes in case of multi-shape versions) upon the action of an external stimulus [53].

If the shape-memory behavior of SMPs is typically recognized as a nonreversible feature, as the external stimulus activates the only changes from a temporary to a permanent shape. This is a so-called “one-way” shape-memory behavior. This type of shape-memory function allows one to satisfy the requirements of a wide variety of applications. By contrast, the development of polymers featuring a reversible shape shifting on the basis of an on-off stimulus are the so called “two-way” shape-memory behavior, which may fulfill the requirements of applications such as sensors and actuators, or as artificial muscles [9, 50-53].

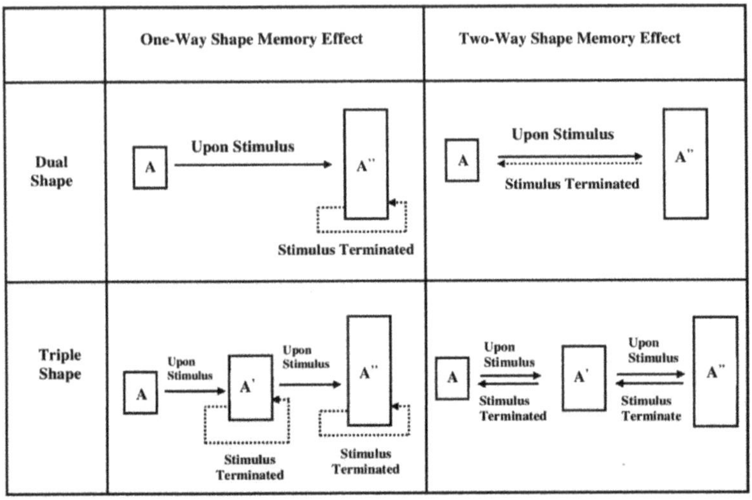

Fig. 7. Illustration of the shape–memory functionality of SMPs. (For one-way SME, once the stimulus is terminated, the temporary shape is retained, while for two-way SME, the temporary shape can be recovered to the initial shape when the stimulus is terminated) (Reproduced with consent from Ref. [3], Elsevier Publication 2014).

2.2.1 One-Way SME and Two-Way SME

Scientists thought that one-way SMEs were not good enough for many applications because they were the most simple and common SMEs of SMPs, and thus they began to explore two-way SMEs of polymers. Research on smart materials with two-way SMEs has attracted considerable attention as these materials are capable of changing their geometries when exposed to an external stimulus.

However, as more and more programming protocols have been devised and reported in the literature, one-way SMPs have been found to be much more flexible, versatile and applicable than the two-way SMPs, because two-way SMEs are difficult to program. One-way SMEs have their original shapes predefined in the manufacturing process, while temporary shapes can be varied based on different shape–memory programming processes [3].

2.2.2 One-Way Reversible SME and Two-Way Reversible SME

Zhou et al. [54] reported a general approach for enabling reversible shifting (two-way) between programmable shapes in semi-crystalline polymeric elastomer, poly(octylene adipate), with uniform chemical composition and without applying a persistent external force. These authors also developed a general strategy for reversible shape-memory (RSM) which can be applied to conventional semicrystalline polymeric elastomers. Strategy integrates three different shape-memory (SM) behaviors: conventional one-way SM, two-way RSM, and one-way RSM. Also the relative contributions of the different strains (one-way SM, two-way RSM, and one-way RSM) to aforementioned shape-memory can be tuned by controlling the respective proportions of chemical crosslinks and crystallites. For instance, the experimental demonstrations of two different RSM concepts are displayed in Fig. 8. The structure of poly(octylene adipate), a semicrystalline elastomer shown in (Fig. 8a), similarly: (i) two-way reversible shape transformation upon heating-cooling cycles (Figs. 8b–e) and (ii) one-way shape reversibility upon heating (Figs. 8f–h). In the first experiment (Fig. 8b), a straight bar of the semicrystalline elastomer was programmed as a hairpin temporary shape by deforming the sample at a temperature of 60°C and then quenching it to 5°C. Subsequent heating to 38° C and cooling to 5°C resulted in a reversible shape transformation of the hairpin between 115° and 155° angles, which occurred spontaneously without applying any external force. The angles of oscillations are highly repeatable. Upon consecutive heating and cooling cycles, the sample shape changed reversibly multiple times without losing memory of both angular shapes (Fig. 8e). Similar reversible behavior was accomplished for a polymer coil (Fig. 8c) and for the folding-unfolding of origami-like polymer sheets (Fig. 8d). Note that the upper temperature, 38°C, corresponds to partial melting of the crystalline phase. In the second experiment (Figs. 8f–h), a one-way reversible SM transformation was evoked in the same polymer by reprogramming the temporary shape at an elevated temperature corresponding to a partial molten state. As shown in Fig. 8f, a sample having a temporary shape of a straight rod was heated to an intermediate temperature of T_i = 34°C (substantially below T_m = 41°C) and manually coiled into a new temporary shape. The new coil was then fixed by quenching to T_0 = 5°C. Subsequent heating from T_0 = 5°C to T_i = 34°C and then to T = 60°C, triggered a one-way transition from coil (second temporary shape) to rod (first temporary shape) and back to coil (permanent shape) (Fig. 1f). In a similar way, an opposite shape transformation was demonstrated, i.e., rod-coil-rod using the same temperature ramp (Fig. 8g).

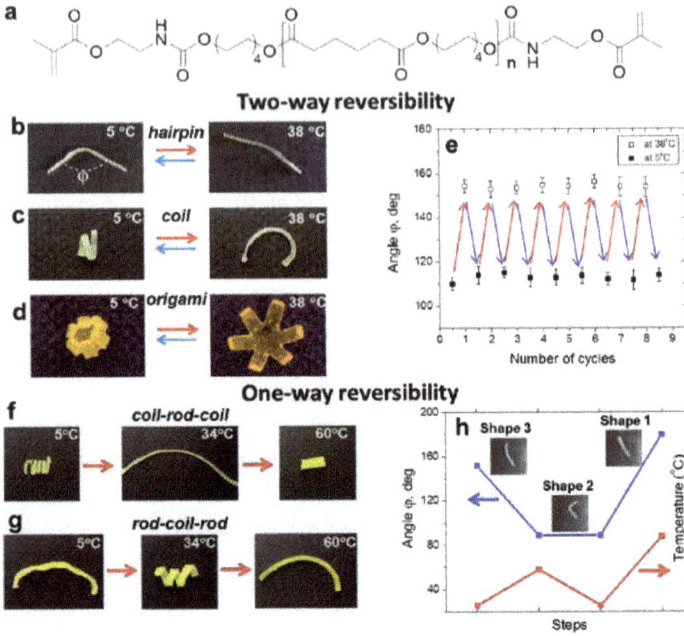

Fig. 8. Two-way and one-way-reversible shape memory. (a) Chemical structure of poly(octylene adipate) (POA) oligomers used for the synthesis of shape-memory elastomers. Their cross-linking density and melting temperature were controlled by varying the degree of polymerization n of the oligomers and copolymerization. (b−e) Examples of two-way reversible shape memory demonstrated by polymer (ν = 200 mol/m^3, T_m = 42°C), for three different shape transformations upon heating to 38°C and subsequent cooling to 5°C. (e) Reversible oscillations of the interior hairpin angle shown in panel (b) during eight consecutive heating−cooling cycles within the 5−38°C temperature range. (f−h) One-way reversible shape transformations were observed during heating from 5 to 60°C. While the two-way transformations may be repeated multiple times, the one-way reversibility occurs only once. (h) Temperature ramp and the corresponding variations of the interior angle of a hairpin sample which was programmed to perform one-way reversible bending. (Adopted with consent from Ref. [54], American Chemical Society Publication 2014).

2.2.3 Triple-Way Shape Memory Effect

Triple-shape polymers, unlike conventional "dual-shape" SMPs which can recover from a temporary shape to a permanent shape, are capable of fixing two temporary shapes and recovering sequentially from the first temporary shape (shape 1), to the second temporary shape (shape 2), and eventually to the permanent shape (shape 3), upon stimulus [22]. Xie et al. [55] reported bilayer polymers that consist of two epoxy dual shape memory polymers which have well separated glass transition temperatures. These bilayer epoxy samples exhibited a triple shape-memory effect (TSME) with shape fixities tailorable by changing the ratio between the two layers. The triple shape fixities of the bilayer epoxy polymers can be explained by the balance of stress between the two layers. Based on this work, it is believed that the following three molecular design criterions should be considered in designing triple-shape memory polymers with optimum TSME:

I. well separated thermal transitions,

II. a strong interface, and

III. an appropriate balance.

Luo and Mather [22] developed a series of triple-shape polymeric composites (TSPCs), using a unique and broadly applicable fiber/matrix composite approach. It is a significant approach as polymer composites afford a large degree of design flexibility in designing triple-shape systems for targeted applications. For example, one can control the transition temperatures by simply selecting different polymers for the fibers and the matrix, or separately functionalize them either chemically or physically to achieve other novel functions such as controlled drug delivery or reversible adhesion. The triple-shape memory effect is demonstrated in Fig. 9 [56].

Jeon et al. [57] studied the shape-memory properties of norbornyl-POSS (POSS-cubic polyhedral oligomeric silsesquioxane, a nanoparticle) copolymers having either cyclohexyl corner groups (CyPOSS) or cyclopentyl corner groups (CpPOSS). The POSS copolymeric composites showed two stages of strain recovery or triple SME, a fast strain-recovery process related to the polynorbornene relaxation in the norbornyl matrix, and a second slow strain recovery process related to the POSS rich domains.

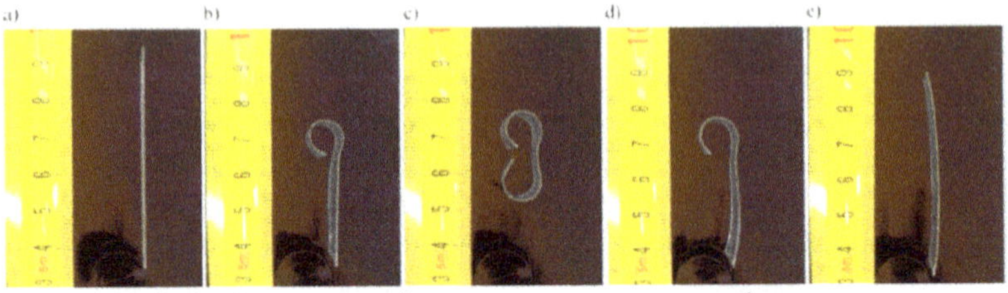

Fig. 9. Demonstration of the triple-shape memory effect for the symmetric blend in practical conditions: (a) original shape 1; (b) shape 2, blended at 100°C and cooled to 65° C; (c) shape 3, blended at 65°C and cooled to 25°C; (d) recovered shape 2 after heating at 70°C; (e) recovered shape 1 after heating to 100°C. (Adopted with consent from Ref. [56], ACS Publication 2014).

2.2.4 Multifunctional SMEs

SMPs which provide more than one functionalities i.e., permeable, optical, biodegradable, switch temperature, mechanical properties and thermal chronic properties or/and which can recover more than two temporary shapes are the multiple shape-memory polymers.

Driven by potential applications, such as intelligent implants or textiles, the question arises whether SME of polymers can be combined with different functions, which are not directly linked to SME. Examples of such functionality combinations are electrical conductivity and SME, change of color and SME, or degradability and SME [53]. Multifunctionality is the targeted combination of material functions, which are not linked with each other. The development of such multifunctional materials is highly interesting challenge for fundamental research. Several approaches have been explored for creating multifunctionality, involving one component polymer systems as well as multimaterial systems, e.g., combining two polymers or a polymer with inorganic nano particles or for an example, if a single macromolecule, in which several functionalities have been integrated to be a polymer network, in which an additional functionality (e.g., hydrolytical degradability) can be obtained by insertion of easily hydrolysable bonds. These can be obtained by the addition of independent functions to SMPs, such as electrical conductivity, thermal conductivity or hydrolytical degradability. It can also be obtained by the integration of various functions in multi-material systems on the nano- or micro-level, such as in composites, multifibers, or multilayer constructs, as well as single material systems [3,53].

Multifunctionality in polymers is often referred to polymers, having several different reactive groups or having several same reactive groups (repeating units) [58]. The polymers having same reactive groups are able to react with ligands, presenting a multiple acceptor unit. In material sciences, the understanding of multifunctionality differs, in which a functionality origins from the bulk material. Thus, the hierarchical organization of different reactive groups in subunits of a polymer, as each group responsible for a certain function, enables as multifunctional materials. Finally, not every function, which is integrated in a polymer, might be required for each application. Therefore, multifunctional polymers cover the aspect of universality, so that they contribute only with the relevant functions to the specific application. Multifunctional materials are realized as composites [59] or hybrid structures of several distinctmaterial phases, in which each phase contributes a different but necessary function. Typical functions include strength, elasticity, degradability, electrical-conductivity or -insulation, thermal conductivity or protection, energy storage, magnetism or magnetic shielding, radiation protection, light emission, self-healing, or actuation [53]. The multifunctional character of such composite materials might be even enhanced by organizing the structured composites in layers, in which each sub-layer contributes with its multifunctionality [60]. Xie et al. [55] prepared shape memory epoxy with a layered structure showing multiple-shape memory effect (MSME). The epoxy with a high T_g was first cured and then an epoxy with a low T_g was cured on top of the first layer. The two layers of epoxy have a strong interface as a result of the reaction among the residual epoxy or amine groups on the surface of the two layers [39]. Recently, He et al. [61] exploited new multiple SME systems based on multicomposite SMPs made of Fe_3O_4-SMP region and a carbon nanotube (CNT)-SMP region separated by the neat T_g type SMP region. As two nanofillers (Fe_3O_4 and CNT) can induce heat at two distinctively different radio frequency ranges (296 kHz and 13.56 MHz), the multi-composite demonstrated well-controlled multiple shape recoveries.

3. Fixity Ratio (R_f) and Recovery Ratio (R_r)

In view of the above functional SMPs, shape-memory performances based on the fixity ratio (the ability to store the temporary shape), the recovery ratio (the ability to recover the original shape), and the switch temperature (the temperature required to switch from the temporary to the original shape) can be easily assessed. Fixity ratio (R_f) and recovery ratio (R_r) are calculated from the relations 1 and 2, respectively.

$$R_f = \frac{\varepsilon_2}{\varepsilon_{s2}} \times 100 \tag{1}$$

$$R_r = \left[1 - \frac{\varepsilon_{r1}}{\varepsilon_2}\right] \times 100 \tag{2}$$

where ε_{s2}, the maximum engineering strain achieved during uniaxial stretching; ε_2, the engineering strain of the temporary shape after unloading; and ε_{r1}, the engineering strain of the recovered original shape.

In this respect, relations 1 and 2 developed for dual-shape memory experiments could be generalized by relations 3 and 4 for multiple-shape memory experiments

$$R_f \text{ (shape i)} = \frac{\varepsilon_i}{\varepsilon_{si}} \times 100 \quad (3)$$

$$R_r \text{ (shape i)} = \left[1 - \frac{\varepsilon_{ri} - \varepsilon_i}{\varepsilon_{i+1} - \varepsilon_i}\right] \times 100 \quad (4)$$

where ε_{si} is the maximum engineering strain achieved during uniaxial stretching of shape i; ε_i is the engineering strain of shape i after unloading, ε_{i+1} is the engineering strain of shape i + 1 after unloading, and ε_{ri} is the engineering strain of recovered shape i [55].

4. Polymer Structures and Their Approaches

Shape-memory behavior can be exploited and demonstrated in various polymer systems that are significantly different in molecular structure and morphology. The polymer segments can be biodegradable or non-biodegradable and natural or synthetic. However, synthetic polymers are preferred. The resulting SMP should be biodegradable and biocompatible polymers, as these are particularly preferred for biomedical applications. In general, these materials degrade by hydrolysis, exposure to water or enzymes under physiological conditions, surface erosion, bulk erosion, or a combination thereof. Non-biodegradable polymers used for medical applications preferably do not include aromatic groups, other than those present in naturally occurring amino acids [40]. Furthermore, natural polymer segments include proteins such as zein, modified zein, casein, gelatin, gluten, serum albumin and collagen, and polysaccharides such as alginate, cellulose, dextran, pullulane and poly(hyaluronic acid) as well as chitin, poly(3-hydroxyalkanoate)s, especially poly(β-hydroxybutyrate), poly(3-hydroxyoctanoate) and poly(3-hydroxy fatty acids).

Examples of synthetic polymers include polyphosphazenes, polyamides, polyester amides, poly(amino acid)s, polyanhydrides, polycarbonates, polyacrylates, polyalkylenes, polyacrylamides, poly(alkylene glycol)s, poly(alkylene oxide)s, poly(alkylene terephthalate)s, poly(ortho ester)s, poly(vinyl ether)s, poly(vinyl ester)s, poly(vinyl halide)s, poly(vinyl pyrrolidone)s, polyesters, polylactides, polyglycolides, polysiloxanes, polyurethanes and copolymers thereof [40].

The shape-memory effect can be obtained in a polymer system containing two kinds of segments, i.e., a soft segment(s) as the switching segment controlling temporary shape and a hard segment(s) as net-point segment determining permanent shape due to entropic elasticity [9, 62,63]. The terms hard segment and soft segment are relative terms, relating to the T_{trans} of the segments. The hard segment(s) has a higher T_{trans} than the soft segment(s). As used herein, the term "segment" refers to a block or sequence of polymer forming part of the shape-memory polymer. The shape-memory polymers can include at least one hard segment and at least one soft segment, or can include at least one kind of soft segment, wherein at least one kind of soft segment is crosslinked, without the presence of a hard segment. The hard segments can be linear oligomers or polymers, and can be cyclic compounds, such as crown ethers, cyclic di, tri, or oligopetides, cyclic oligo(ester amides), diisocynates and diols [64] and the soft segments are either polyethers or polyesters [63].

The physical interaction between the hard segments can be based on charge transfer complexes, hydrogen bonds or other interactions, since some segments have melting temperatures that are higher than the degradation temperature. In these cases, there is no melting or glass transition temperature for the segment. A non-thermal mechanism, such as a solvent, is required to change the segment bonding.

Lee et al. [64] reported that the polyurethanes (PU) synthesized from diphenylmethane diisocyanate and polyol using 1,4-butanediol as chain extender and phase-separated structures consisting of a hard domain and a soft domain. Due to the thermodynamic immiscibility, hard segments could bind themselves via hydrogen bonding and crystallization, making the PU solid below its melting temperature. Reversible phase transformation of the soft segment is reported to be responsible for the shape-memory effect. Thus, the shape-memory effect can be controlled by mole ratio between hard and soft segments and polymerization process as required for a target application [40,64].

Practically all polymerization methods including addition, condensation, free radical, ring opening [3], chemical crosslinking [39], radiation crosslinking [65], thermally induced crosslinking [66], sol-gel crosslinking [22], movable crosslinking [11,12], photochemical reaction [21,39] etc. can be employed to synthesize SMPs. In addition to these, there are many approaches developed for the synthesis of SMPs to improve the properties for target applications. The detailed synthetic approaches are discussed below:

4.1 Homopolymers

4.1.1 Poly(N-isoproprylacrylamide)

Poly(N-isopoprylacrylamide), from the family of poly(N-substituted acrylamide), is a synthetic temperature-sensitive polymer. The structure of PNIPAAm is shown in Fig. 10. The fact that it is soluble in water at room temperature makes it the most studied and widely used material for drug delivery systems. This happens owing to its hydrophilic interactions, due to the predominance of hydrogens. Poly(N-isopoprylacrylamide) in its hydrated form acts like an inert material, however, above 32°C the polymeric chain becomes hydrophobic due to dehydration, and precipitation occurs. At the lower critical solution temperature (LCST, 32°C), the solution becomes opaque and turns into a gel with a transition temperature very similar to the body temperature, with predominance of hydrophobic interactions [67].

This polymer, incorporated in therapeutic systems, when injected in the human body turns into a gel at the body temperature, forming a small deposit and providing a controlled delivery of the drug. The LCST can be adjusted through the copolymerization of the hydrophobic monomers or through controlling molecular weight of the polymer [27].

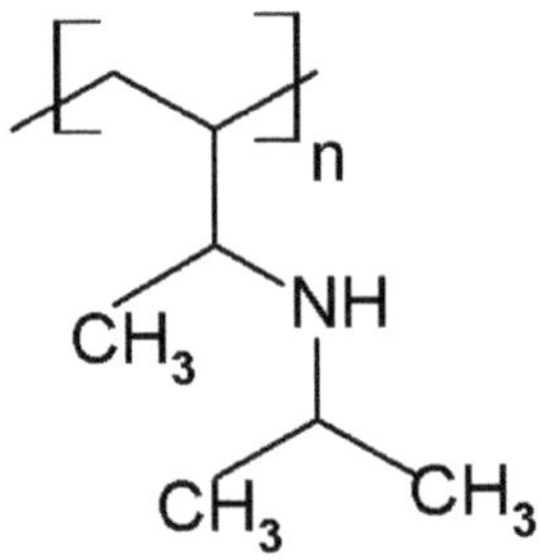

Fig. 10 Structure of PNIPAAm.

4.1.2 Poly(methyl vinyl ether)

Poly(methyl vinyl ether) (PMVE) has a transition temperature of exactly 37°C, which makes it very interesting for biomedical application. The structure of PMVE is shown in Fig. 11. and this is synthesized by cationic polymerization in an inert condition. Nucleophiles like alcohol or amino groups cannot be tolerated during the synthesis, which limits the potentiality of PMVE.

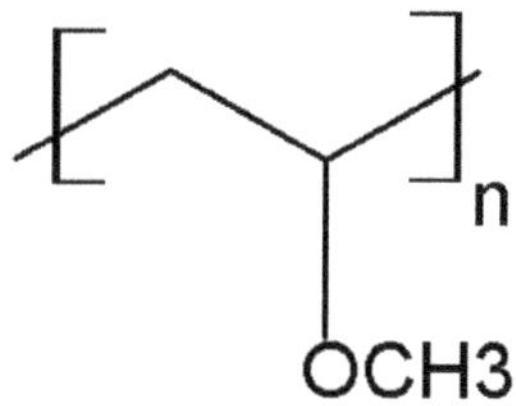

Fig. 11.Structure of PMVE.

4.1.3 Poly(N-vinyl caprolactam)

Poly(N-vinyl caprolactam) has not been studied as intensively as PMVE and PNIPAAm, but it possesses very interesting properties suitable for medical and biotechnological applications, e.g., solubility in water and organic solvents, biocompatibility, high absorption ability and a transition temperature within the settings of these applications (33°C) [68].

4.1.4 Poly(N-ethyl oxazoline)

Poly(N-ethyl oxazoline) has a transition temperature of around 62°C, which is too high for any drug delivery application. However, Schmaljohann [68] recently prepared a double thermo-responsive system by graft polymerisation of ethyl oxazoline onto a modified PNIPAAm backbone. Currently, these systems are explored for their potential in drug delivery, because they tend to aggregate micelles above the LCST. Unfortunately, the poly(oxazoline) chemistry has the disadvantage that it is not very tolerant against unprotected functionalities.

4.2 Interpenetrating Networks

Interpenetrating network technology is one of the useful routes for the preparation of SMPs. Interpenetrating networks are defined as networks where two components are crosslinked, but not covalently bonded to each other. The original shape is determined by the network with the highest crosslink density and mechanical strength. IPN can include at least one physically crosslinked polymer network (thermoplastic polymer) and at least one covalently crosslinked polymer network (thermosetting polymer) that cannot be separated by any physical methods. In that again semi-IPNs can be defined as network having two independent components, where one component is a crosslinked polymer (a polymer network) and the other component is a non-crosslinked polymer (a homopolymer or copolymer), wherein the components cannot be separated by physical methods. SIPNs is one of the methods to increase both thermal and mechanical properties of SMPs. Lorenzo et al. [69] reported that IPN of chitosan, and poly(N-isopropylacrylamide) showed enhanced loading capacity and controlled release properties. Zhang and Peppas [70] reported the synthesis and characterization of poly(methacrylic acid) (PMAA)/poly(N-isopoprylacrilamide) IPNs stimuli to both pH and temperature. Ma et al. [71] demonstrated that synthesized semi-IPN of sodium carboxymethylcellulose (CMC)/poly(N-isopropylacrylamide) crosslinked with clay exhibited excellent mechanical properties with high tensile strength and elongation at break (Fig. 12). Guo and Gao [72] reported that the synthesized polyampholyte hydrogels of carboxymethylated chitosan and poly(N-isopropylacrylamide) with N,N'-methylenebisacrylamide (BIS) as the crosslinking agent, showed pH and temperature responsive functions. These polymers have been extensively studied in the biomedical field as these two factors (pH and temperature) can be easily controlled and are applicable both *in vitro* and *in vivo* conditions. An interpenetrating network of poly(acrylic acid) and polyacrylamide is one of the examples for a system with upper critical solution temperature behavior within the biomedical setting. The transition temperature is at 25°C. The UCST behavior is caused by the cooperative effects coming from the hydrogen bonding between acrylic acid and acrylamide units [16].

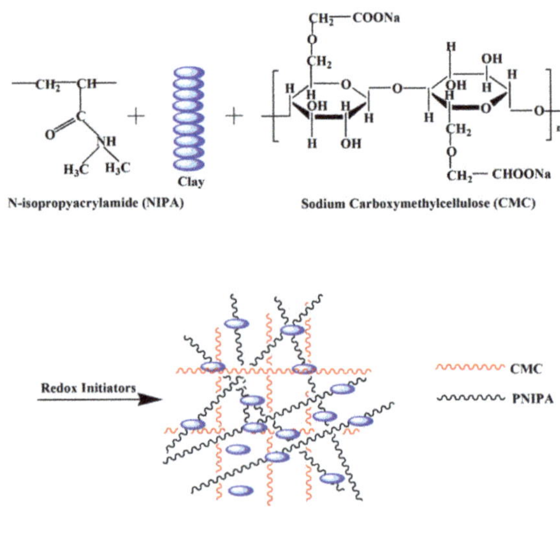

Fig. 12. Scheme of the formation process of CMC/PNIPA/Clay hydrogel.(Reproduced with consent from Ref. [71], Elsevier Publication 2007).

4.3 Copolymers

Copolymers are widely used as SMPs in view of their advantages of having combined properties of two or more repeating units. Berg et al. [34] addressed the variety of distinct chemical approaches that are being explored currently for SMPs. The copolymers can be in the form of linear diblock, triblock, tetrablock, or multiblock copolymers, branch or graft co-polymers. Peponi et al. [73] reported that the synthesized triblock PCL-PLLA copolymer for SMPs showed potential biomedical applications. Suchao-in and Chirachanchai [74] reported a graft copolymer consisting of polystyrene-b-poly(ethylene-co-butylene)-b-polystyrene (m-SEBS) as the main chain while polybutylene succinate (PBS) as the grafted side chains. As the melting temperatures of poly(ethylene-co-butylene) and PBS are 55-65° C and 105-115°C, respectively, the resulting polymer demonstrated triple-SM behavior based on its two crystalline segments [35]. Gradient copolymers, a lesser-studied cousin of block copolymers with many of the same characteristics, lead to interesting properties when they are used as SMPs. Luo et al. [76] designed a "V-shaped" gradient copolymer of styrene and methyl acrylate by radical polymerization. The "triblock-gradient" copolymer underwent microphase separation similar to a triblock copolymer, but exhibited a broad T_g due to change in composition along the length of the polymer chain [35]. After 3 cycles, dual-SM fixities and recoveries using the entire broad T_g for actuation were >99% and >98%, respectively, indicating the potential for using this type of architecture in biomedical applications [35].

Hu et al. [75] reported in their review that synthesized stimuli-responsive tertiary amine methacrylate-based block copolymers having supramolecular self-assembly can typically exihibit pH responsive features due to the protonation of tertiary amine moieties in acidic condition and deprotonating in alkaline condition having potential application in the biomedical field as pH responsive SMPs.

4.4 Polyrotaxanes (PR)

This is a novel type of crosslinked polymer networs in which polymer chains neither crosslinked by covalent bonds nor by attractive interactions as in traditional physical network, but are interlocked by ***movable crosslinks***. As ring like molecules can rotate and slide along the polymer chains, these matrices demonstrate excellent swelling capacity and mechanical strength. Creation of this fully recoverable high strain SMP may facilitate further innovation into the recent generation of smart devices for biomedical applications. Wang et al. [11] synthesized the SMP network (shown in Fig. 13) using a α-cyclodextrin in PR as crosslinker, methylacrylate monomer and methylmethacrylate as co-monomers and azobisisobutylnitrile (AIBN) as initiator which showed an outstanding shape-memory effect and excellent mechanical strength even at high strain. Imran et al. [12]

reported the formation of thermosensitive and water soluble shape-memory polymers with PR as crosslinker and N-isopropyl acrylamide as monomer. Chiang and Chu [77] reported the photoresponsive shape-memory polymers with hybrid alginate polymeric hydrogel synthesized by Ca^{+2} mediated crosslinking through the reaction with a mixture of β-cyclodextrin-grafted alginate (β-CD-Alg) and diazobenzene modified PEG (Az_2-PEG). Both these thermosensitive and photoresponsive polymers exhibited excellent shape-memory effects, which could find enormous biomedical applications in the near future.

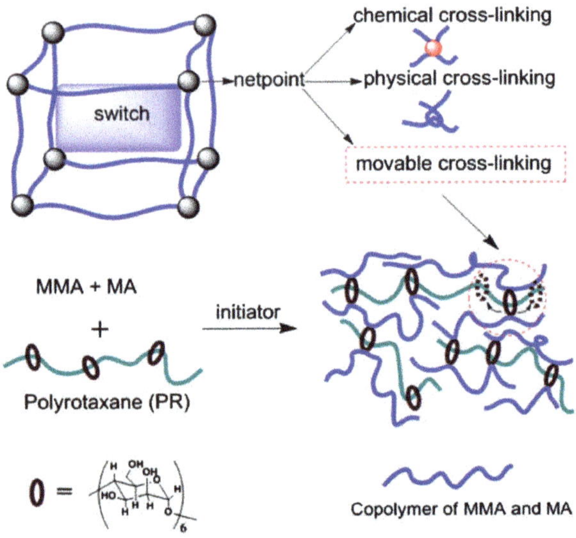

Fig. 13. Illustration of SMP network with movable crosslinks.(Adopted with consent from Ref. [11], Royal Society of Chemistry Publication 2014).

4.5 Blends

Blending technique is one of the approaches for synthesizing SMPs. This technique, helps to tune the switch temperature and mechanical properties of the SMPs [78]. Jeong et al. [79] blended thermoplastic shape-memory polyurethanes (SMPU) with poly(vinyl chloride) to vary the switch temperature and improve the mechanical strength of SMPU. The poly(vinyl chloride) was miscible with the soft segment of the SMPU. The switch temperature of the blends could be changed smoothly with different compositions.

Polylactide, a biodegradable SMP, has the disadvantages of high brittleness and low toughness. Zhang et al. [80] toughened polylactide using a polyamide elastomer from polyamide-12 and poly(tetramethyleneoxide). The mechanical properties of the polylactide were markedly improved. Song et al. [81] reported biocompatible SMP polymer blend foams composed of thermoplastic polyurethane (TPU) and polylactic acid, which will find application as biomaterials.

Blending of crystalline and amorphous polymers is a facile strategy to create new SMPs [78]. The resulting blended SMPs are of great industrial interest compared to covalently crosslinked SMPs as they can be prepared easily and quickly by blending two polymers. The crystalline polymer acts as the crosslinking structure and the amorphous polymer provides the switch. The crystalline polymers may include poly(vinylidene fluoride), polylactide, poly(hydroxyl butyrate), poly(ethylene glycol), polyethylene, poly(ethylene-co-vinyl acetate), poly(vinyl chloride), poly(vinylidene chloride), and copolymers of poly(vinylidene chloride) and poly(vinyl chloride). The amorphous polymers include poly(vinyl acetate), poly(methyl acrylate), poly(ethyl acrylate), atactic poly(methyl methacrylate), isotactic poly(methyl methacrylate) and other poly(alkylmethacrylate)s. Mather and coworkers [82] have made significant contributions in this area. Kolesov and Radusch [83] blended and crosslinked linear high density polyethylene, ethylene-1-octene copolymers and short-chain branched polyethylene to prepare SMPs with multiple crystallization and melting behavior. Dual-shape and triple-shape memory was observed after two- and three-step programming of the blends. Similarly, Cuevas et al. [84] blended and crosslinked two semicrystalline polymers, poly(cyclooctene) and polyethylene, to create two well-separated crystalline domains in a single polymer, for demonstrating the properties of multiple-shape recovery.

4.6 Epoxy Based Polymers

Epoxy based materials are used extensively in demanding applications due to their excellent chemical resistance, superior thermal and mechanical properties and good adhesion properties. However, their development as shape-memory (SM) material, has remained limited, possibly due to their relatively low failure strains and high moduli. In the case of epoxies (i.e., covalently crosslinked networks), the transformation temperature (T_{trans}) of their shape-memory effect coincides with their glass transition temperature (T_g). Their permanent shape is dictated by the crosslink points formed during the curing reaction, whereas their permanent shape results from a mechanical deformation performed on the cured samples above T_{trans} (¼T_g). Above T_g, SM epoxies exhibit an elastic behavior. Therefore, in response to mechanical deformation, the more mobile polymer chains maintain a state of maximum entropy and minimum internal energy by allowing conformational changes to occur. This translates to a macroscopic deformation of the epoxy networks, which can be fixed by subsequent cooling while maintaining the applied force. The work thereby performed on the sample (alternatively, the energy input in the system) can be stored as strain energy when decreasing the temperature below T_g (i.e., polymer vitrification). Only subsequent heating above T_g will enable for the stored strain energy to be released as the mobility of the polymer chains between crosslinks increases. The SM epoxy will thereby recover its initial, permanent shape provided that the external force applied during the deformation stage is removed. Obviously, the strength and energy that will be stored as a result of the deformation and cooling depend on the mechanical properties of the materials. For this reason, various SM epoxies are synthesized in order to provide polymers with varying structural and thermo-mechanical properties. Several epoxies based on formulations consisting of a mixture of an aromatic and an aliphatic diepoxide with an aliphatic diamine curing agent were produced [85]. Xie and Rousseau [86,87] reported that epoxy systems give guidelines for material improvement for their use as high performance SMPs.

4.7 Polymer Brushes

Another approach in developing a biomedical SMP is the fabrication of polymer brushes. Ionov et al. [88] demonstrated the fabrication of nanostructuredstimuli responsive polymer brushes using microtubules as templates and reported the ''grafting from'' of thermo-responsive poly(N-isopropylacrylamide) brushes by atom transfer radical polymerization (ATRP) initiated on microtubules. The outer surface of microtubules containing reactive amino groups provides the possibility of chemical modification with fluorescent and other functional groups. Schematic diagram illustrating this approach is presented in Fig. 14.

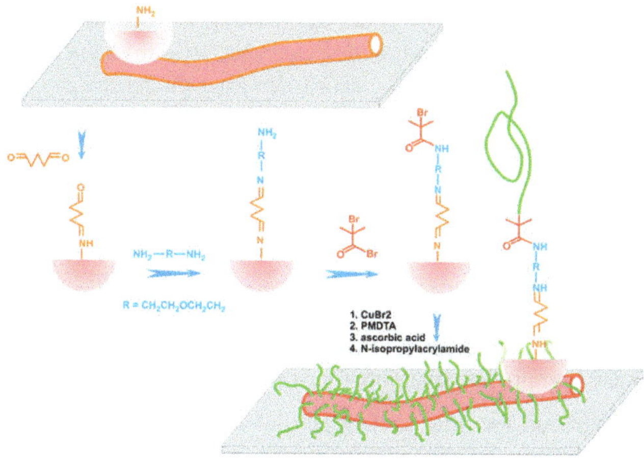

Fig. 14. Schematic diagram of microtubule modification and polymer grafting. The polymer chains were grown from microtubules that were adsorbed on a hydrophobic glass substrate and sequentially crosslinked with glutaraldehyde, amino-functionalised using 2,2'-(ethylenedioxy)-bis(ethylamine) and modified by 2-bromo-2-methylpropanoyl bromide. (Adopted with consent from Ref. [88], Royal Society of Chemistry Publication 2009).

Kitano et al. [89] synthesized 2-(2-methoxyethoxy)ethyl methacrylate (MDM), and constructed polymer brush (PMDM) on a monolayer of colloidal gold. The resulting brush confirmed gold colloids on the surface, which are sensitive to temperature. The polymeric brush which was prepared by atom transfer radical polymerization using 2-(2'-bromoisobutyroyloxy)ethyl disulfide as initiator (DT-PMDM), preheated with $NaBH_4$, did not absorb lysozyme significantly below the LCST of the polymer chain, whereas it substantially absorbed the protein above the LCST, signifying that these brushes are suitable for various biomedical applications.

5. Applications of Shape-Memory Polymers

There have been a number of interesting biomedical applications proposed for stimuli-responsive polymer systems [90] as discussed below:

5.1 Drug Delivery

As the name suggests this method enables the administration of a pharmaceutical drug to achieve a therapeutic effect in humans or animals. However, there are many obstacles in achieving successful drug delivery. These obstacles include the drug's low solubility, environmental or enzymatic degradation, fast clearance rates from the body, non-specific toxicity and inability to cross the biological barriers, and the release of expected amount of drug at the desired time. In order to overcome these obstacles, drug delivery carriers are being used and most of them are based on polymers [27]. Consequently, shape-memory polymeric carriers are useful to deliver drugs. These carriers allow the delivery of optimum concentration of drugs at the right time by releasing the drugs in response to an external stimulus. For example, the polymer chains of a carrier may expand as a result of the increase in temperature, thus enabling the drug to diffuse out from the carrier. For instance, thermo-sensitive polymers were used to solubilize some hydrophobic drugs, such as paclitaxel (an anticancer drug, which is not water soluble) and to produce excellent formulations with low solubility drugs as a carrier for drug deliver [18].

For the oral delivery of any drug, certain physiological aspects of the body and more specifically the gastrointestinal tract are important. For instance, an obvious pH change occurs along the gastrointestinal tract; however, more subtle changes occur within the various tissues in the body. Chronic wounds have a pH between 7.4 and 5.4, and cancer tissue has been reported to be acidic in extracellular environment. The same is true for the different cellular compartments in the body. Since variations in pH occur within the body, this property can be exploited for a direct response to a certain tissue or cellular compartment by a pH response shape-memory polymer. Also, local pH changes in response to specific substrates can be generated and exploited for modulating drug release [23].

Levotiroxine, a synthetic hormone, is usually administered orally to treat hypothyroidism and gout, since it stimulates the lipid metabolism and induces lipolysis. In this context, Azarbayjani et al. [91] have prepared smart polymeric nanofibers from a blend of Polyvinyl alcohol (PVA) and PNIPAAm. Through *in vitro* analysis, it was concluded that these nanofibers had delivered 97% of the drug, and were able to promote a sustained delivery of levotiroxine through diffusion and erosion mechanisms. This maintains the appropriate levels of the drug for a long time and promotes the accumulation of active substances in the superficial skin layers, and thereby decreases the side effects of systemic absorption. However, the nanofibers developed with only PVA had delivered 65% of the drug [18].

5.2 Cancer Therapy

Drug molecules that are conjugated to a polymer are usually inactive. These conjugates are therefore known as prodrugs. Use of prodrugs is particularly advantageous for cytotoxic drugs, e.g. the incorporation of a targeting system in cancer chemotherapy may avoid or minimize adverse reactions that may lead to nonspecific toxicity. However, only an optimum release of drug at the site of action gives these prodrugs the full advantage. The triggering, targeting and controlling of the drug release according to variations in pH can be achieved in two ways. Firstly, the extracellular tissue can be targeted: tumor tissue has an extracellular pH between 6.5 and 7.2, which is slightly lower than the normal pH of 7.4. Secondly, the lysosomes may be targeted: after cellular uptake, the drug conjugate reaches the lysosome which has a pH of 4.5–5.0, and in this case hydrolytic enzymes such as cathepsin B may also be utilized to release the drug. It has been established that nanocarriers can concentrate preferentially on solid tumors by virtue of the enhanced permeation and retention (EPR) mechanism. Once accumulated at the tumor site, they act as a local drug storage depending upon the makeup of the carrier. In this case, pH-sensitive nanoparticles that are capable of boosting the delivery of the anticancer drug (paclitaxel) to tumor cells are used. A representative poly(β-amino ester) (PbAE) with biodegradable and pH sensitive properties was used to formulate the delivery system. Chawla and Amiji [92] have developed and characterized nanoparticles using poly(ε-caprolactone) (PCL) to increase the local concentration of tamoxifen in estrogen receptor (ER)-positive breast cancer. Similarly, Lee et al. [93], reported poly(L-histidine)-co-PEG in combination with poly-L-lactic acid-co-poly(ethylene glycol) (PLLA-co-PEG) and adriamycin (ADR) as drug and was subjected for extracellular tumor targeting.

5.3 Tissue Engineering

It is an interdisciplinary field, and works on the principles of engineering and the life sciences towards the development of biological substitutes to restore or improve tissue function. It aims to regenerate or replace biologically damaged or diseased tissue or generate replacement organs for a wide range of medical conditions, such as heart diseases, diabetes, cirrohosis, osteoarthritis, spinal cord injury and disfiguration. This requires the use of a biocompatible material/scaffold, usually natural materials like proteins or synthetic polymers, which have the ability to convey both nutrients and growth factors to encapsulated cells. Synthetic polymers used in tissue engineering should be non-toxic, readily available and relatively inexpensive to produce. Thermo-responsive polymers in tissue engineering are commonly used in two situations: as substrates that enable the cell growth and proliferation and as injectable gels, for *in situ* formationof the scaffold. In the first application, the thermo responsive ability of the polymers is used to regulate the cells' attachment and detachment from a surface. In fact, in one study, the polymer surface was even reusable for repeated cell culture. The second application involves the encapsulation of cells in a 3-dimensional (3D) structure in the body. Compared to the *in vitro* formation of the construct, the *in situ* formation of cell/scaffold allows the delivery of encapsulated cells, nutrients and growth factors to the defects of any shape using minimally invasive techniques. The basic idea of *in situ* formation is shown in Fig. 15. Specifically, the thermo responsive polymer is mixed at room temperature with the cells and then injected into the body. Upon injection, due to the temperature increase (to 37°C) that is above the polymer's LCST, the polymer forms a physical gel. The cells are encapsulated within the 3D structure of the gel [27]. An extensive summary of polymeric scaffolds was presented by Agrawal and Ray [94], in which various scaffolds made of synthetic biodegradable polymers such as poly(lactic acid) (PLA), poly(ε-caprolactone) poly(glycolic acid) (PGA) and their copolymer (PLGA) were investigated. PLA is considered as a scaffold material for the support of cell growth; however, it was found that this material was not chemically reactive enough. To overcome this problem, many authors [95-100] proposed surface modification by introducing reactive groups. Based on these, many other polymeric scaffolds have also been developed for tissue engineering applications, such as breast reconstruction [95] as well as the replacement and regeneration of damaged bone [96] and cartilage [97,98]. For example, polyanhydrides used as successful scaffolds for orthopedic implants [99,100] and tyrosine-derived polycarbonates have produced interesting results when they used as scaffolds in tissue engineering. As demonstrated by Choueka et al. [101], these polymers exhibited an intimate contact with the bone. In general, tissue engineering requires synthetic materials to display carefully tailored bulk and surface properties, and are specifically designed to function as scaffolds to promote tissue growth and organization by providing a three-dimensional framework with favorable cell responses. More specifically, we believe that SMPs can provide new challenges by exhibiting the appropriate and required matching of their mechanical and micro-mechanical properties to those of hosting and surrounding cells and tissues.

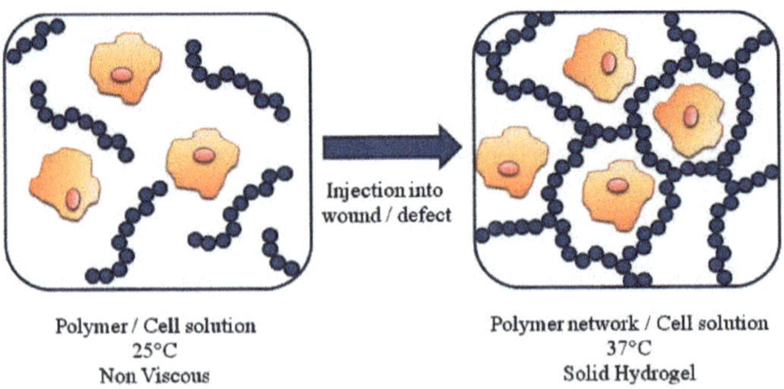

Fig. 15. In situ formation of a scaffold in tissue engineering.(Adopted with consent from Ref. [27], MDPI Publication 2011).

Phase sensitive polymers have major applications in the lysozyme release. Singh and Singh [102] reported a phase sensitive system, which was prepared by adding lysozyme to poly(D,L-lactic acid) (PLA)-triacetin solutions. Alhough, phase sensitive smart polymeric formulations have wide applications for the controlled release of several proteins one of the problems is the burst releasing during a few hours following injection into the body. This could be due to the time lag between the injection of the delivery system and the formation of the gel depot. Burst release of lysozyme and insulin was modulated in benzyl benzoate/benzyl alcohol solvent systems and polymer concentration. Increase in the proportion of the hydrophilic solvent led to an increase in the burst release as addition of hydrophilic solvent increases the affinity between the water influx rates. These polymeric systems are also known to display a high initial release of the drug followed by a more sustained release profile. The solvents used in phase sensitive smart polymeric systems should be nontoxic and biocompatible to avoid severe tissue irritation or necrosis at the site of administration [42].

Gene therapy aims at the treatment of many genetic diseases as it is a technique for correcting defective genes that are responsible for genetic disorders. Specifically, the delivery of the appropriate therapeutic gene (DNA) into the cells that will replace, repair or regulate the defective gene (that causes the disease) is a vital step for gene therapy. DNA is a negatively charged, hydrophilic molecule and its delivery into the nucleus of the cell which requires it to pass through the negatively charged and hydrophobic cell membrane is not feasible. Consequently, gene delivery carriers (also called vectors or vehicles) have been developed. Nature's way to carry genes is viruses and these were the first carriers used for gene delivery. However viruses have many disadvantages, the most severe of which is the immune response and for this reason, non-viral carriers were developed. Many of these are polymer-based, because polymers are cheaper and safer than viruses and also easier to tailor compared to other gene delivery carriers like liposome. Interestingly, thermo responsive polymers have been used to enhance the transfection efficiency by changing the temperature either during the complexion and/or during the incubation or transfection period [27]. In particular, Ward and Georgiou [103] reported polyethyleneimmine (PEI) with grafted PNIPAAm, chitosan grafted with PNIPAAm, linear and branched N-isopropylacrylamide (NIPAAm), dimethylaminoethylmethacrylate (DMAEMA) and PEI, and PEG with grafted PEI chains. The complexion and transfection temperature was changed to enhance the transfection efficiency. In other studies only the incubation or complexion temperatures were varied using random terpolymers of poly(NIPAAm-co-DMAEMA-co-butylmethacrylate) or PNIPAAm copolymers, while both complexion and incubation temperature were varied using a polyarginine polymer conjugated with PNIPAAm. In an interesting study Zhou et al. [104] used [poly(N,N-dimethylaminopropylacrylamide)-b-PNIPAAm]-star polymers as thermoresponsive polymers in a different manner. The complexion was undertaken at room temperature which was below the polymer LCST and then the complex was deposited on a surface above the LCST. On that surface, the cells were allowed to incubate at 37°C. The result was increased transfection to cells cultured on the surface compared to cells grown on a surface where DNA with no polymer was deposited.

5.4 Articles and Devices for Biomedical Applications

Shape-memory polymers find enormous applications for medical devices [105]. For example a laser-activated device employed for the removal of blood clots. Fig. 16 depicts the removal of blood clot from the blood vessel. The device is inserted by minimally invasive surgery into the blood vessel, and upon laser activation, the shape-memory material coils into its permanent shape, enabling the mechanical removal of the thrombus (blood clot). Another example of a medical challenge to be addressed is obesity, which is one of the major health problems in developed countries. In most cases, overeating is the key problem, which can be circumvented by methods for curbing appetite one of its solutions can be biodegradable intragastric implants that inflate after an approximate predetermined time and provide the patient with a feeling of safety after the intake of only a small amount of food [106,107]. Shape-memory foams have been proposed as a measuring device to survey the shape of a human ear canal, and prepared hearing aid which can be fitted properly [105]. The material is commercially available polyurethane foam with a T_g switching transition. The foam shows full recovery after 83% compression.

Another application of shape-memory polymers is in stents for the prevention of strokes [108]. Here, coils of a composite consisting of tantal and a polyetherurethane ($T_{trans} = T_g$; $T_g = 33°C$) have been studied. Tantal acted as radio-opaque filler for diagnostic detection. The filling does not affect the shape-recovery behavior but lowers T_g and the maximum recovery stress, whereas polyurethane-based shape-memory polymers are used *in-vivo*, having biocompatibility and cytotoxicity. In an *in vivo* study, commercially available aromatic shape-memory polyetherurethanes show similar biocompatibility and cytotoxicity compared with non shape-memory polyurethanes [109].

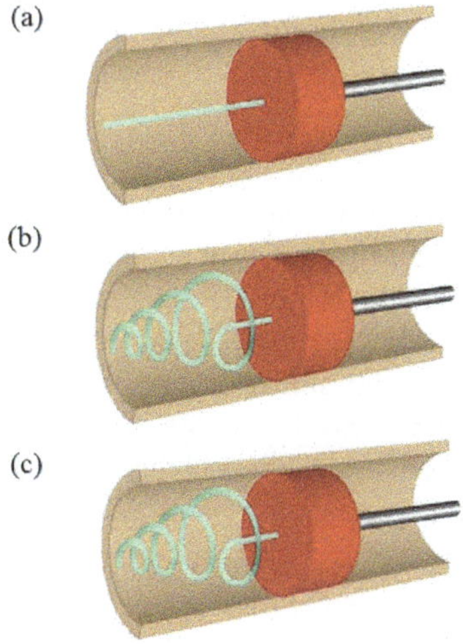

Fig. 16. Depiction of removal of a clot in a blood vessel using the laser-activated shape-memory polymer microactuator coupled to an optical fiber: (a) In its temporary straight rod form, the microactuator is delivered through a catheter distal to the blood clot; (b) The microactuator is then transformed into its permanent corkscrew form by laser heating; (c) The deployed microactuator is retracted to capture the thrombus. (Reproduced with consent from Ref. [110], The Optical Society of America Publication 2005).

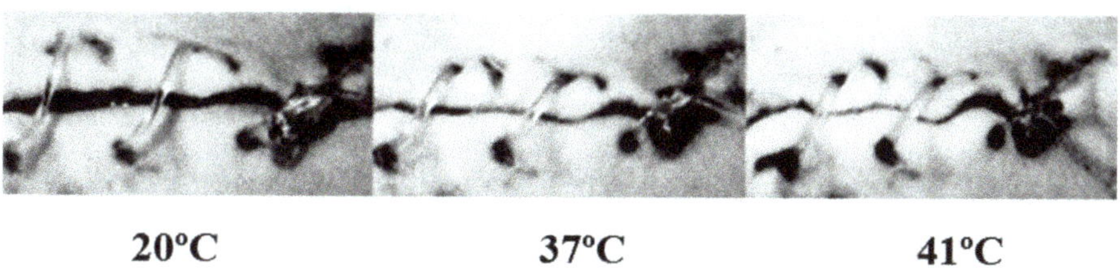

Fig. 17. Shows the shrinkage of the fiber with increasing temperature for early recovery of wounds using shape-memory sutures.(Reproduced with consent from Ref. [63], The Royal Society of Chemistry Publication 2007).

The polymeric compositions can be used to manufacture articles for their use in biomedical applications. For example, sutures (Fig. 17), orthodontic materials, bone screws, nails, plates, catheters, tubes, films, orthopedic braces, splints, tape for preparing casts, contact lenses, drug delivery devices, implants, thermal indicators etc. have been important applications.

Another preferred use is, in the area of catheters. A catheter can be rigid below body temperature for ease of insertion, but after warming to body temperature, can soften to provide comfort to the patient. The SMP compositions are preferably prepared from biocompatible and biodegradable polymers for most of the applications, because the biodegradable polymers degrade at a controlled rate depending on the composition and crosslinking of the polymer. Degradable polymeric implants eliminate the need for implant retrieval and can be used simultaneously to deliver therapeutic agents. The polymer compositions can be formed into the shape of an implant which can be implanted within the body to serve a mechanical function. The polymer compositions can be combined with fillers, reinforcement materials, radio imaging materials, excipients or other materials as needed for a particular implant application [40]. Biodegradable shape-memory block copolymers can be used as self-expandable stents due to transparency and high mechanical strength [111]. The SM polymeric gels can also be used in biomedical science depending on the nature of the applications [112]. Ahmad et al. [113] reported that polyurethane-based shape-memory polymer actuators can be used as artificial bandages where gradient pressure is required between the ankle and knee for the treatment of leg ulcers. Due to the high refractive index of the shape-memory gel (SMG), it was employed to develop a Plano-Convex Lens as an alternative to contact lenses, which are made of N,N'-dimethyl acrylamide. A promising application of polymeric hydrogels is in the biomedical field of soft contact lenses [112]. Biobased poly(propylene sebacate) (BPE) hasbeen used as shape-memory polymer with tunable switching temperature as *in vitro* fibroblast response and degradation demonstrated that the SMPs made of BPEs were potentially biocompatible and biodegradable, rendering them as potential candidates for biomedical devices such as different stents [114].

5.5 Adaptable Biological Devices

Nguyen et al. [115] fabricated tunable shrink-induced honeycomb microwell arrays for uniform embryoid bodies (EBs) from a single cell suspension. By printing various microwell patterns onto pre-stressed polystyrene sheets and employing heat induced shrinking, high aspect micromolds were generated. Notably, they achieved rounded bottom polydimethylsiloxane (PDMS) wells, which are not easily achievable with standard micro-fabrication methods, but using SMPs the achievement of spherical EBs are critical. Furthermore, by simply controlling the size of the microwells and the concentration of the cell suspension, they were able to control the initial size of the cell aggregate, thus influencing the lineage commitment. In addition, these microwells are easily adaptable and scalable to most standard well plates, and they are easily integrated into commercial liquid handling [3].

Shape-memory polymer based **Micro-Electro Mechanical (MEM)** systems have been proposed by several investigators for rapid and accurate blind placement and release of medical devices in the body and these are called as Bio-MEMs. Similarly, micro pumps for drug delivery have been designed from shape-memory alloys for fine control of fluid rates. These devices hold a lot of untapped potential due to relatively low stiffness of SMPs. This issue was triggered by the efforts to develop composite SMPs [21] that could provide the biocompatibility of polymers as well as the stiffness/conductivity of metals. Liu et al. [63] and Ratna and Kocsis [116] have provided a comprehensive review that cover a large number of biomedical applications based on SMPs as listed in Table 2.

6. Future Prospects, Challenges and Limitations

Shape-memory polymers play a key role in the area of biomedical applications [104]. Polymers bring numerous possibilities as novel SMPs such as multiple-SMPs, two-way SMPs, spatially controlled SMPs, electro-active SMPs, photoactive SMPs and self-healing SMPs. While the field is progressing rapidly and many exciting achievements have been made, some of the challenges are still awaiting to be overcome [39]. There are many smart polymers discovered currently for drug delivery, diagnostic and tissue engineering applications, but they still have to pass the challenges of a clinical trial. Many smart polymers have not undergone a clinical trial. Some of the reasons for this are the potential cellular toxicity of the smart polymers, especially for applications involving intracellular delivery. Further, though many of the smart polymeric carriers are most effective in reaching their cellular target, but they may not readily excreted via the kidney, after delivering the drug. This is because of the high molecular weight polymers used, and so the polymers which are not biodegradable and have high molecular weight tend to accumulate in the body. This is another important reason why these polymers did not undergo any clinical trial testing. Another reason is that an injection of a bio-molecular drug without a carrier is already a well-accepted method. But for the treatment of cancer a smart polymer drug delivery system can offer features such as special protection of the drug during circulation, physical targeting and/or active targeting cancer cells with ligands. But they may not be successful in entering actual clinical testing unless they offer a unique possibility that every last cancer cell can be killed.

Table 2. Potential biomedical applications of SMPs [39].

Applications	References
Aneurysm occlusion devices	[108,117-119]
Bio-Micro-electro mechanical (MEM) systems	[21]
Bone defect fillers	[120,121]
Cardiac valve repair	[41,122]
Cells manipulating and capturing	[123,124]
Clot removal devices	[14,125,126]
Controlled drug release	[127-134]
Embolic devices	[117]
Endoscopic surgery suture	[135-139]
Hair treatment	[140,141]
Kidney dialysis needles	[21,142]
Micro tweezers in medicine	[143]
Orthodontics, orthopedic cast, orthopedics morphix suture anchor	[21,144-152]
Pharyngeal mucosa reconstruction	[153]
Physiological monitoring	[154]
Post-surgical treatment of mitral insufficiency	[155,156]
Study of cell proliferation	[115,157-158]
Neuronal probe	[159]
Microfluidic devices	[160]
Treatment of esophageal stenosis	[161-163]
Vascular stents	[164-168]
Surgery inside living cells, Endoscopic surgery	[21,169]
Wound dressing	[170]
Self-healing	[171-180]
Neuroprosthetics	[21]

Driven by their usefulness in the area of self-actuating biomedical devices, there is increasing growth in research, focused on the development of biodegradable and/or drug-eluting SMPs. Challenges of SMPs still exist, mainly with how to incorporate the shape-memory properties, biodegradability and drug-eluting properties into one material to fulfill the multi-dimensional requirements in modulus, triggering/fixing condition, degree of fixing/recovery, biocompatibility/biodegradability, drug-eluting kinetics and many other aspects [63,90]. Given that the SMPs have superior strain recovery in comparison to shape-memory alloys (SMAs), the first area of improvement is to increase stress recovery. A reduction in strain recovery for increased stress recovery is desirable in a number of applications. A second area for material chemistry improvements would be to design chemical structures that can more easily and cheaply be manufactured. A thermoplastic SMP that could be easily molded and machined and then subsequently treated (e.g., heat, radiation, chemical) to lock (e.g., crosslink) into the primary shape would be desirable. A third development that would open up new application areas would be a material that could be remotely triggered to change from a bio-stable to a biodegradable chemical structure [21].

The application of SMPs in medicine is limited. This is also due to a lack of commercial suppliers of medical-grade materials. As more small companies drive worldwide regulatory agencies to approve specific devices, the attraction of profit is likely to initiate the commercial availability of medical grade SMPs. So, business is one of the hurdles for the early applications of SMPs in the biomedical field [21]. Finally, the future of SMPs and their limitation depend on the creativity of material chemists and engineers [63,90].

7. Conclusions

In this chapter, we have made an attempt to encapsulate the most recent advances of SMPs employed in biomedical applications, such as tissue engineering (regenerative medicine), drug delivery, gene therapy, cancer therapy, adaptable biological devices or articles and so on.

The unique attributes of shape-memory effect, biocompatibility and other properties of shape-memory polymers make them a worthy technology for numerous potential self-deployable medical products. All SMPs generally contain two networks, which are called the permanent networks and the temporary networks. The permanent networks are either covalent crosslinks or physical crosslinks, and the "bonds" exist above transition temperature (T_{trans}) determining permanent shape due to entropic elasticity. The temporary networks typically rely upon vitrification, crystallization, or some other physical interactions, such as hydrogen bonding and ionic bonding, and can be novel high strain SMP network (modified polyrotaxanes) with movable crosslinks which are different from conventional chemically crosslinked or physically crosslinked networks.

Generally, all the existing SMPs fall into four major categories according to the types of stimulus applied to induce the SME, namely thermo-responsive SMP, pH responsive SMP, dual stimuli responsive SMP (respond simultaneously to both stimuli) and other types of stimuli responsive SMPs include light, electric field and magnetic field induced, phase sensitive and water activated SMPs. Their structural approaches related to morphological changes have also been explained in detail, with a prime aim to explore biomedical applications.

SMPs have advantages include use of diverse external stimuli and trigger of the shape recovery; highly flexible programming; a broad range of structural designs; tunable properties; and they are well-suited to human senses/tissues. These properties are extremely important for biomedical applications. Some of the major drawbacks of SMPs include relatively low modulus that results in their lower recovery stress/forces; long response time; lower cycle life and relatively weak material stability. On the other hand, a small amount of crosslinking agent can significantly improve the mechanical strength and recovery stress of SMP, and their composites have further enhanced and broadened the applications of shape-memory polymers. The shape-memory effect in SMPs, as conceptually integrates different shape-memory protocols. They are: one-way SME; two-way SME; one way reversible; two-way reversible; triple-way shape-memory effect and multifunctional shape-memory effect.

In conclusion, the development of SMPs suitable for biomedical applications remains to be a challenging topic. We have suitably discussed the various processing techniques. The advantages of SMPs including drawbacks, applications pertaining to biomedical field, future prospects and challenges have been discussed adequately. At the end, we have also made efforts to provide valuable hints for an integrated solution of SMPs and their design to meet the demands of biomedical applications.

Acknowledgements

The authors wish to thank the UGC, New Delhi for providing the financial support under CPEPA [Grant No. 8-2/2008(NS/PE)] and UPE-FAR-I [Grant No.14-3/2012(NS/PE)] programs.

References

[1] K.S.S. Kumar, A.K. Khatwa, C.P.R. Nair, High transition temperature shape memory polymers (SMPs) by telechelic oligomer approach, Reactive Functional Polymers 78 (2014) 7-13. http://dx.doi.org/10.1016/j.reactfunctpolym.2014.02.008

[2] Y. Liu, Y. Li, H. Chen, G. Yang, X. Zheng, S. Zhou, Water-induced shape-memory poly(D,L-lactide)/microcrystalline cellulose composites, Carbohydrate Polymers 104 (2014) 101–108. http://dx.doi.org/10.1016/j.carbpol.2014.01.031

[3] J. Hu, Y. Zhu, H. Huang, J. Lu, Recent advances in shape–memory polymers: structure, mechanism, functionality, modeling and applications, Progress in Polymer Science 37 (2012) 1720–1763. http://dx.doi.org/10.1016/j.progpolymsci.2012.06.001

[4] J. Hu, S. Chen, A review of actively moving polymers in textile applications, Journal of Materials Chemistry 20 (2010) 3346–3355. http://dx.doi.org/10.1039/b922872a

[5] M.R. Aguilar, C. Elvira, A. Gallardo, B. Vázquez, J.S. Román, Topics in tissue engineering, in: N. Ashammakhi, R. Reis, E. Chiellini (Eds.), Smart Polymers and their Applications as Biomaterials, Spain, 2007, pp. 1-27.

[6] J. Xu, J. Song, Thermal responsive shape memory polymers for biomedical applications, in: R.F. Rezai (Ed.), Biomedical Engineering–Frontiers and Challenges, In Tech CC BY-NC-SA 3.0 license, USA, 2011, pp. 125-142. http://dx.doi.org/10.5772/19256

[7] B. Yan, S. Gua, Y. Zhang, Polylactide-based thermoplastic shape memory polymer nanocomposites, European Polymer Journal 49 (2013) 366-378. http://dx.doi.org/10.1016/j.eurpolymj.2012.09.026

[8] Inamuddin, R.K. Jain, S. Hussain, M. Naushad, Poly (3,4-ethylenedioxythiophene): polystyrene sulfonate zirconium(IV) phosphate compsiteionomeric membrane for artifical muscle application.RSC Adv 5 (2015) 84526-84534.

[9] S. Pandini, F. Baldi, K. Paderni, M. Messori, M. Toselli, F. Pilati, A. Gianoncelli, M. Brisotto, E. Bontempi, T. Riccò, One-way and two-way shape memory behaviour of semi-crystalline networks based on sol-gel cross-linked poly(e-caprolactone), Polymer 54 (2013) 4253-4265. http://dx.doi.org/10.1016/j.polymer.2013.06.016

[10] J.Y. Sun, X. Zhao, W.R.K. Illeperuma, O. Chaudhuri, K.H. Oh, D.J. Mooney, J.J. Vlassak, Z. Suo, Highly stretchable and tough hydrogels, Nature 489 (2012) 133-136. http://dx.doi.org/10.1038/nature11409

[11] Y. Wang, X. Li, Y. Pan, Z. Zheng, X. Ding, Y. Peng, High-strain shape memory polymers with movable cross-links constructed by interlocked slide-ring structure, Royal Society of Chemistry Advances 4 (2014) 17156 -17160. http://dx.doi.org/10.1039/c4ra00165f

[12] D. Pathania, G. Sharma, M. Naushad, V. Priya, A biopolymerbasedhybridcationexchanger pectincerium(IV) iodate:Synthesis, characterization andanalyticalapplications. Des & Water Treat 57 (2016) 468-475.

[13] S. Hayashi, S. Kondo, P. Kapadia, E. Ushioda, Room-temperature-functional shape-memory polymers, Plastics Engineering 51 (1995) 29-31.

[14] D.J. Maitland, M.F. Metzger, D. Schumann, A. Lee, T.S. Wilson, Photothermal properties of shape memory polymer micro-actuators for treating stroke, Lasers in Surgery and Medicine 30 (2002) 1-11. http://dx.doi.org/10.1002/lsm.10007

[15] W.M. Huang, B. Yang, Y. Zhao, Z. Ding, Thermo-moisture responsive polyurethane shape-memory polymer and composites: A review, Journal of Material Chemistry 20 (2010) 3367–3381. http://dx.doi.org/10.1039/b922943d

[16] A. Gandhi, A. Paul, S.O. Sen, K.K. Sen, Studies on thermoresponsive polymers: Phase behaviour, drug delivery and biomedical applications, Asian Journal of Pharmaceutical sciences 10 (2015) 99-107. http://dx.doi.org/10.1016/j.ajps.2014.08.010

[17] R. Xiao, J. Guo, D.L. Safranski, T.D. Nguyen, Solvent-driven temperature memory and multiple shape memory effects, Soft Matter 11 (2015) 3977-3985. http://dx.doi.org/10.1039/C5SM00543D

[18] H. Almeida, M. Helena Amaral, P. Lobao, Temperature and pH stimuli-responsive polymers and their applications in controlled and self-regulated drug delivery, Journal of Applied Pharmaceutical Science 02 (2012) 1-10.

[19] J. Leng, X. Lan, Y. Liu, S. Du, Shape-memory polymers and their composites: stimulus methods and applications, Progress in Materials Science 56 (2011) 1077–1135. http://dx.doi.org/10.1016/j.pmatsci.2011.03.001

[20] F.E. Feninat, G. Laroche, M. Fiset, D. Mantovani, Shape memory materials for biomedical applications, Advanced Engineering Materials 4 (2002) 91-104. http://dx.doi.org/10.1002/1527-2648(200203)4:3<91::AID-ADEM91>3.0.CO;2-B

[21] W. Small, P. Singhal, T.S. Wilsona, D.J. Maitland, Biomedical applications of thermally activated shape memory polymers, Journal of Material Chemistry 20 (2010) 3356–3366. http://dx.doi.org/10.1039/b923717h

[22] X. Luo, P.T. Mather, Triple-shape polymeric composites (TSPCs), Advanced Functional Materials 20 (2010) 2649–2656. http://dx.doi.org/10.1002/adfm.201000052

[23] P. Bawa, V. Pillay, Y.E. Choonara, L.C. du Toit, Stimuli-responsive polymers and their applications in drug delivery, Biomedical Materials 4 (2009) 1-15. http://dx.doi.org/10.1088/1748-6041/4/2/022001

[24] M. Lemanowicz, A. Gierczycki, W. Kuznik, R. Sancewicz, P. Imiela, Determination of lower critical solution temperature of thermosensitive flocculants, Minerals Engineering 69 (2014) 170–176. http://dx.doi.org/10.1016/j.mineng.2014.07.022

[25] J. Li, C.L. Lewis, D.L. Chen, M. Anthamatten, Dynamic mechanical behavior of photo-cross-linked shape-memory elastomers, Macromolecules 44 (2011) 5336–5343. http://dx.doi.org/10.1021/ma2004019

[26] T.Y. Liu, S.H. Hu, D.M. Liu, S.Y. Chen, I.W. Chen, Biomedical nanoparticle carriers with combined thermal and magnetic responses, Nano Today 4 (2009) 52-65. http://dx.doi.org/10.1016/j.nantod.2008.10.011

[27] M.A. Ward, T.K. Georgiou, Thermoresponsive polymers for biomedical applications, Polymers 3 (2011) 1215-1242. http://dx.doi.org/10.3390/polym3031215

[28] R.P. Shaikh, V. Pillay, Y.E. Choonara, L.C. Toit, V.M.K. Ndesendo, P. Bawa, S. Cooppan, A review of multi-responsive membranous systems for rate-modulated drug delivery, AAPS PharmSciTech 2 (2010) 441-459. http://dx.doi.org/10.1208/s12249-010-9403-2

[29] Y. Hiruta, T. Funatsu, M. Matsuura, J. Wang, E. Ayano, H. Kanazawa, pH/temperature-responsive fluorescence polymer probe with pH-controlled cellular uptake, Sensors and Actuators B: Chemical 207 (2015) 724–731. http://dx.doi.org/10.1016/j.snb.2014.10.065

[30] M.F. Leung, J. Zhu, P. Li, F.W. Harris, Novel synthesis and properties of smart core-shell microgels, Macromolecular Symposia 226 (2005) 177-185. http://dx.doi.org/10.1002/masy.200550817

[31] D. Kuckling, H.J.P. Adler, K.F. Arndt, L. Ling, W.D. Habicher. Temperature and pH dependent solubility of novel poly(N-isopropylacrylamide) copolymers, Macromolecular Chemistry and Physics 201 (2000) 273-280. http://dx.doi.org/10.1002/(SICI)1521-3935(20000201)201:2<273::AID-MACP273>3.0.CO;2-E

[32] G. Sun, X.Z. Zhang, C.C. Chu, Formulation and characterization of chitosan-based hydrogel films having both temperature and pH sensitivity, Journal of Material Science: Materials in Medicine 18 (2007) 1563–1577. http://dx.doi.org/10.1007/s10856-007-3030-9

[33] X. Wu, W.M. Huang, Y. Zhao, Z. Ding, C. Tang, J. Zhang, Mechanisms of the shape memory effect in polymeric materials, Polymer 5 (2013) 1169-1202. http://dx.doi.org/10.3390/polym5041169

[34] A. Lendlein, H. Jiang, O. Junger, R. Lange, Light-induced shape-memory polymers, Nature 434 (2005) 879-882. http://dx.doi.org/10.1038/nature03496

[35] G.J. Berg, M.K. McBride, C. Wang, C.N. Bowman, New directions in the chemistry of shape memory polymers, Polymer 55 (2014) 5849-5872. http://dx.doi.org/10.1016/j.polymer.2014.07.052

[36] H. Finkelmann, E. Nishikawa, G.G. Pereira, M. Warner, A new opto-mechanical effect in solids, Physical Review letters 87 (2001) 015501-1- 015501-4. http://dx.doi.org/10.1103/PhysRevLett.87.015501

[37] S. Chatani, C. J. Kloxin, C.N. Bowman, The power of light in polymer science: photochemical processes to maniplulate polymer formation, structure, and properties, Polymer Chemistry 5 (2014) 2187-2201. http://dx.doi.org/10.1039/C3PY01334K

[38] S.Q. Wang, D. Kaneko, M. Okajima, K. Yasaki, S. Tateyama, T. Kaneko, Hyperbranched polycoumarates with photofunctional multiple shape memory, Smart Materials 52 (2013) 11143-11148.

[39] H. Meng, G. Li, A review of stimuli-responsive shape memory polymer composites, Polymer 54 (2013) 2199-2221. http://dx.doi.org/10.1016/j.polymer.2013.02.023

[40] R.S. Langer, A. Lendlein, A. Schmidt and H. Grablowitz, U.S. Patent 6,160,084. A. (2000)

[41] R. Mohr, K. Kratz, T. Weigel, M.L. Gabor, M. Moneke, A. Lendlein, Initiation of shape-memory effect by inductive heating of magnetic nanoparticles in

thermoplastic polymers, Proceedings of the National Academy of Sciences 103 (2006) 3540-3445. http://dx.doi.org/10.1073/pnas.0600079103

[42] A. Mahajan, G. Aggarwal, Smart polymers: Innovations in novel drug delivery, International Journal of Drug Development and Research 3 (2011) 16-30.

[43] W.M. Huang, B. Yang, L. An, C. Li, Y.S. Chan, Water-driven programmable polyurethane shape memory polymer: Demonstration and mechanism, Applied Physics Letters 86 (2005) 114105-1-114105-3. http://dx.doi.org/10.1063/1.1880448

[44] M.C. Chen, H.W. Tsai, Y. Chang, W.Y. Lai, F.L. Mi, C.T. Liu, H.S. Wong, H.W. Sung, Rapidly self-expandable polymeric stents with a shape-memory property, Biomacromolecules 8 (2007) 2774-2780. http://dx.doi.org/10.1021/bm7004615

[45] H. Luo, J. Hu, Y. Zhu, S. Zhang, Y. Fan, G. Ye, Achieving shape memory: Reversible behaviors of cellulose-PU blends in wet–dry cycles, Journal of Applied Polymer Science 125 (2012) 657-665. http://dx.doi.org/10.1002/app.36292

[46] H. Luo, J. Hu, Y. Zhu, Tunable shape recovery of polymeric nano-composites, Materials Letters 65 (2011) 3583-3585. http://dx.doi.org/10.1016/j.matlet.2011.08.006

[47] Y. Liu, H. Lv, X. Lan, J. Leng, S. Du, Review of electro-active shape-memory polymer composite, Composites Science and Technology 69 (2009) 2064–2068. http://dx.doi.org/10.1016/j.compscitech.2008.08.016

[48] E. Smela, Conjugated polymer actuators for biomedical applications, Advanced Materials 15 (2003) 481-494. http://dx.doi.org/10.1002/adma.200390113

[49] G.G. Wallace, G.M. Spinks, P.R. Teasdale, Conductive electroactive polymers: Intelligent material systems, Technomic Publication Company, Lancaster, 1997.

[50] H.H. Qin, P.T. Mather, Combined one-way and two-way shape memory in a glass-forming nematic network, Macromolecules 42 (2009) 273–280. http://dx.doi.org/10.1021/ma8022926

[51] J. Zotzmann, M. Behl, D. Hofmann, A. Lendlein, Reversible triple-shape effect of polymer networks containing polypentadecalactone and poly(epsilon-caprolactone)-segments, Advanced Materials 22 (2010) 3424–3429. http://dx.doi.org/10.1002/adma.200904202

[52] J.K. Kocsis, S. Kéki, Biodegradable polyester-based shape memory polymers: Concepts of (supra) molecular architecturing, Express Polymer Letters 8 (2014) 397–412. http://dx.doi.org/10.3144/expresspolymlett.2014.44

[53] M. Behl, M.Y. Razzaq, A. Lendlein, Multifunctional shape-memory polymers, Advanced Materials 22 (2010) 3388–3410. http://dx.doi.org/10.1002/adma.200904447

[54] J. Zhou, S.A. Turner, S.M. Brosnan, Q. Li, J.M.Y. Carrillo, D. Nykypanchuk, O. Gang, V.S. Ashby, A.V. Dobrynin, S.S. Sheiko, Shape shifting: reversible shape memory in semicrystalline elastomers, Macromolecules 47 (2014) 1768-1776. http://dx.doi.org/10.1021/ma4023185

[55] T. Xie, X. Xiao, Y.T. Cheng, Revealing triple-shape memory effect by polymer bilayers, Macromolecular Rapid Communications 30 (2009) 1823–1827. http://dx.doi.org/10.1002/marc.200900409

[56] C. Samuel, S. Barrau, J.M. Lefebvre, J.M. Raquez, P. Dubois, Designing multiple-shape memory polymers with miscible polymer blends: evidence and origins of a triple-shape memory effect for miscible PLLA/PMMA blends, Macromolecules 47 (2014) 6791-6803. http://dx.doi.org/10.1021/ma500846x

[57] H. Jeon, P. Mather, T. Haddad, Shape memory and nanostructure in poly(norbornyl-POSS) copolymers, Polymer International 49 (2000) 453-457. http://dx.doi.org/10.1002/(SICI)1097-0126(200005)49:5<453::AID-PI332>3.0.CO;2-H

[58] K. Matyjaszewski, N.V. Tsarevsky, Nanostructured functional materials prepared by atom transfer radical polymerization, Nature Chemistry 1 (2009) 276-288. http://dx.doi.org/10.1038/nchem.257

[59] L. Schadler, Nanocomposites: Model interfaces, Nature Materials 6 (2007) 257-258. http://dx.doi.org/10.1038/nmat1873

[60] A.C. Balazs, T. Emrick, T.P. Russell, Nanoparticle polymer composites: Where two small worlds meet, Science 314 (2006) 1107-1110. http://dx.doi.org/10.1126/science.1130557

[61] Z. He, N. Satarkar, T. Xie, Y.-T. Cheng, J.Z. Hilt, Remote controlled multishape polymer nanocomposites with selective radiofrequency actuations, Advanced Materials 23 (2011) 3192–3196. http://dx.doi.org/10.1002/adma.201100646

[62] A. Lendlein, S. Kelch, Shape-memory polymers, Angewandte Chemie International Edition 41 (2002) 2034-2057. http://dx.doi.org/10.1002/1521-3773(20020617)41:12<2034::AID-ANIE2034>3.0.CO;2-M

[63] C. Liu, H. Qin, P.T. Mather, Review of progress in shape-memory polymers, Journal of Material Chemistry 17 (2007) 1543–1558. http://dx.doi.org/10.1039/b615954k

[64] B.S. Lee, B.C. Chun, Y.C. Chung, K.II Sul, J.W. Cho, Structure and thermomechanical properties of polyurethane block copolymers with shape memory effect, Macromolecules 34 (2001) 6431–6437. http://dx.doi.org/10.1021/ma001842l

[65] A.A. Basfar, S. Lotfy, Radiation-crosslinking of shape memory polymers based on poly(vinyl alcohol) in the presence of carbon nanotubes, Radiation Physics and Chemistry 106 (2015) 376–384. http://dx.doi.org/10.1016/j.radphyschem.2014.08.024

[66] O. M'barki, A. Hanafia, D. Bouver, C. Faur, R. Seacousse, U. Delabre, C. Blot, P. Guenoun, A. Deratani, D. Quemener, C. Pochat-Bohatier, Greener method to prepare porous polymer membrane by combining thermally induced phase separation and crosslinking of poly(vinyl alcohol) in water, Journal of Membrane Science 458 (2014) 225–235. http://dx.doi.org/10.1016/j.memsci.2013.12.013

[67] Y. Chen, X.L. Tang, B.T. Chen, G. Qiu. Low temperature plasma vapor treatment of thermo-sensitive poly(N-isopropylacrylamide) and its application, Applied Surface Science 268 (2013) 332– 336. http://dx.doi.org/10.1016/j.apsusc.2012.12.089

[68] D. Schmaljohann, Thermo and pH-responsive polymers in drug delivery, Advanced Drug Delivery Reviews 58 (2006) 1655–1670. http://dx.doi.org/10.1016/j.addr.2006.09.020

[69] C.A. Lorenzo, A. Concheiro, A.S. Dubovik, N.V. Grinberg, T.V. Burova, V.Y. Grinberg, Temperature-sensitive chitosan-poly(N-isopropylacrylamide) interpenetrated networks with enhanced loading capacity and controlled release properties, Journal of Controlled Release 102 (2005) 629-641. http://dx.doi.org/10.1016/j.jconrel.2004.10.021

[70] J. Zhang, N.A. Peppas, Synthesis and characterization of pH- and temperature-sensitive poly(methacrylicacid)/poly(N-isopropylacrylamide) interpenetrating polymeric networks, Macromolecules 33 (2000) 102-107. http://dx.doi.org/10.1021/ma991398q

[71] J. Ma, Y. Xu, B. Fan, B. Liang, Macromolecular nanotechnology preparation and characterization of sodium carboxymethylcellulose/poly(N-

isopropylacrylamide)/clay semi-IPN nanocomposite hydrogels, European Polymer Journal 43 (2007) 2221–2228. http://dx.doi.org/10.1016/j.eurpolymj.2007.02.026

[72] B.L. Guo, Q.Y. Gao, Preparation and properties of a pH/temperature-responsive carboxymethyl chitosan/poly(N-isopropylacrylamide) semi-IPN hydrogel for oral delivery of drugs, Carbohydrate Research 342 (2007) 2416–2422. http://dx.doi.org/10.1016/j.carres.2007.07.007

[73] L. Peponi, I.N. Baena, A. Sonseca, E. Gimenez, A.M. Fernandez, J.M. Kennya, Biobased polymers and related materials, synthesis and characterization of PCL–PLLA polyurethane with shape memory behavior, European Polymer Journal 49 (2013) 893–903. http://dx.doi.org/10.1016/j.eurpolymj.2012.11.001

[74] K. Suchao-in, S. Chirachanchai, "Grafting to" as a novel and simple approach for triple-shape memory polymers, Applied Materials & Interfaces 5 (2013) 6850-6853J. http://dx.doi.org/10.1021/am402214j

[75] Hu, G. Zhang, Z. Ge, S. Liu, Stimuli-responsive tertiary amine methacrylate-based block copolymers: Synthesis, supramolecular self-assembly and functional applications, Progress in Polymer Science 39 (2014) 1096–1143. http://dx.doi.org/10.1016/j.progpolymsci.2013.10.006

[76] Y. Luo, Y. Guo, X. Gao, B.-G. Li, T. Xie, A general approach towards thermoplastic multishape-memory polymers via sequence structure design, Advanced Materials 25 (2013) 743-748. http://dx.doi.org/10.1002/adma.201202884

[77] C.Y. Chiang, C.C. Chu, Synthesis of photoresponsive hybrid alginate hydrogel with photo-controlled release behavior, Carbohydrate Polymers 119 (2015) 18–25. http://dx.doi.org/10.1016/j.carbpol.2014.11.043

[78] Q. Meng, J. Hu, A review of shape memory polymer composites and blends, Composites: Part A 40 (2009) 1661–1672. http://dx.doi.org/10.1016/j.compositesa.2009.08.011

[79] H.M. Jeong, J.H. Song, S.Y. Lee, B.K. Kim, Miscibility and shape memory property of poly(vinyl chloride)/thermoplastic polyurethane blends, Journal of Materials Science 36 (2001) 5457-5463. http://dx.doi.org/10.1023/A:1012481631570

[80] W. Zhang, L. Chen, Y. Zhang, Surprising shape-memory effect of polylactide resulted from toughening by polyamide elastomer, Polymer 50 (2009) 1311–1315. http://dx.doi.org/10.1016/j.polymer.2009.01.032

[81] J.J. Song, I. Srivastava, J. Kowalski, H.E. Naguib, Fabrication and characterization of a foamed polylactic acid (PLA)/ thermoplastic polyurethane (TPU) shape memory polymer (SMP) blend for biomedical and clinical applications, Proc. SPIE: Behavior and Mechanics of Multifunctional Materials and Composites 9058 (2014) 90580B

[82] P.T. Mather, C. Liu and C.J. Campo, U. S. Patent 7,371,799. (2007)

[83] I.S. Kolesov, H.J. Radusch, Multiple shape-memory behavior and thermal-mechanicalproperties of peroxide cross-linked blends of linear and short-chain branched polyethylenes, Express Polymer Letters 2 (2008) 461-473. http://dx.doi.org/10.3144/expresspolymlett.2008.56

[84] J.M. Cuevas, R. Rubio, L. German, J.M. Laza, J.L. Vilas, M. Rodriguez, L.M. Leon, Triple-shape memory effect of covalently crosslinked polyalkenamer based semicrystalline polymer blends, Soft Matter 8 (2012) 4928-4935. http://dx.doi.org/10.1039/c2sm07481h

[85] M. Souri, Y.C. Lu, A. Erol, S.S. Pulla, H.E. Karaca, Characterization of unconstraint and constraint shape recoveries of an epoxy based shape memory polymer, Polymer Testing 41 (2015) 231-238. http://dx.doi.org/10.1016/j.polymertesting.2014.11.006

[86] T. Xie, I.A. Rousseau, Facile tailoring of thermal transition temperatures of epoxy shape memory polymers, Polymer 50 (2009) 1852-1856. http://dx.doi.org/10.1016/j.polymer.2009.02.035

[87] I.A. Rousseau, T. Xie, Shape memory epoxy: Composition, structure, properties and shape memory performances, Journal of Materials Chemistry 20 (2010) 3431–3441. http://dx.doi.org/10.1039/b923394f

[88] L. Ionov, V. Bocharova, S. Diez, Biotemplated synthesis of stimuli-responsive nanopatterned polymer brushes on microtubules, Soft Matter 5 (2009) 67–71. http://dx.doi.org/10.1039/B813295J

[89] H. Kitano, H. Kago, K. Matsuura, Temperature-responsive polymer brush constructed on a colloidal gold monolayer, Journal of Colloid and Interface Science 331 (2009) 343–350. http://dx.doi.org/10.1016/j.jcis.2008.11.058

[90] A.S. Hoffman, Stimuli-responsive polymers: Biomedical applications and challenges for clinical translation, Advanced Drug Delivery Reviews 65 (2013) 10–16. http://dx.doi.org/10.1016/j.addr.2012.11.004

[91] A.F. Azarbayjani, J.R. Venugopal, S. Ramakrishna, P.F.C. Lim, Y.W. Chan, S.Y. Chan, Smart polymeric nanofibers for topical delivery of Levothyroxine, Journal of Pharmacy & Pharmaceutical Sciences 13 (2010) 400-410. http://dx.doi.org/10.18433/J3TS3G

[92] J.S. Chawla, M.M. Amiji, Biodegradable poly(e-caprolactone) nanoparticles for tumor-targeted delivery of tamoxifen, International Journal of Pharmaceutics 249 (2002) 127-138. http://dx.doi.org/10.1016/S0378-5173(02)00483-0

[93] E.S. Lee, K. Na, Y.H. Bae, Polymeric micelle for tumor pH and folate-mediated targeting Journal of Controlled Release 91 (2003) 103–113. http://dx.doi.org/10.1016/S0168-3659(03)00239-6

[94] C.M. Agrawal, R.B. Ray, Biodegradable polymeric scaffolds for musculoskeletal tissue engineering, Journal of Biomedical Materials Research 55 (2001) 141-150. http://dx.doi.org/10.1002/1097-4636(200105)55:2<141::AID-JBM1000>3.0.CO;2-J

[95] C.W. Patrick Jr., P.B. Chauvin, G.L. Robb, Tissue engineered adipose, in : C. W. Patrick, A.G. Mikos, L.V. McIntire (Eds.), Frontiers in Tissue Engineering, Elsevier Science, New York, 1998, pp. 369-382. http://dx.doi.org/10.1016/B978-008042689-1/50021-2

[96] C.M. Agrawal, J. Bert, J.D. Heckman, B.D. Boyan, Protein release kinetics of a biodegradable implant for fracture non-unions, Biomaterials 16 (1995) 1255-1260. http://dx.doi.org/10.1016/0142-9612(95)98133-Y

[97] K.A. Athanasiou, D. Korvick, R.C. Schenck, Biodegradable implants for the treatment of osteochondral defects in a goat model, Tissue Engineering 3 (1997) 363-373. http://dx.doi.org/10.1089/ten.1997.3.363

[98] G. V. Navokovic, I. Martin, B. Obradovic, S. Treppo, A.J. Grodzinsky, R. Langer, L.E. Freed, Bioreactor cultivation conditions modulate the composition and mechanical properties of tissue-engineered cartilageJournal of Orthopaedic Research 17 (1999) 130-138.

[99] S.E. Ibim, K.E. Uhrich, R. Bronson, S.F. El-Amin, R. Langer, C.T. Laurencin, Poly(anhydride-co-imides): in vivo biocompatibility in a rat model, Biomaterials 19 (1998) 941-951. http://dx.doi.org/10.1016/S0142-9612(98)00019-2

[100] S.E. Ibim, K.E. Uhrich, M. Attawia, V.R. Shastri, S.F. El-Amin, R. Bronson, R. Langer, C.T. Laurencin, Preliminary in vivo report on the osteocompatibility of poly(anhydride-co-imides) evaluated in a tibial model, Journal of Biomedical

Material Research 43 (1998) 374-379. http://dx.doi.org/10.1002/(SICI)1097-4636(199824)43:4<374::AID-JBM5>3.0.CO;2-5

[101] J. Choueka, J.L. Koval, H. Alexander, K.S. James, L.A. Hooper, J. Kohn, Canine bone response to tyrosine-derived polycarbonates and poly(L-lactic acid), Journal of Biomedical Material Research 31 (1996) 35-41. http://dx.doi.org/10.1002/(SICI)1097-4636(199605)31:1<35::AID-JBM5>3.0.CO;2-R

[102] S. Singh, J. Singh, Controlled release of a model protein lyzozyme from phase sensitive smartpolymer systems, International Journal of Pharmaceutics 271 (2004) 189-196. http://dx.doi.org/10.1016/j.ijpharm.2003.11.010

[103] M.A. Ward, T.K. Georgiou, Thermoresponsive terpolymers based on methacrylate monomers: Effect of architecture and composition, Journal of Polymer Science Part A 48 (2010) 775-783. http://dx.doi.org/10.1002/pola.23825

[104] Y.M. Zhou, A. Ishikawa, R. Okahashi, K. Uchida, Y. Nemoto, M. Nakayama, Y. Nakayama, Deposition transfection technology using a DNA complex with a thermoresponsive cationic star polymer, Journal of Controlled Release 123 (2007) 239-246. http://dx.doi.org/10.1016/j.jconrel.2007.08.026

[105] M. Behl, A. Lendlein, Shape-memory polymers, Materials Today 10 (2007) 20-28. http://dx.doi.org/10.1016/S1369-7021(07)70047-0

[106] A. Lendlein and R.S. Langer, W.O. 2004073690 A1. (2004)

[107] D. Marco, W.O. 2006092789 A2. (2006)

[108] J.M. Hampikian, B.C. Heaton, F.C. Tong, Z. Zhang, C.P. Wong, Mechanical and radiographic properties of a shape memory polymer composite for intracranial aneurysm, Materials Science and Engineering C 26 (2006) 1373-1379. http://dx.doi.org/10.1016/j.msec.2005.08.026

[109] S. Farèet, V. Valtulina, P. Petrini, E. Alessandrini, G. Pietrocola, M.C. Tanzi, P. Speziale, L. Visai, In vitro interaction of human fibroblasts and platelets with a shape-memory polyurethane, Journal of Biomedical Materials Research Part A 73 (2005) 1-11.

[110] W. Small IV, T.S. Wilson, W. J. Benett, J.M. Loge, D.J. Maitland, Laser-activated shape memory polymer intravascular thrombectomy device, Optics Express 13 (2005) 8204-8213 http://dx.doi.org/10.1364/OPEX.13.008204

[111] L. Xue, S. Dai, Z. Li, Biodegradable shape-memory block co-polymers for fast self-expandable stents, Biomaterials 31 (2010) 8132-8140. http://dx.doi.org/10.1016/j.biomaterials.2010.07.043

[112] M.H. Kabir, T. Hazama, Y. Watanabe, J. Gong, K. Murase, T. Sunada, H. Furukawa, Smart hydrogel with shape memory for biomedical applications, Journal of the Taiwan Institute of Chemical Engineers 45 (2014) 3134–3138. http://dx.doi.org/10.1016/j.jtice.2014.09.035

[113] M. Ahmad, J. Luo, M. Mohsen, Feasibility study of polyurethane shape-memory polymer actuators for pressure bandage application, Science and Technology of Advanced Materials 13 (2012) 015006-015013. http://dx.doi.org/10.1088/1468-6996/13/1/015006

[114] B. Guo, Y. Chen, Y. Lei, L. Zhang, W.Y. Zhou, A.B.M. Rabie, J. Zhao, Biobased poly(propylene sebacate) as shape memory polymer with tunable switching temperature for potential biomedical applications, Biomacromolecules 12 (2011) 1312–1321. http://dx.doi.org/10.1021/bm2000378

[115] D. Nguyen, S. Sa, J.D. Pegan, B. Rich, G.X. Xiang, K.E. McCloskey, J.O. Manilay, M. Khine, Tunable shrink-induced honeycomb microwell arrays for uniform embryoid bodies, Lab on a Chip 9 (2009) 3338–3344. http://dx.doi.org/10.1039/b914091c

[116] D. Ratna, J. K. Kocsis, Recent advances in shape memory polymers and composites: a review, Journal of Materials Science 43 (2008) 254–269. http://dx.doi.org/10.1007/s10853-007-2176-7

[117] D.J. Maitland, W. Small, J.M. Ortega, P.R. Buckely, J. Rodriguez, J. Hartman, Prototype laser-activated shape memory polymer foam device for embolic treatment of aneurysms, Journal of Biomedical Optics 12 (2007) 030504-1-030504-3. http://dx.doi.org/10.1117/1.2743983

[118] J.M. Ortega, J. Hartman, J. Rodriguez, D.J. Maitland, Post-treatment hemodynamics of a basilar aneurysm and bifurcation, Annals of Biomedical Engineering 36 (2008) 1531-1546. http://dx.doi.org/10.1007/s10439-008-9535-7

[119] J. Rodriguez, Y. Yu, M. Miller, T. Wilson, J. Hartman, F. Clubb, Opacification of shape memory polymer foam designed for treatment of intracranial aneurysms, Annals of Biomedical Engineering 40 (2012) 883-897. http://dx.doi.org/10.1007/s10439-011-0468-1

[120] I.A. Rousseau, E.J. Berger, J.N. Owens and H.G. Kia, U.S. Patent 0,249,682 A1. (2010)

[121] J. Cui, K. Kratz, A. Lendlein, Adjusting shape-memory properties of amorphous polyether urethanes and radio-opaque composites thereof by variation of physical parameters during programming, Smart Materials and Structures 19 (2010) 065019-065029. http://dx.doi.org/10.1088/0964-1726/19/6/065019

[122] P.R. Buckley, G.H. McKinley, T.S. Wilson, W. Small IV, W. J. Benett, J.P. Bearinger, M.W. McElfresh, D.J. Maitland, IEEE Transactions on Biomedical Engineering 53 (2006) 2075-2083. http://dx.doi.org/10.1109/TBME.2006.877113

[123] L. Song, W. Hu, G. Wang, G. Niu, H. Zhang, H. Cao, K. Wang, H. Yang, S. Zhu, Tailored (meth)acrylate shape-memory polymer networks for ophthalmic applications, Macromolecular Bioscience 10 (2010) 1194-1202. http://dx.doi.org/10.1002/mabi.201000028

[124] B. Simpson, G. Nunnery, R. Tannenbaum, K. Kalaitzidou, Capture/release ability of thermo-responsive polymer particles, Journal of Materials Chemistry 20 (2010) 3496-3501. http://dx.doi.org/10.1039/b922972h

[125] W. Small, T.S. Wilson, P.R. Buckley, W.J. Benett, J.M. Loge, J. Hartman, D.J. Maitland, Prototype fabrication and preliminary in vitro testing of a shape memory endovascular thrombectomy device, IEEE Transactions on Biomedical Engineering 54 (2007) 1657-1666. http://dx.doi.org/10.1109/TBME.2007.892921

[126] J. Hartman, W. Small IV, T.S. Wilson, J. Brock, P.R. Buckley, W.J. Benett, J.M. Loge, D.J. Maitland, Embolectomy in a rabbit acute arterial occlusion model using a novel electromechanical extraction device, American Journal of Neuroradiology 28 (2007) 872-874.

[127] C. Wischke, A.T. Neffe, S. Steuer, A. Lendlein, Evaluation of a degradable shape-memory polymer network as matrix for controlled drug release, Journal of Controlled Release 138 (2009) 243-250. http://dx.doi.org/10.1016/j.jconrel.2009.05.027

[128] G. Li, G. Fei, H. Xia, J. Han, Y. Zhao, Spatial and temporal control of shape memory polymers and simultaneous drug release using high intensity focused ultrasound, Journal of Materials Chemistry 22 (2012) 7692-7696. http://dx.doi.org/10.1039/c2jm30848g

[129] C. Wischke, A. Lendlein, Shape-memory polymers as drug carriers—a multifunctional system, Pharmaceutical Research 27 (2010) 527-529. http://dx.doi.org/10.1007/s11095-010-0062-5

[130] C. Wischke, A.T. Neffe, A. Lendlein, Shape-Memory Polymers in Advances in Polymer Science, Springer, New York, 2010.

[131] J.J. Wykrzykowska, Y. Onuma, P.W. Serruys, Advances in stent drug delivery: The future is in bioabsorbable stents, Expert Opinion on Drug Delivery 6 (2009) 113-126. http://dx.doi.org/10.1517/17425240802668495

[132] A.T. Neffe, B.D. Hanh, S. Steuer, A. Lendlein, Polymer networks combining controlled drug release, biodegradation, and shape memory capability, Advanced Materials 21 (2009) 3394-3398. http://dx.doi.org/10.1002/adma.200802333

[133] C. Wischke, A.T. Neffe, S. Steuerc, A. Lendlein, Comparing techniques for drug loading of shape-memory polymer networks-effect on their functionalities, European Journal of Pharmaceutical Sciences 41 (2010) 136-147. http://dx.doi.org/10.1016/j.ejps.2010.06.003

[134] K. Nagahama, Y. Ueda, T. Ouchi, Y. Ohya, Biodegradable shape-memory polymers exhibiting sharp thermal transitions and controlled drug release, Biomacromolecules 10 (2009) 1789-1794. http://dx.doi.org/10.1021/bm9002078

[135] A. Lendlein, R. Langer, Biodegradable, elastic shape-memory polymers for potential biomedical applications, Science 96 (2002) 1673-1676. http://dx.doi.org/10.1126/science.1066102

[136] R.S. Langer and A. Lendlein, World Patent W.O. 2003088818 A2. (2003)

[137] A. Lendlein, S. Kelch, Shape-memory polymers as stimuli-sensitive implant materials, Clinical Hemorphology and Microcirculation 32 (2005) 105-116.

[138] A. Lendlein, K. Kratz, S. Kelch, Smart implant materials, Medical Device Technology 16 (2005) 12-14.

[139] Smart surgery future materials (through Textile Technology Index) (2004) 33-34.

[140] A. Lendlein and U. Ridder, U.S. Patent 0088135 A1. (2007)

[141] A. Lendlein, G. Lang, K. Knitz, T. Krause, A. Kalbfleisch, J. Allwohn, J. Burghaus, G. Sendelbach, A. Beyer, D. Mattinger, S. Uhl, S. Birkel, A. Duchscherer, T.A.A. Hasson, H.-Y. Jiang, S. Morsheim and Y. Ghazlan, U.S. Patent 0244353 A1. (2005)

[142] J. Ortega, W. Small, T. Wilson, W. Benett, J. Loge, D. Maitland, A shape memory polymer dialysis needle adapter for the reduction of hemodynamic stress within arteriovenous grafts, IEEE Transactions on Biomedical Engineering 54 (2007) 1722-1724. http://dx.doi.org/10.1109/TBME.2007.892927

[143] Y. Yu, T. Ikeda, Photodeformable polymers: A new kind of promising smart material for micro- and nano-applications, Macromolecular Chemistry and Physics 206 (2005) 1705-1708. http://dx.doi.org/10.1002/macp.200500318

[144] Y.C. Jung, J.W. Cho, Application of shape memory polyurethane in orthodontic, Journal of Materials Science: Materials in Medicine 21 (2010) 2881-2886. http://dx.doi.org/10.1007/s10856-008-3538-7

[145] P.T. Mather, C. Liu and C.J. Burstone, World Patent 2006071520. (2006)

[146] S. Yasuo, U.S. Patent 19980189973. (1998)

[147] C. Liu, P.T. Mather, C. Burstone, Proceedings of the annual technical conference, Society of Plastics Engineers, 64th Brookfield, CT, USA, 2006, pp. 1356-1360.

[148] Q.H. Meng, J.L. Hu, B.H. Liu, Y. Zhu, A low-temperature thermoplastic anti-bacterial medical orthotic material made of shape memory polyurethane ionomer: influence of ionic group, Journal of Biomaterials Science, Polymer Edition 20 (2009) 199-218. http://dx.doi.org/10.1163/156856209X404497

[149] Q.H. Meng, J.L. Hu, Y. Zhu, Properties of shape memory polyurethane used as a low-temperature thermoplastic biomedical orthotic material: influence of hard segment content, Journal of Biomaterials Science, Polymer Edition 19 (2008) 1437-1454. http://dx.doi.org/10.1163/156856208786140355

[150] T. Ware, G. Ellson, A. Kwasnik, S. Drewicz, K. Gall, W. Voit, Tough shape-memory polymer-fiber composites, Journal of Reinforced Plastics and Composites 30 (2011) 371-380. http://dx.doi.org/10.1177/0731684410395418

[151] C.M. Yakacki, T.D. Nguyen, R. Likos, R. Lamell, D. Guigou, K. Gall, Impact of shape-memory programming on mechanically-driven recovery in polymers, Polymer 52 (2012) 4947-4954. http://dx.doi.org/10.1016/j.polymer.2011.08.027

[152] C.M. Yakacki, J. Griffis, M. Poukalova, K. Gall, Bearing area: A new indication for suture anchor pullout strength, Journal of Orthopaedic Research 27 (2009) 1048-1054. http://dx.doi.org/10.1002/jor.20856

[153] D. Rickert, GMS Current Topics in Otorhinolaryngology, Head and Neck Surgery 8 (2009) PMC3199816.

[154] S. Pasche, S. Angeloni, R. Ischer, M. Liley, J. Luprano, G. Voirin, Wearable biosensors for monitoring wound healing, Advances in Science and Technology 57 (2008) 80-87. http://dx.doi.org/10.4028/www.scientific.net/AST.57.80

[155] P. Lafont, A.D. Lantada and L. Yustos, Spain, P200603149. (2006)

[156] V. Lorenzo, A.D. Lantada, P. Lafont, H.L. Yustos, C. Fonseca, J. Acosta, Physical ageing of a PU-based shape memory polymer: Influence on their applicability to the development of medical devices, Materials & Design 30 (2009) 2431-2434. http://dx.doi.org/10.1016/j.matdes.2008.10.023

[157] K. Ishida, R. Hortensius, X. Luo, P.T. Mather, Soft bacterial polyester-based shape memory nanocomposites featuring reconfigurable nanostructure, Journal of Polymer Science Part B: Polymer Physics 50 (2012) 387-393. http://dx.doi.org/10.1002/polb.23021

[158] K.A. Davis, K.A. Burke, P.T. Mather, J.H. Henderson, Dynamic cell behavior on shape memory polymer substrates, Biomaterials 32 (2011) 2285-2293. http://dx.doi.org/10.1016/j.biomaterials.2010.12.006

[159] M. Ebara, K. Uto, N. Idota, J.M. Hoffman, T. Aoyagi, Shape-memory surface with dynamically tunable nano-geometry activated by body heat, Advanced Materials 24 (2012) 273-278. http://dx.doi.org/10.1002/adma.201102181

[160] A.A. Sharp, H.V. Panchawagh, A. Ortega, R. Artale, S. R. Burns, D.S. Finch, K. Gall, R.L. Mahajan, D. Restrepo, Toward a self-deploying shape memory polymer neuronal electrode, Journal of Neural Engineering 3 (2006) L23-30. http://dx.doi.org/10.1088/1741-2560/3/4/L02

[161] X. Yu, L. Wang, M. Huang, T. Gong, W. Li, Y. Cao, D. Ji, P. Wang, J. Wang , S. Zhou, A shape memory stent of poly(e-caprolactone-co-DL-lactide) copolymer for potential treatment of esophageal stenosis, Journal of Materials Science: Materials in Medicine 23 (2012) 581-589. http://dx.doi.org/10.1007/s10856-011-4475-4

[162] T.G. Leong, C.L. Randall, B.R. Benson, N. Bassik, G.M. Stern, D.H. Gracias, Tetherless thermobiochemically actuated microgrippers, Proceedings of the National Academy of Sciences 106 (2009) 703-708. http://dx.doi.org/10.1073/pnas.0807698106

[163] V.J. Neiman, S. Varghese, Synthetic bio-actuators and their applications in biomedicine, Smart Structures and Systems 7 (2011) 185-198. http://dx.doi.org/10.12989/sss.2011.7.3.185

[164] C. Yakacki, R. Shandas, C. Lanning, B. Rech, A. Eckstein, K. Gall, Unconstrained recovery characterization of shape-memory polymer networks for cardiovascular applications, Biomaterials 28 (2007) 2255-2263. http://dx.doi.org/10.1016/j.biomaterials.2007.01.030

[165] G.M. Baer, W. Small IV, T.S. Wilson, W.J. Benett, D.L. Matthews, J. Hartman, D. J. Maitland, Fabrication and in vitro deployment of a laser-activated shape memory polymer vascular stent, Biomedical Engineering Online 6 (2007) 43-1-43-8. http://dx.doi.org/10.1186/1475-925X-6-43

[166] S.H. Ajili, N.G. Ebrahimi, M. Soleimani, Polyurethane/polycaprolactane blend with shape memory effect as a proposed material for cardiovascular implants, Acta Biomaterialia 5 (2009) 1519-1530. http://dx.doi.org/10.1016/j.actbio.2008.12.014

[167] G.M. Baer, T.S. Wilson, W. Small, J. Hartman, W.J. Benett, D.L. Matthews, D.J. Maitland, Thermomechanical properties, collapse pressure, and expansion of shape memory polymer neurovascular stent prototypes, Journal of Biomedical Materials Research Part B Applied Biomaterials 90 (2009) 421-429. http://dx.doi.org/10.1002/jbm.b.31301

[168] L. Xue, S. Dai, Z. Li, Synthesis and characterization of elastic star shape-memory polymers as self-expandable drug-eluting stents, Journal of Materials Chemistry 22 (2012) 7403-7411 http://dx.doi.org/10.1039/c2jm15918j

[169] L. Sun, W.M. Huang, Thermo/moisture responsive shape-memory polymer for possible surgery/operation inside living cells in future, Materials and Design 31 (2010) 2684-2689. http://dx.doi.org/10.1016/j.matdes.2009.11.036

[170] V.J. Cornelius, N. Majcen, M.J. Snowden, J.C. Mitchell, B. Voncina, Smart materials IV, in: H.V. Nicolas (Ed.), Preparation of SMART wound dressings based on colloidal microgels and textile fibres, Proceedings of the SPIE, SPIE Publications, Bellingham, Washington, 2007, pp. 6413.

[171] G. Li, M. John, A self-healing smart syntactic foam under multiple impacts, Composites Science and Technology 68 (2008) 3337-3343. http://dx.doi.org/10.1016/j.compscitech.2008.09.009

[172] J. Nji, G. Li, A biomimic shape memory polymer based self-healing particulate composite, Polymer 51 (2010) 6021-6029. http://dx.doi.org/10.1016/j.polymer.2010.10.021

[173] G. Li, N. Uppu, Shape memory polymer based self-healing syntactic foam: 3-D confined thermomechanical characterization, Composites Science and Technology 70 (2010) 1419-1427. http://dx.doi.org/10.1016/j.compscitech.2010.04.026

[174] J. Nji, G. Li, Damage healing ability of a shape-memory-polymer-based particulate composite with small thermoplastic contents, Smart Materials and Structures 21 (2012) 025011-1-025011-10. http://dx.doi.org/10.1088/0964-1726/21/2/025011

[175] G.Z. Voyiadjis, A. Shojaei, G. Li, P.I. Kattan, A theory of anisotropic healing and damage mechanics of materials, Proceedings of the Royal Society A 468 (2012) 163-183. http://dx.doi.org/10.1098/rspa.2011.0326

[176] G. Li, H. Meng, J. Hu, Healable thermoset polymer composite embedded with stimuli-responsive fibres, Journal of Royal Society Interface 9 (2012) 3279-3287. http://dx.doi.org/10.1098/rsif.2012.0409

[177] J. Nji, G. Li, A self-healing 3D woven fabric reinforced shape memory polymer composite for impact mitigation, Smart Materials and Structures 19 (2010) 035007-1-035007-9. http://dx.doi.org/10.1088/0964-1726/19/3/035007

[178] G.Z. Voyiadjis, A. Shojaei, G. Li, A thermodynamic consistent damage and healing model for self healing materials, International Journal of Plasticity 27 (2011) 1025-1044. http://dx.doi.org/10.1016/j.ijplas.2010.11.002

[179] M. John, G. Li, Self-healing of sandwich structures with a grid stiffened shape memory polymer syntactic foam core, Smart Materials and Structures 19 (2010) 075013-1-075013-12. http://dx.doi.org/10.1088/0964-1726/19/7/075013

[180] X. Xiao, T. Xie, Y.T. Cheng, Self-healable graphene polymer composites, Journal of Materials Chemistry 20 (2010) 3508-3514. http://dx.doi.org/10.1039/c0jm00307g

CHAPTER 2

Application of Bioactive Composite Green Polymer for the Development of Artificial Organs

Arijit Nath, Tiyasha Kanjilal, Chiranjib Bhattacharjee
Chemical Engineering Department, Jadavpur University, Kolkata 700032, West Bengal, India
cbhattacharyya@chemical.jdvu.ac.in, c.bhatta@gmail.com
Phone: +91 98364 02118, Telefax: +91 33 2414 6203

Abstract

Presently, the concept of 'Green Chemistry for Sustainable Development' is a guiding source for development of next generation materials, known as composite green polymers. Green composite polymers are used for development of hard tissue implants, such as, artificial bone, spine instrumentation, knee hip replacement and bone cement etc. Moreover green composite polymers are used for soft tissue replacements and implantations, such as, ureter prosthesis, catheters, vascular grafts, tendons and ligaments instead of conventional metal alloy and ceramic. The present chapter represents various applications of green composite polymers in medical technology with special interest on development of artificial organs.

Keywords

Composite green Polymer, Artificial organs, Hard tissue applications; Soft cell and cartilage; Prosthetic limbs

Contents

1. Introduction

In the present century incredible technological innovations and improvements are bringing new hope and opportunities to the healthcare sector. These evolutions improve the methodologies of determination of diseases, heart failure, synthesis of serological items, biomaterials, replacement of organs, generation of tissue and so on. It is the multi-disciplinary approach associated with physics, chemistry, biology and computer engineering. Recently, advances in the development of composite materials enable major improvements in medical technology, more particularly, design of biomedical, modern prosthetic and orthopedic devices [1].

2. Composite polymer

Presently, instead of conventional polymer, composite polymer has drawn considerable attention. Different materials are combined judiciously to produce another material which has a more useful structure and high mechanical properties, i.e. low weight and high specific strength, specific stiffness, fractures resistance, corrosion resistance, fatigue strength. Moreover, such composite are popular because of their low cost and their simple fabrication procedure [2]. Composite materials can be defined as materials which consist of two or more chemically different materials or phases (matrix phase and dispersed phase), separated by a proper interface. Matrix is the primary phase, usually uniform, more ductile and soft. On the other hand, a dispersed phase is usually stronger than a matrix phase, embedded in the matrix phase in a discontinuous form [2-3]. In the composite, two main constituents remain recognizable unlike in the blend [4]. Depending upon the feedstock ingredients many different types of polymers are available [5]. Different types of polymers, such as polyester, vinyl ester, epoxy, phenolic, polyimide, polyamide, polypropylene, polyether ether ketone and others are prevalently used for matrix preparation [6]. There are various methods, such as hand lay-up, bag molding process, pultrusion, filament winding, preformed molding compounds, resin transfer molding, injection molding, reaction injection molding, reinforced reaction injection molding, spray-up etc. which have been developed to reduce the resin content of the final product. The main disadvantages of polymer matrix composites are high thermal expansion coefficient and low thermal resistance [7-8].

3. Composite green polymer

Polymers are originally developed for common engineering applications instead of tissue engineering in human bodies. There are also various shortcomings of polymers for their intended medical applications. One of the major problems of polymer materials for tissue implantation is that they are much stiffer than human cortical bone [6, 9]. According to the load sharing principle of a composite theory, if a stiff material is placed in bone, the bone will be subjected to reduce mechanical environment and consequently the bone will resorb [10]. For example, bone resorption in the proximal femur leads to aseptic loosening of the prosthesis (a very common problem) [11]. Elastic characteristics of the implant provide physiologically acceptable stress in the femur. Synthetic polymer, such as polyethylene for total hip replacement has been successfully used for a long time. After a long clinical investigations, some synthetic polymers, such as polypropylene, polyethylene, polyurethane, poly(-vinyl chloride), poly(-tetrafluoroethylene), polyamides, poly(-etheretherketone), poly(-ethylene terephthalate), polyacetal, polycarbonate and polysulfone are considered as biocompatible [12-13]. Recently, the emergence of tissue and biomedical engineering has led to the invention of biodegradable polymers, such as poly(-glycolic acid), poly(-lactic acid), poly(-hydroxybutyrate), poly(e-caprolactone) which are generally used to construct degradable scaffolds and various types of cells [14].

In the present century, due to scarcity of oil and gas resources and with increasing interest on green chemistry, synthesis of bio-composite materials are facing great challenges [15-16]. Natural polymers are generally derived from renewable plant or animal sources, which have low mechanical properties. Table 1 presents an overview of properties of different types of synthetic, natural and green polymers.

Table 1. Mechanical properties of different types of synthetic, natural and green polymers.

Polymer type	Density ($kg.m^{-3}$)	Young's modulus (GPa)	Tensile strength (MPa)	Elongation (%)
Carbon HS	1800–1840	225–260	4400–4800	0.0
E-glass	2550–2600	72–85	1900–2050	1.8–4.8
Coir	1150–1220	4–6	135–240	15–35
Cotton	1520–1560	7–12	350–800	5–12
Flax	1420–1520	75–90	750–940	1.2–1.8
Hemp	1470–1520	55–70	550–920	1.4–1.7
Jute	1440–1520	35–60	400–860	1.7–2
Kenaf	1435–1500	60–66	195–666	1.3–5.5
Ramie	1450–1550	38–44	500–680	2–2.2

Sisal	1400–1450	10–25	550–790	4–6
Bamboo	600–1100	11–32	140–800	2.5–3.7
poly(-lactic acid)	1210–1250	0.35–3.83	21–60	2.5–6
poly(-ydroxybutyrate)	1180–1260	3.5–4	24–40	5–8
Starch	1000–1390	0.125–0.85	5–6	31–44
Epoxies	1110–1400	2.35–3.08	45–89.6	2–10
Polyester	1040–1400	2.07–4.41	41.4–89.6	2–2.6
Polypropylene	890–910	0.9–1.55	28–41	100–600

Without any contradiction, composite green polymers provide high mechanical properties, good abrasion resistance and so on. To synthesize green composite polymers, various kinds of natural fibers were employed with biodegradable resins, such as poly(-lactic acid), poly(-hydroxybutyrate), polysaccharides of plant origin; cellulose and alginate, animal origin; chitin and of microbe origin; hyaluronate and modified starch [17-19]. Green composite polymers are generally used for fabricating short life-span products, sporting equipment and biomedical items. In the present chapter medical applications of different types of composite green polymers have been discussed.

4. Application of green composite polymer in medical technology

Over the years, a wide variety of polymer composite materials have been used for fabricating various transplant organs [20]. A wide variety of polymer composite materials were investigated and developed with the progress of medical technology. The green polymer composite materials have been classified as vital/avital composites and avital/avital composites. The avital/avital composites are further divided into non resorbable, partially resorbable and fully resorbable composite biomaterials (Fig. 1).

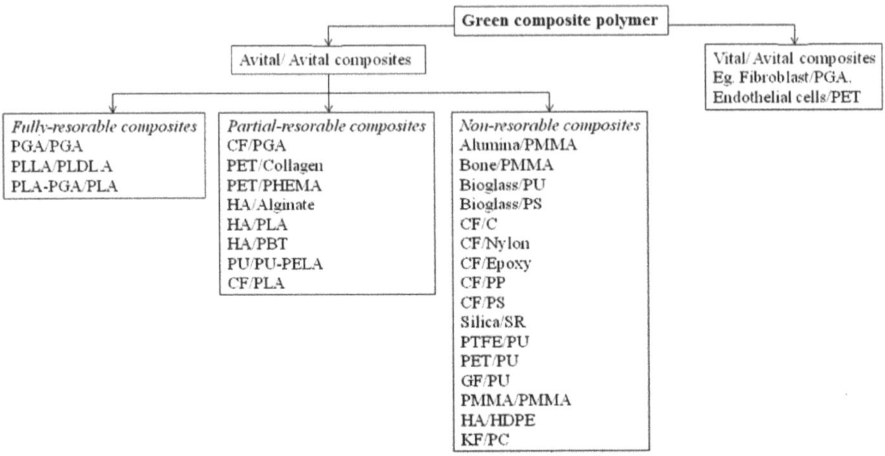

Figure 1. Classification of composite biomaterials used in medical purposes. PGA: Poly(-glycolic acid); PLLA: Poly(L-lactic acid); PLDLA: Poly(L-DL-lactic acid); PLA: Poly(-lactic acid); CF: Carbon fibers; PET: Poly(-ethylene terepthalate); HA: Hydroxyapatite; PBT: Poly(-butylene terephthalate); PU: Polyurethane; PELA: Block copolymer of lactic acid and polyethylene glycol; PMMA: Poly(-methylmethacrylate); PS: Polysulfone; PP: Polypropylene; SR: Silicone rubber; PTFE: Poly(-tetrafluroethylene); PU: Polyurethane; GF: Glass fibers; HDPE: High density polyethylene; KF : Kevlar fiber.

The non resorbable composites are particularly promising for long-term implants, such as total joint replacements, spine rods, bone cement, fusion cages, plates, discs, dental posts and hernia patches because they are not degraded in vivo (inside the body) environment. They are also proposed for short term purposes, such as bone plates, rods, screws and ligaments. Contradictorily the resorbable polymer composites loose their mechanical integrity in vivo. They are particularly promising for short-term uses. They are generally used for making bone plates, screws, rods, pins, ligaments, vascular grafts, tendons and artificial skin [21-23]. In Fig. 2, various applications of different green polymer composites in human prosthesis are depicted.

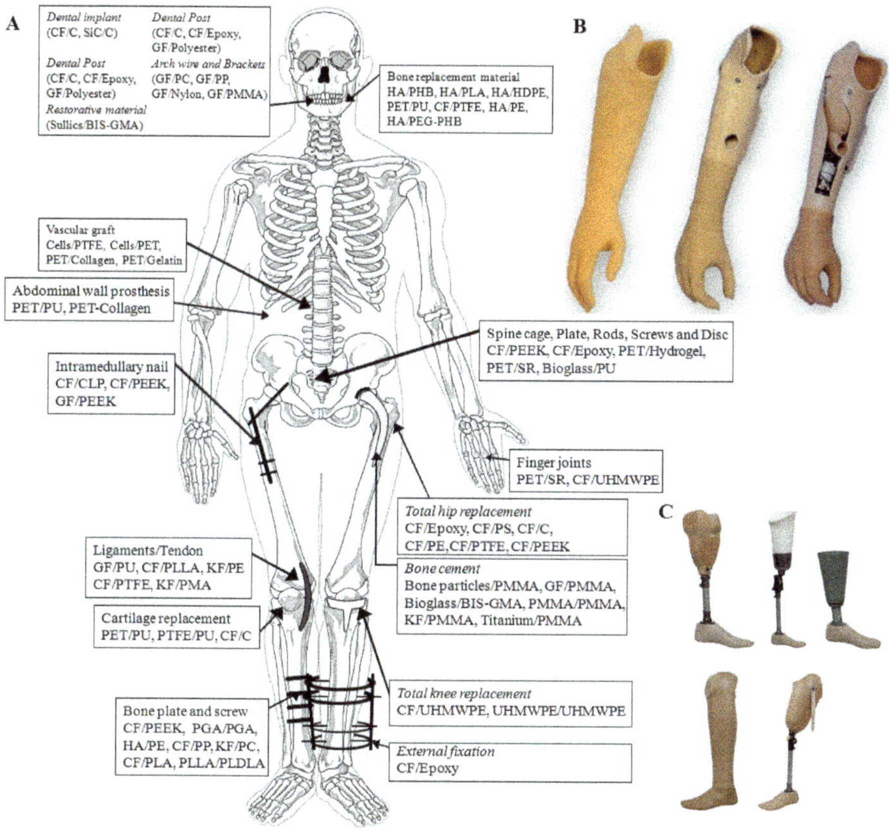

Figure 2. Various applications of different green polymer composite in human prosthesis, A. Artificial organ and bone; [BIS-GMA: Bis-phenol A glycidyl methacrylate; C: Carbon; CF: Carbon fibers; GF: Glass fibers; HA: Hydroxyapatite; HDPE: High density polyethylene; KF: Kevlar fiber; LCP: Liquid crystalline polymer; LDPE: Low density polyethylene; MMA: Methylmethacrylate; PA: Polyacetal; PBT: Poly(-butylene terephthalate); PC: Polycarbonate; PCL: Polycaprolactone; PE: Polyethylene; PEA: Polyethylacrylate; PEEK: Poly(-etherether ketone); PEG: Poly(-ethylene glycol); PELA: Block copolymer of lactic acid and polyethylene glycol; PET: Poly(-ethylene terepthalate); PGA: Poly(-glycolic acid); PHB: Poly(-hydroxybutyrate); PHEMA: Poly(HEMA) or poly(2-hydroxyethyl methacrylate); PLA: Poly(-lactic acid); PLLA: Poly(L-lactic acid); PLDLA : Poly(L-DL-lactic acid); PMA: Poly(-methylacrylate); PMMA: Poly(-methylmethacrylate); PP: Polypropylene; PS: Polysulfone; PTFE: Poly(-tetrafluroethylene); PU: Polyurethane; PVC: Poly(-

vinylchloride); SR: Silicone rubber; THFM: Tetrahydrofurfuryl methacrylate; UHMWPE: Ultra high molecular weight polyethylene], B. Artificial hand, C. Artificial leg.

4.1. Tissue replacement

Conventional tissues are classified as hard and soft tissues. Blood vessels, skin, cartilage and ligaments are few examples of soft tissues, whereas bone and tooth are considered as hard tissues. As the name suggests, in general, the soft tissues are less strong (tensile strength) and stiffer (elastic modulus) than hard tissues. Considering the structural and mechanical compatibility with soft tissues, composite polymers are often used for soft tissue applications. It has been reported that proper matching with the host tissues implant produces a desired tissue remodeling. The composite polymer materials have several orthopedic applications because of their low elastic modulus and high strength [24-25]. The principal advantage of an implant, fabricated from composite materials is that its properties and design can be changed and tailored to suit the physiological and mechanical conditions of the host tissues. Therefore, composite materials have greater potential for structural biocompatibility than the homogenous monolithic materials [26]. Another advantage of polymer composite biomaterials include, absence of releasing metal ions, such as nickel or chromium which may cause loosening of the implant, patient discomfort, allergic skin reactions, low fracture toughness of ceramic materials, corrosion and fatigue failure of metal alloys [27, 28]. Moreover for polymer composite materials, the radio transparency can be adjusted. The subsequent sections present detail information about hard and soft tissues replacement by polymer composite.

4.1.1. Hard tissue applications

4.1.1.1. Bone fracture repair

Skeletal bone is composed with collagen fibers with hydroxyapatite nano-crystalls precipitate, which provide the supporting structure for the body [29]. Bone fractures can be treated different ways, such as internal fixation and external fixation. The internal fixation requires opening of the fracture site whereas for external fixation is not required to open the fracture site. For the external fixation approach the bone fragments are detained in alignment by various ways, such as splints, braces, casts and external fixators. The casting materials or plaster bandages are used to build splints, casts or braces [30]. It is essential that the casting material should be a composite material made of woven cotton fabrics (woven gauze) and plaster of paris matrix (calcium sulphate), fabrics of glass and polyester fibers. Presently, casts made of glass or polyester fiber fabrics and water-activated polyurethanes are popular. It is desirable that an ideal cast material should be easy to handle, conformable to anatomical shape, light weight, strong, stiff, radiolucent, water proof and easy to remove. Moreover, cast material should be permeable for ventilation without which the patient's skin may be scorched or weakened [31]. A typical external fixation system comprises of pins or Kirschner wires, which are pierced through the bone and apprehended under high tension by screws to the external frame [32]. The external fixators are developed with high rigidity and strength to ensure stability, oriented at different angles across the bone. External fixators are prepared by carbon fibers/epoxy composite because of their lightweight, sufficient stiffness and strength [33]. Furthermore, the assessment of the bone union made of polymer composites by radiography has become easy because radiolucency of polymer composites is always acceptable and they do not cause artifacts in the radiographs. Bone external fixation is also employed for bone lengthening purposes. With the internal fixation approach, the bone fragments are positioned together by different ways using implants, such as intramedullary nails, wires, pins, screws and plates. The conventional implants are developed with stainless steel, Co-Cr or Ti alloys. Surgical wires and pins are generally used for simplest implantation, such as joining small fragments of bones. Often they are used to provide additional stability in fractures of long bones (humerus, femur, tibia, radius, ulna, fibula and so on). Proper implant design and implantation technique ensure the desired physico-mechanical outcome of the fixation, avoid devascularization and additional tissue trauma at the fracture site [34]. The non-resorbable composite plates are fabricated from either thermoplastic composite or thermoset polymer composites [35-38]. Due to the toxic effects of monomers in partially cured epoxy composite materials, investigations on these materials have gradually decreased. As a result, good quality thermoplastic composites, such as carbon fiber/poly(-methyl methacrylate) [39], carbon fiber/polypropylene [40], carbon fiber/polysulfone [41-43, 22], carbon fiber/polyethylene [44], carbon fiber/nylon, carbon fiber/polybutylene terephthalate [45] and carbon fiber/poly(-etheretherketone) [46-50, 20-21] were developed and successfully utilized as bone plates. Unlike metal alloy plates and thermoset composites, the thermoplastic composites are considered free from complications associated with unused monomers those can be bent or contoured to the shape of the bone during surgery. Among all, the carbon fiber/poly(-etheretherketone) is considerable biocompatible and has high strength, fatigue resistance [51-52] and biological inertness with no mutagenicity or carcinogenicity, good resistance to radiation (sterilization technique) and hydrolysis degradation [53-54]. Due to disadvantages of short carbon fiber reinforced poly(-etheretherketone) composites, Mayer with his co-worker developed knitted carbon fiber/poly(-etheretherketone) composite bone plates using commingled yarns of carbon and poly(-etheretherketone) fibers [52, 55]. The development of braided carbon fiber/poly(-etheretherketone) composite bone plates using a new technique was reported by Ramakrishna [20]. Initially micro-braided yarns by combining carbon and poly(-etheretherketone) fibers were made and subsequently micro-braided yarns were again braided into flat fabrics of desired dimensions. Carbon fiber/carbon and carbon fiber/poly(-etheretherketone) composite screws were developed for osteo-synthethesis [56-57, 49]. In 1997 Peter with his co-workers developed the squeeze casting method for fabrication of complex shaped components with fiber contents as high as 62% by volume [49]. It was reported that a combination of polymer composite plates and screws can overcome the corrosion problem. Non-resorbable polymer composites were developed in such a way that they were labile in-vivo condition (no modification in plate stiffness with implantation time). During the bone healing progresses, it is expected that the bone is subjected to gradual increase of stress to reduce the stress-shielding effect. This may be justified by the fact that the plate loses its rigidity in-vivo condition. Presently, resorbable polymers are drawn considerable attention for bone plate development [58]. As the bio-resorbable polymers loose most of their mechanical properties in few weeks, few researchers proposed fully resorbable composites with poly(-lactic acid) fibers and calcium phosphate based glass fibers [59-60]. The advantages of resorbable composite prostheses are that they do not need to be removed. However, the applications of resorbable materials are restricted where the loads are moderated [61]. To improve mechanical properties, the resorbable polymers are modified with variety of non- resorbable materials including carbon [62-66] and polyamide [67-68] fibers and they are classified as partially resorbable composites.

Intramedullary nails or rods are generally used to staple long bone fractures, such as femoral neck or inter-trochanteric bone. It is introduced into the intramedullary cavity of the bone and fixed in position using screws or with a friction fit approach. Compared to the plate fixation, torsional resistance of the intramedullary nails is much less than that of plates. However, the intramedullary nail fixation is better than plate fixation to resist bending, since it is located in the center of the bone. Stainless steel is one of the widely used materials for intramedullary nails. Recently, Lin et al. proposed short glass fibers/poly(-etheretherketone) composite material for intramedullary application [69]. Few researchers developed unidirectional carbon fiber reinforced liquid crystalline polymer composite intramedullary rod, which is biologically inert, with higher flexural strength than the yield strength of stainless steel and elastic modulus close to the bone [70].

4.1.1.2. Spine instrumentation

The human body spine serves two distinct and conflicting (apparently) roles. It provides a strong, yet mobile, central axis onto which the whole skeleton distributes the load, it protects the spinal cord and the roots of delicate nerves, which connect the brain to periphery [13]. Over the years, numbers of spine related disorders, including metastasis of vertebral body and disc, facet degeneration, disc herniation, stenosis and structural abnormalities, such as scoliosis, kyphosis and spondylolistheses are reported. A variety of reasons such as birth deformities, tumorous lesions (metastasis), aging, mechanical loads caused by sports and work can lead to spine disorders. Two techniques, namely spinal fusion and disc replacement are used for spine disorder treatment. Depending on the patient condition and prognosis, these methods are used alone or in combination. Various methods are employed in spinal fusion which signifies surgical immobilization of joint between two vertebrae. One such method involves the surgical removal of the affected portions of vertebrae and replacement of the gap with synthetic bone graft. Homologous or autologous bone grafts are limited do to the risk of shortage of donor bone sites (with risk of hepatitis and AIDS in case of autologous donors), infection and postoperative resorption and collapse of the graft. Few researchers developed bioglass/polyurethane composite material for vertebral body bone grafting [71-73]. Another approach of spine fusion is to use special vertebral prostheses, such as cages, baskets and threaded inserts, made of metals or bioceramics [74-76]. In some cases stainless steel or titanium made rods, screws and plates are used in conjunction with prostheses to provide the necessary stabilization. There are some disadvantages of these implants, such as poor form fit of these implants and complicated postoperative judgment with X-rays, computed tomography (CT), magnetic resonance imaging (MRI), collapse of instrumented spine, injury of neurological structures and blood vessels. Moreover, composite cages, plates and screws made from carbon fiber/poly(-etheretherketone) and carbon fiber/polysulfone were developed for lumbar interbody fusion, stabilizing the replacement body and spine [76-77, 52, 49]. The composite cage has an advantage because its elastic modulus is similar to the bones which promote the maximum bone growth into the cage. Moreover, composite cages are radiolucent, therefore it does not hamper radiographic evaluation of the bone fusion. It was reported that fusion alters the biomechanics of the spine and increase stresses at junction between fused and unfused segments. This promotes further disc degeneration, which seems to contradict the primary purpose of its use for the patient. Such arguments have given rise to intervertebral disc prostheses. On the other hand, an artificial disc is used for problems related to intervertebral discs. Some researchers proposed metal balls and injectable silicone elastomers or hydrogels to replace the nucleus after discectomy [78]. Presently composites are proposed for disc prostheses either alone or in combinations [11]. However, there are no artificial discs that are acceptable for long term applications. In few cases, an adjustable stainless steel rod, known as Harrington spinal distraction rod is used for stabilizing the curvature. Schmitt-Thomas with co-workers tried to develop a polymer composite rod using braided, unidirectional carbon fibers and biocompatible epoxy resin which could overcome the problems of metal alloys such as corrosion [79].

4.1.1.3. Joint replacements

Joints allow the movement of body and its parts. All the joints have two opposing articular surfaces and are protected by a thin articular cartilage layer. Joints are lubricated by elastic-viscous synovial fluid. Osteoarthritis is one of the regular causes for joint deterioration and sometimes hypertrophic changes in the bone and cartilage of joints. Moreover, joints are damaged upon exposure to severe metabolic disorder and mechanical injury. For a long period, a number of artificial joints were developed and applied for replacing various joints in the body. Considering the complexity of joint function, extent of loading and severity of the physiological environment, joint replacement is most demanding for all types of implant applications in the human body and also they are placed in the body permanently.

4.1.1.3.1. Knee replacement

The knee joint has a more complicated biomechanics of movements and geometry than hip joints. The incidence of knee injuries and degeneration are higher than most other joint displacements. A typical knee replacement signifies the replacement of femoral and tibial components. The materials used for femoral components are generally Co-Cr and Ti alloys, whereas the ultra high molecular weight polyethylene, supported by a metallic tibial tray is used for the replacement of tibial component. It has been reported that ultra high molecular weight polyethylene undergoes cold deformation, which leads to sinking of prosthesis [80-81]. To reduce its cold flow (creep) deformation, carbon fibers were frequently used to reinforce ultra high molecular weight polyethylene [82]. The reinforcement enhances the tensile yield strength, creep resistance, stiffness and fatigue strength of ultra high molecular weight polyethylene. Contradictory results have been reported on the effect of carbon fibers on wear characteristics of ultra high molecular weight polyethylene. Early studies reported that wear characteristics are reduced because of the use of carbon fibers. However, some researchers reported that the composite wear rates have increased with respect to unreinforced ultra high molecular weight polyethylene [83].

4.1.1.3.2. Hip replacement

Total hip replacement is another common artificial joint replacement procedure in the human body [84]. Several types of hip replacements have been designed by changing the geometry and material of acetabular cups and femoral stems as well as fixation methods.

A typical hip replacement consists of a cup type acetabular section and a femoral module whose head is designed to adjust into the acetabular cup. Stainless steel, Co-Cr and Ti alloys for femoral shaft and neck, and Co-Cr alloy or ceramics, such as alumina and zirconia are used for the head or the ball. To decrease the friction and eliminate metallic wear on particles, Charnley used a stainless steel femoral module with a mating acetabular component, made of poly(-tetrafluroethylene) for the acetabular components [85]. The advantages of poly(-tetrafluroethylene) are a low friction coefficient, high thermal stability, hydrophobic nature, stablity in most types of chemical environments and inertness in human body. It was reported that poly(-tetrafluroethylene) acetabular cups for total hip substitute prostheses shows unfavorable high distortion and wear. The wear debris promotes extensive tissue reaction and even arrangement of granuloma. This attributes to its high wear under high stresses during sliding and low compressive strength and stiffness. Therefore, poly(-tetrafluroethylene) is not commonly used for load bearing applications. Contradictorily, acetabular cup made of ultra high molecular weight polyethylene were developed and used successfully. The ultra high molecular weight polyethylene cups are usually supported with a metal backing which might be the cause of creep deformation, plastic distortion and high wear or erosion. To improve the quality of ultra high molecular weight polyethylene cups with respect to creep resistance, stiffness and strength, reinforcing ultra high molecular weight polyethylene was made with carbon fibers [44, 85-86]. Recently certain designs have been developed with dense alumina or zirconia ball and matching acetabular cup. The potential advantages of ceramic materials are low coefficient of friction and wear rate, high hardness and compressive strength, and good biological acceptance. It was reported that stiffness mismatch stimulates abnormal stresses in the bone which leads to bone resoprtion and eventual aseptic loosening of prosthesis [12, 87-92]. Materials suitable for implant design are very limited because for this design high strength material is prerequisite. The implant properties of polymer composites can be modified by selecting material ingredients and controlling their composition which is useful in reducing the development of high stress regions, and more uniform and efficient transfer of stress (spatially or locally) from the stem to the bone. Compared to metals, they are more elastic and also offer modulus and strength. It was reported that composite implants made by carbon fiber/carbon or carbon fiber/polysulfone have high static and fatigue strengths which promote faster bone bonding compared to conventional implants. Moreover, they are stable with no release of soluble compounds [87, 53]. In 2000, Simoes and Marques reported composite stems by braided hybrid carbon-glass fiber and epoxy resin [93]. Chang with his co-workers developed carbon fiber/epoxy stems by imposing 120 layers of unidirectional plies in a pre-determined orientation [94]. It was reported that carbon fiber/poly(-etheretherketone) composite stems possess a mechanical behavior similar with femur [95-96, 49, 75]. Both in-vitro and in-vivo animal studies corroborated the mechanical stability of carbon fiber/poly(-etheretherketone) up to 6 months. Porous surface coatings have been developed from various materials including bioactive polymers which promote the growth of bone tissues.

4.1.1.3.3. Other joint replacements

Various joint (ankle, toe, shoulder, elbow, wrist and finger joints) replacements have been performed by Co-Cr and Ti alloys, silicon rubber, high density poly(-ethylene) and ultra high molecular weight polyethylene. To provide higher strength and creep resistance some designs have been fabricated from carbon fiber/ultra high molecular weight polyethylene instead of ultra high molecular weight polyethylene. In order to improve the flexural properties and tear strength, sometimes silicon rubber is reinforced with poly(-ethylene terephthalate) fabrics. Goldner and Urbaniak have reported that the composite prosthesis reduces joint pain, increase function and improve the stability of the joint [97].

4.1.1.3.4. Bone cement

Both proper fixation of bones and joints as well as its design are important. One of the common methods for fixing artificial joints to bones is press-fitting the joint prosthesis onto the bone using a grouting material, known as, bone cement. The most widely used bone cement is acrylic bone cement, i.e., poly(methyl methacrylate), which is self-polymerizing and contains solid poly(methyl-methacrylate) powder and liquid methyl-methacrylate monomer [98]. Strength characteristics of cement are important for its clinical success because it transfers the load from prosthesis to bone [99]. Moreover problem related with mechanical failure and instability is another main issue [100-101]. It has been reported that bone cement, monomers are released into the blood stream through exothermic reaction that may induce local bone necrosis [102]. To increase the quality of bone cement, researchers endeavored to improve the mechanical properties of it by reinforcing it with Kevlar, carbon [103], stainless steel and Ti alloy wires, and polymer fibers, such as ultra high molecular weight polyethylene [104-107] and poly(methyl methacrylate) [108]. Fiber reinforcement reduces the peak temperature during polymerization of the cement which reduces the tissue necrosis [105]. Sometimes poly(methyl methacrylate) bone cement is mixed with bone particles or surface-reactive glass powders in order to combine immediate mechanical fixation of surface-active glasses (Bioglass) with the bone [109] and poly(methyl methacrylate) with chemical bonding of bone particles [110-112]. Although, there is experimental evidence on its adequate mechanical performance reinforced cements have not yet been established in current clinical practice due to some limitations, such as the addition of fibers increases the apparent viscosity of bone cement. Moreover uniform allocation of fibers in the bone cement is hard. Gerhart and co-workers developed a partially resorbable bone cement, which is a composite of a gelatin matrix and tricalcium phosphate particles [113]. It was proposed that this composite bone cement provides immediate structural support and subsequent resorption of resorbable component. As a result, it facilitates bone development and direct bonding by the host bone. Contradictorily, it was reported that poly(methyl methacrylate) bone cement cannot provide bonding by the host bone even though it provides immediate structural support. Moreover poly(methyl methacrylate) is weak for accumulation of fatigue damage because its strength is considerably higher than partially resorbable bone cement. It is necessary to mention that partially resorbable bone cement may be limited by a propensity for particle migration, away from the implant site.

4.1.1.3.5. Bone replacement (synthetic bone graft) materials

Synthetic bone grafts are necessary for treatment and replacement of fractured bones [114]. The bone graft materials must be adequately stiff and strong and also potential to make bond with residual bones. To improve the mechanical properties, such as load bearing applications, reinforced polyethylene was developed using hydroxyapatite particles which are bioactive and biocompartable [115-120]. This composite has a bio-compatible elastic modulus and strain. It has several disadvantages when the volume fraction of hydroxyapatite increases up to 50%. It was reported that over 40% volume fraction of hydroxyapatite particulate, the composite becomes brittle. Moreover, the bioactivity of the composite is less than its original one because the available surface area of hydroxyapatite is reduced and the rate of bone bonding of hydroxyapatite is slow. To increase the interface between hydroxyapatite particles and the bone tissues, partially resorbable composites were developed with polyethylene glycol, polybutylene terephthalate [121], poly(-lactic acid) [122-126], poly(-hydroxybutyrate) [115, 126], alginate and gelatin [127].

4.1.1.3.6. Dental applications

Teeth consist of two components, i.e. crown and root, are non-homogenous, anisotropic and unsymmetrical. Dental treatment is one of the most common medical treatments, ranging from filling cavities (also called 'dental caries') to replace fractured or decayed teeth. Different biomaterials are used for dental treatments (Cavity lining, cavity filling, luting, endodontic, bridge and crown, prosthetic, preventive) depending on its ability to resemble the physical, mechanical and esthetic properties of the original tooth structure. Dental restorative materials are used to fill the tooth cavities and sometimes to mask discoloration or to correct contour and alignment deficiencies of tooth. Several materials, such as amalgam, alumina, zirconia, gold, acrylic resins and silicate cements are commonly used for restoring decayed teeth. Due to some reasons, gold amalgam (for cosmetic reasons), silver-mercury amalgam fillings (long-term toxicity), acrylic resins and silicate cements (poor mechanical properties) are not acceptable in all cases. Dental composite resins are generally used to restore posterior as well as anterior teeth because they are translucent with a refractive index, matching with enamel. The dental composite resin consists of quartz, barium glass and colloidal silica as fillers, and bis-phenol A glycidyl methacrylate as the matrix polymer. Urethane dimethacryate resin is also applied in many cases instead of bis-phenol A glycidyl methacrylate. Glass filler decreases the shrinkage upon polymerization of resin and divergence the thermal expansion coefficient between teeth and composite resin. Glass filler offer high strength, stiffness and good wear resistance for dental composite resins [128]. Strong bonding between the resin and fillers is achieved by the use of silane-coupling agents [129]. In situations where the amount of coronal tooth structure remains small (also referred as pulp less tooth), a dental post or a cast dowel is used to reinforce the remaining tooth structure, on which the core and crown are built [130-131]. Cores are usually formed from composite resins or amalgam or precious cast or nonprecious alloys in combination with a metal post [132]. Traditionally posts are made of stainless steel, Ni-Cr, Au-Pt or Ti alloys, based on the assumption that the post should be rigid. Failures include corrosion of posts, bending or crack of posts, loss of retention, root fracture and core fracture. In recent years this old basic tenet is strongly questioned and it is recommended that modulus mismatch between post and dentine should be reduced in order to reduce the occurrence of root fractures (root fracture frequency is 2-4%) and failure of restorations. Newer posts made of zirconia, short glass fiber reinforced polyester and unidirectional carbon fiber reinforced epoxy composite are introduced in medical technology [133, 134]. These new posts are adequately rigid, resistant to fatigue and corrosion [135]. Ramakrishna with his co-workers developed a bio-comparable dental post with braided carbon fiber/epoxy composite technology [136-138]. This composite material has a high stiffness in the coronal region and its stiffness gradually decreases to a value comparable with dentine at the apical end. Moreover to overcome dental root fracture, graded stiffness post abolishes stress concentration in dentine and reduces post/dentine interfacial shear stresses by mean of decreasing the probability of post loosening from dentine. In an extreme case, the condemned or damaged tooth is extracted and it is replaced with a dental implant, i.e. an artificial tooth. The dental implant are made of a wide range of materials, such as ceramics (alumina, zirconia, glass and carbon), metals (Co-Cr-Mo alloys, silver, Ti alloys, stainless steel, platinum), polymers (polysulfone, ultra high molecular weight polyethylene, poly(-tetrafluroethylene), poly(-methyl methacrylate) and poly(-ethylene terephthalate)) and composites (SiC/carbon and carbon fiber/carbon) [139-143]. Sometimes a bridge, a partial denture (false teeth) is used to replace one or more teeth completely. The high cost and time consuming preparation of current gold bridges resulted in the development of relatively inexpensive and easy to use biocompatible materials, such as Kevlar fiber/poly(methyl methacrylate) [144], glass fibers/poly(methyl methacrylate) [145], carbon fiber/poly(methyl methacrylate) [146], ultra high molecular weight polyethylene/ poly(methyl methacrylate) [147] and composite bridges and dentures [148]. Orthodontic arch wires (approximately 0.5 mm diameter) are used to correct the alignment of teeth. Sometime a splint is used to correct traumatized teeth. Conventionally arch wires are developed by stainless steel and Ni-Ti (beta titanium) alloys. Several researchers have proposed glass fibers/polycarbonate, glass fibers/polypropylene, glass fibers/Nylon and glass fibers/poly(-methyl methacrylate) composite materials for arch wires [149-151]. The advantages of composite arch wires are aesthetics, possibility of varying stiffness without changing component dimensions [152].

4.1.2. Soft tissue applications

Different types of implants have been used for treatment of soft tissue deformities or defects. Depending upon intended applications, soft tissue implants fill the space created by some defects, provide mechanical support or serve as a scaffold for tissue growth.

4.1.2.1. Ureter prosthesis

Ureter prostheses made of polyethylene, poly(vinyl chloride), poly(-tetrafluroethylene), nylon and silicon rubber were developed and used without success because of the difficulty of adjusting a fluid-tight prosthesis to the living system, blockage of passage by calcification deposits from urine and constant danger of microbial infection. Contradictorily, polyester fiber reinforced glycol methacrylate gel prostheses with a fabric backing was successfully used [153, 154]. It has been reported that fabric backing promotes effortless conjugation of prosthesis on to the mucous membrane without any irritation and the hydrophilic nature of the gel facilitates maintaining a clear inner space.

4.1.2.2. Tendons and ligaments

Tendons and ligaments are the load-bearing soft tissue. Treatment and tendon replacement are performed in two stages, i.e., in a first operation, the tendon is replaced by a gliding implant that promotes the formation of a new tendon sheath and in a second operation, the gliding implant is replaced by tendon graft [155]. Synthetic biopolymers, including polypropylene, polyurethane, ultra high molecular weight polyethylene, poly(-tetrafluroethylene), poly(-ethylene terephthalate), Kevlar 49, carbon and reconstituted collagen fibers in multifilament form or braided form have been used for artificial tendon and ligament synthesis [156-166]. To improve the handling properties and reduce particle migration, prostheses are coated with polymers, such as silicon rubber, poly(2-hydroxyethyl methacrylate) and poly(-lactic acid). In 1991, Pradas and Calleja reported that combining flexible polymer, such as poly(-ethylacrylate) or pol(-ymethylacrylate) with crimped Kevlar-49 fibers, the stress-strain behavior of artificial ligaments can be improved to a certain extent [167]. Few scientists have developed a ligament prosthesis by reinforcing a hydrogel matrix (poly(HEMA) or poly(2-hydroxyethyl methacrylate)) with helically wound rigid poly(-ethylene terephthalate) fibers to improve both static and dynamic mechanical behaviors [168-170].

4.1.2.3. Vascular grafts

Arterial blood vessels consisting of collagen and elastin fibers, smooth muscle, endothelium and ground substance have complex multilayered structures. Vascular grafts are used to replace diseased or blocked segments of natural cardiovascular systems. Generally, vascular grafts are developed by woven or knitted fabric tubes of poly(-ethylene terephthalate) material or extruded porous wall tubes of poly(-tetrafluroethylene) and polyurethane materials. In some cases, vascular grafts are impregnated with gelatin or collagen to improve the dimensional stability of grafts and to seal the pores. The composite graft has isocompliant and anisotropic behaviors, similar with the natural artery. Poly(-ethylene terephthalate) (Dacron) made vascular grafts (woven or knitted fabric tubes) are successfully used for the replacement of large diameter blood vessels (12-38 mm diameter). It was reported that although absorptions of platelet and protein onto poly(-ethylene terephthalate) are low, the mentioned material is thrombogenic. Poly(-ethylene terephthalate) vascular grafts are seeded with endothelial cells to reduce the thrombogenic activity and to improve patency of prosthetic. Generally, a medium diameter (6-12 mm) vascular graft has been developed from expanded polytetrafluroethylene (e-poly-(tetrafluroethylene) or Gore Tex). The mechanical behavior of porous non-woven microstructure of e-poly-(tetrafluroethylene) vascular grafts have been found more similar to that of host blood vessels compared to non-porous (solid) vascular grafts. Moreover, the inner (luminal) surface of e-poly(-tetrafluroethylene) graft promotes the development of neointima (newly formed endothelial tissue lining) that avoids many complications, such as formation of emboli (dislodged blood clot) and thrombi (blood clot). Few researchers have developed composite grafts, comprised of polyurethane (Lycra trade name) fibers in a matrix, mixture of PELA (block copolymer of lactic acid and polyethylene glycol) and polyurethane [171-173].

4.1.3. Prosthetic limbs

Legs are not only the most important organs responsible for walking, running and so on but they also carry the load of the entire body. Leg bone fractures as well as injury are common. Primarily, artificial legs are designed by wood or metallic materials to restore the ability to walk for the amputees. Due to poor durability, caused by corrosion and moisture, and limited by their weight, these materials are becoming less acceptable [174]. Several designs of artificial limbs by different thermoset polymer composites, reinforced with glass, carbon, or Kevlar fibers are now widely used in these systems [175]. A typical artificial leg system consists of three different parts, such as socket, shaft and foot. Artificial sockets can be divided into two categories, i.e. indirect and direct sockets. The indirect socket is developed by wrapping several layers of knitted fabrics or woven on the customized plastic mold, vacuuming the fabrics enclosed in a plastic bag and impregnating the vacuumed fabrics with polyester resin [176]. A direct socket is developed directly on the stump of a sick, without using any kind of mold. The benefit of direct socket fabrication compared to indirect socket is that it can diminish the problem related with fitting between the stump and socket. Direct sockets are developed using a combination of braided or knitted carbon or glass fiber fabrics and water-curable (water-activated) resins. Without any contradiction, the knitted fabric reinforced sockets are flexible and more conformable to the patient's stump, whereas the braided fabric reinforced sockets are strong and stiff [177]. Often shaft or stem is developed by filament wound or laminated woven/braided fabric carbon fiber reinforced epoxy composites. It provides structural support and force share to mimic the skeleton and whole body [178]. In many designs, the foot unit consists of forefoot and heel components, which are made of laminated carbon fiber/epoxy composites [179]. As these composites are light weighted, they are responsible to conserve energy during movement of the body, which help people to feel more placate compared to conventional wood or polymer made artificial feet [180].

4.1.4. Soft cell and cartilage

Hernia is defined as irregular protrusion of tissue, organ or a portion of an organ through an abnormal break in the surrounding cavity's connective or muscular tissue wall. Various materials, such as poly(-tetrafluroethylene), poly(-ethylene terephthalate), polypropylene, nylon, carbon, stainless steel and tantalum in the form of fabrics or meshes are used to repair hernias [181]. The fabrics or meshes promote the growth of tissue which provides the stability to the prosthesis. Recently, Werkmeister et al. have developed poly(-ethylene terephthalate) fabrics coated with collagen and polyurethane materials suitable for repairing hernia and abdominal wall (liver, gallbladder, spleen, stomach, pancreas, intestine, and kidney in abdominal cavity) abnormality [182]. The green composite material is more biologically compatible and exhibits better mechanical properties than uncoated fabrics which promote the tissue generation. Presently, composite prostheses are used for restoring the conductive hearing loss from otosclerosis. Otology prostheses are generally made of poly(methyl methacrylate), polyethylene, polytetrafluroethylene, silicon rubber and carbon fiber/ polytetrafluroethylene composites. Researchers developed polyethylene/polyurethane flexible composite materials as tympanic membrane replacements [183-184]. Moreover, composite green polymers are used for prosthetic sphincters for gastrointestinal tracts, and tracheal and urethral prostheses.

5. Case study

Cherian and co-workers have developed crystalline polymer composites from renewable cellulosic sources. Pineapple leaf fiber was used with polyurethane to improved mechanical properties and biodegrability. Effects of high pressure defibrillation and chemical purification were studied to investigate the pineapple leaf fiber morphology from micro to nanoscale. X-ray diffraction and microscopy techniques were used to study the structure of prepared nanofibers and composites. Degradable polyurethane was used for preparation of film, nanofibril mats and nanocomposite. Microscopy studies suggested that used individual processes (high pressure defibrillation and chemical purification) lead to a unique morphology of interconnected web-like structure of pineapple leaf cellulose fibers. It was reported that chemical treatments resulted the distinguished pineapple leaf microfibres and network structure of pineapple leaf nanofibres was developed by further steam treatment. Whereas high pressure defibrillation contributed a unique morphology of the interconnected web-like structure of pineapple leaf nanofibers. The pineapple leaf nanofibers having a width in the range of 5–15 nm due to steam explosion after high pressure chemical treatment, which are the cementing materials around the fibre bundles. After high pressure and chemical treatment the average diameter of the resulted fibers was about 10–15 μm. Compare to other conventional methods, percentage yield and aspect ratio of the nanofiber, synthesized by this technique were found to be very high. Composite polymer films were prepared by solvent casting with polyurethane. Nonwoven nanofibril mats were prepared by pineapple leaf cellulose fibers. Furthermore, nanocomposite with high tensile strength and high strain-to-failure was synthesized by solvent casting, based on dry crystalline cellulose and hydrophilic polyurethane. The addition of 5 wt% of cellulose nanofibrils to polyurethane increased the stiffness by 2600% and the strength nearly 300%. The cellulose composite material was used for developing heart valve and vascular valve. In a rat model the biological capability of developed pineapple leaf nanocellulose–polyurethane valve has been tested. It was proven that the developed pineapple leaf nanocellulose–polyurethane valve has good biological durability, stability, fatigue resistance and haemodynamics. Pineapple leaf fiber derived nanocellulose embedded polyurethane has been utilized for fabrication of vascular grafts with a wall thickness of 0.7–1.0 mm. It has been reported that nanocellulose–polyurethane vascular grafts withstood hydraulic pressures up to 300 kPa and elongation at break of 800–1200% [185].

Reference

[1] F.R.F. Gibson, A review of recent research on mechanics of multifunctional composite materials and structures. Compos Struct. 92 (2010) 2793–2810. *http://dx.doi.org/10.1016/j.compstruct.2010.05.003*

[2] T. Huber, J. Müssig, O. Curnow, S. Pang, S. Bickerton, M.P. Staiger, A critical review of all-cellulose composites. J. Mater. Sci. 2011, 47, 1171-1186. *http://dx.doi.org/10.1007/s10853-011-5774-3*

[3] O. Faruk, A.K. Bledzki, H.P. Fink, M. Sain, Biocomposites reinforced with natural fibers (2000–2010). Prog. Polymer Sci. 37(11) (2012) 1552-1596. *http://dx.doi.org/10.1016/j.progpolymsci.2012.04.003*

[4] A. Baptista, F. Isabel, J.P. Borges, Cellulose-based composite systems for biomedical applications, Biomass based Biocomposites. UK: Smithers Rapra Technology. (2013) 47-60.

[5] M. Wang, Developing bioactive composite materials for tissue replacement. Biomaterials. 24(13) (2003) 2133–2151. *http://dx.doi.org/10.1016/S0142-9612(03)00037-1*

[6] P.M.M. Dicker, P.F. Duckworth, A.B. Baker, G. Francois, M.K. Hazzard, P.M. Weaver, Green composites: A review of material attributes and complementary applications. Compos. Part A - Appl. S. 56 (2014) 280–289.

[7] S. Ramakrishna, J. Mayer, E. Wintermantel, K.W. Leong, Biomedical applications of polymer-composite materials: a review. Comp. Sci. Tech. 61(9) (2001) 1189-1224. *http://dx.doi.org/10.1016/S0266-3538(00)00241-4*

[8] H.P.S.A. Khalil, A.H. Bhat, A.F.I. Yusra, Green composites from sustainable cellulose nanofibrils: A review. Carbohydr. Polym. 87 (2012) 963– 979. *http://dx.doi.org/10.1016/j.carbpol.2011.08.078*

[9] K.G. Satyanarayana, G.G.C. Arizaga, F. Wypych, Biodegradable composites based on lignocellulosic fibers - An overview. Progr. Polym. Sci. 34(9) (2009) 982–1021. *http://dx.doi.org/10.1016/j.progpolymsci.2008.12.002*

[10] I. Van de Weyenberg, T. Chi Truong, B. Vangrimde, I. Verpoest, Improving the properties of UD flax fibre reinforced composites by applying an alkaline fiber treatment. Compos. Part A-Appl. S. 37 (2006) 1368–1376.

[11] L. Ambrosio, P.A.Netti, S.Iannace, S.J. Huang, L. Nicolais, Composite hydrogels for intervertebral disc prostheses. J. Mater.Sci: Mater. M. 7(5) (1996) 251-254. *http://dx.doi.org/10.1007/BF00058561*

[12] H.C. Amstutz, P. Campbell, N. Kossovsky, I.C. Clartk, Mechanism and clinical significance of wear debris-induced osteolysis. Clin. Orthop. Relat. Res. 276 (1992) 7-18. *http://dx.doi.org/10.1097/00003086-199203000-00003*

[13] G.B.J. Andersson, Intervertebral disk, in: V. Wright, E.L.Radin (Eds.), Mechanics of human joints: physiology, pathophysiology, and treatment. Marcel Dekker Inc., New York (1993) pp. 293-311.

[14] Layth Mohammed, M. N. M. Ansari, Grace Pua, Mohammad Jawaid, M. Saiful Islam, A Review on Natural Fiber Reinforced Polymer Composite and Its Applications. Int. J. Polymer Sci. 2015 (2015), 1-15. *http://dx.doi.org/10.1155/2015/243947*

[15] F.P. La Mantia, M. Morreale, Green composites: A brief review. Compos. Part. A-Appl. S. 42(6) (2011) 579–588.

[16] A.K. Mohanty, M.Misra, L.T. Drzal, Sustainable bio-composites from renewable resources: Opportunities and challenges in the green materials world. J. Polym. Sci. Environ. 10(1) (2002) 19–26. *http://dx.doi.org/10.1023/A:1021013921916*

[17] H.P.S. Abdul Khalil, H. Ismail, M.N. Ahmad, A. Ariffin, K. Hassan, Conventional agro-composites from chemically modified fibres. Ind. Crop. Prod. 26(3) (2007) 315-353. *http://dx.doi.org/10.1016/j.indcrop.2007.03.010*

[18] H.P.S. Abdul Khalil, R.N. Kumar, S.M. Asri, N.A. Nik Fuaad, M.N. Ahmad, Hybrid thermoplastic pre-preg oil palm frond fibers (OPF) reinforced in polyester composites. Polym-Plast. Technol. 46(1) (2007) 43–50.

[19] H.P.S. Abdul Khalil, S. Hanida, S.C.W. Kang, N.A. Nik Fuaad, Agro-hybrid composite: The effects on mechanical and physical properties of oil palm fiber (EFB)/glass hybrid reinforced polyester composites. Polym-Plast. Technol. 46(1) (2007)203–218.

[20] S. Ramakrishna, Biomedical applications of polymer composite materials,in: C.S. Hong, C.S. Kim (Eds.), Proc. of the second Asian-Australasian Conference on Composite Materials (ACCM-2000). Advanced Institute of Science and Technology, Korea, (2000) pp. A431-A436.

[21] L.M. Wenz, K. Merritt, S.A. Brown, A. Moet, A.D. Steffee, In vitro biocompatibility of polyetheretherketone and polysulfone composites. J. Biomed. Mater. Res. 24(2) (1990) 207-215. *http://dx.doi.org/10.1002/jbm.820240207*

[22] G.B. McKenna, G.W. Bradley, H.K. Dunn, W.O.Statton, Mechanical properties of some fiber reinforced polymer composites after implantation as fracture fixation plates. Biomaters. 1(4) (1980) 189-192. *http://dx.doi.org/10.1016/0142-9612(80)90015-0*

[23] L.A. Latour Jr., J. Black, Development of FRP composite structural biomaterials: Ultimate strength of the fiber/matrix interfacial bond in in vivo simulated environments. J. Biomed. Mater. Res. 26(5) (1992) 593-606. *http://dx.doi.org/10.1002/jbm.820260504*

[24] B. Harris, The mechanical behavior of composite materials, The Mechanical Properties of Biological Materials, Cambridge University Press, Cambridge, (1980) pp. 37-74.

[25] S.R. Lee, H.M. Park, H.T. Lim, K.Y. Kang, L. Li, W.J. Cho, C.S. Ha, Microstructure, tensile properties, and biodegradability of aliphatic polyester/clay nanocomposites. Polymer. 43(8) (2002) 2495–2500. *http://dx.doi.org/10.1016/S0032-3861(02)00012-5*

[26] T.W. Lin, A.A. Corvelli, C.G. Frondoza, J.C. Roberts, D.S. Hungerford, Glass peek composite promotes proliferation and osteo-calcin production of human osteoblastic cells. J. Biomed. Mater. Res. 36(2) (1997) 137-144. *http://dx.doi.org/10.1002/(SICI)1097-4636(199708)36:2<137::AID-JBM1>3.0.CO;2-L*

[27] G.W. Hastings, Biomedical applications of CFRPs, in: E.Fitzer (Ed.), Carbon Fibre and their Composites. Springer-Verlag, Berlin, (1988) pp. 261-271.

[28] K. Tayton, J. Bradley, How stiff should semi-rigid fixation of the human tibia be?, J. Bone. Joint Surg. Br. 65(3) (1983) 312-315.

[29] A.E. Profio, Biomedical Engineering, John Wiley & Sons Inc. New York, 1993.

[30] J. Black, Orthopaedic Biomaterials in Research and Practice, Churchill Livingstone. New York, 1988.

[31] D. Philips, Characterization and development of 3D-knitted composites, PhD thesis, Katholieke University Leuven, Belgium, 1999.

[32] K.P. Baidya, S. Ramakrishna, M. Rahman, A. Ritchie, Advanced textile composite ring for ilizarov external fixator system. Proc Inst Mech Eng H. 215(1) (2001) 11-23.
http://dx.doi.org/10.1243/0954411011533490

[33] K.P. Baidya, S. Ramakrishna, M. Rahman, A. Ritchie, Quantitative radiographic analysis of fiber reinforced polymer composites. J. Biomater. Appl. 15(3) (2001) 279-289.
http://dx.doi.org/10.1106/BKLQ-E2YG-D2LA-RG3R

[34] E.Y.S. Chao, H.T. Aro, Biomechanics of fracture fixation, in: V.C. Mow, W.C. Hayes (Eds). Basic Orthopaedic Biomechanics, Lippincott-Raven, Philadelphia (1997) pp. 317-350.

[35] M.S. Ali, T.A. French, G.W. Hastings, T. Rae, N. Rushton, E.R. Ross, C.H. Wynn-Jones, Carbon fibre composite bone plates. Development, evaluation and early clinical experience. J Bone Joint Surg Br. 1990 72(4) 586-591.

[36] J.S. Bradley, Carbon fiber reinforced plastics for orthopedic implants, in: G.W. Hastings, D.F. Williams, (Eds.), Mechanical properties of biomaterials, Wiley, Chichester, New York, (1980) pp. 379-386.

[37] G.B. McKenna, G.W. Bradley, H.K. Dunn, W.O. Statton, Mechanical properties of some fiber reinforced polymer composites after implantation as fracture fixation plates. Biomaterials. 1: (1980) 189-192.
http://dx.doi.org/10.1016/0142-9612(80)90015-0

[38] J. Suwanprateeb, K.E. Tanner, S. Turner, W. Bonfield, Influence of sterilization by gamma irradiation and of thermal annealing on creep of hydroxyapatite-reinforced polyethylene composites. J. Biomed. Mater. Res. 39(1) (1998) 16–22
http://dx.doi.org/10.1002/(SICI)1097-4636(199801)39:1<16::AID-JBM3>3.0.CO;2-L

[39] S.L.Y. Woo, W.H. Akeson, B. Levenetz, R.D. Coutts, J.V. Matthews, D. Amiel, Potential application of graphite fiber and methyl methacrylate resin composites as internal fixation plates. J. Biomed. Mater. Res. 8(5) (1974) 321-338. *http://dx.doi.org/10.1002/jbm.820080513*

[40] P. Christel, J.L. Leray, L. Sedel, E. Morel, Mechanical evaluation and tissue compatibility of materials for composite bone plates, in: G.W. Hastings, D.F. Williams (Eds.), Mechanical Properties of Biomaterials,Wiley, New York, (1980) pp. 367-377.

[41] L. Claes, W. Hutter, R. Weiss, Mechanical properties of carbon fiber reinforced polysulfone plates for internal fixation, in: P. Christel, A.Meunier, A.J.C. Lee (Eds.), Biological and Bio-mechanical Performance of Biomaterials,The Netherlands: Elsevier Science Publishers, Amsterdam, (1997) pp. 81-86.

[42] M.S. Hunt, Development of carbon fiber/polysulfone orthopaedic implants, Mat. Des. 8(2) (1987) 113-119. *http://dx.doi.org/10.1016/0261-3069(87)90117-8*

[43] W. Huttner, G.Keuscher, M.Nietert, Carbon fiber reinforced polysulfone thermoplastic composites,in: P. Ducheyne, G. Van der Perre, A.E. Aubert (Eds.), Biomaterials and Biomechanics,The Netherlands: Elsevier Science Publishers, Amsterdam, (1984) pp. 167-172.

[44] N. Rushton, T. Rae, The intra-articular response to particulate carbon fiber reinforced high density polyethylene and its constituents: an experimental study in mice. Biomaterials. 5(6) (1984) 352-356. *http://dx.doi.org/10.1016/0142-9612(84)90034-6*

[45] N. Gillett, S.A. Brown, J.H. Dumbleton, R.P. Pool, The use of short carbon fiber reinforced thermoplastic plates for fracture fixation. Biomaterials. 6 (1985) 113-121. *http://dx.doi.org/10.1016/0142-9612(85)90074-2*

[46] K.A. Jockisch, S.A. Brown, T.W. Bauer, K. Merritt, Biological response to chopped-carbon-fiber-reinforced peek, J. Biomed. Mater. Res. 26(2) (1992) 133-146. *http://dx.doi.org/10.1002/jbm.820260202*

[47] K.B. Kwarteng, C. Stark, Carbon fiber reinforced PEEK (APC-2/ AS4) composites for orthopedics implants. SAMPE Quarterly. 21(2) (1990) 10-14.

[48] J. Mayer, K. Ruffieux, B. Koch, E. Wintermantel, T. Schulten, A. Hatebur, The double die technique (DDT): biomaterials processing for adaptable high fatigue

resistance thermoplastic carbon fiber osteosynthesis plates, J. Biomed. Eng. App. Bas. C. 5 (1993) 778-783.

[49] T. Peter, R. Tognini, J. Mayer, E. Wintermantel, Homoelastic, anisotropic osteosynthesis system by net-shape processing of endless carbon fiber reinforced polyetheretherketone (PEEK),in: J.C.H. Goh, A. Nather (Eds.), Proc. of 9th Conference on Biomedical Engineering, Singapore, National University of Singapore, (1997) pp. 317-319.

[50] D.F. Williams, Consensus and definitions in biomaterials,in: C. de Putter, K. de Lange, K. de Groot, A.J.C. Lee (Eds.), Advances in Biomaterials, Elsevier Science, Amsterdam, (1988) pp. 11-16.

[51] F.N. Cogswell, Thermoplastic Aromatic Polymer Composites, Butterworth Heinemann, UK, 1992.
http://dx.doi.org/10.1533/9781845699253

[52] J. Mayer, Gestrickeaus Kohlenstoffasern fur biokompatibleVerbundwerkstoffe, dargestellt an einerhomoelastischen Osteosyntheseplatte, PhD thesis, ETH Zurich, Switzerland, 1994.

[53] P. Christel, L.Claes, S.A. Brown, Carbon reinforced composites in orthopedic surgery, in: M. Szycher (Eds.), High Performance Biomaterials: A Comprehensive Guide to Medical and Pharmaceutical Applications, Technomic, Lancaster, (1991) pp. 499¬518.

[54] C. Morrison, R. Macnair, C. MacDonald, A. Wykman, I. Goldie, M.H. Grant, In vitro biocompatibility testing of polymers for orthopedic implants using cultured fibroblasts and osteoblasts, Biomaterials. 16(13) (1995) 987-992.
http://dx.doi.org/10.1016/0142-9612(95)94906-2

[55] J. Mayer, E. Wintermantel, Thermoforming process for knitted fabric reinforced thermoplastics. New manufacturing techniques for load bearing anisotropic implants. In: Bhattacharyya D, (Eds.) Composite Sheet Forming. Amsterdam: Elsevier Science, (1997) pp. 403-40.
http://dx.doi.org/10.1016/S0927-0108(97)80012-4

[56] S. Blazewicz, J. Chlopek, A. Litak, C. Wajler, E. Staszkow, Experimental study of mechanical properties of composite carbon screws, Biomaterials. 18(5) (1997) 437-439.
http://dx.doi.org/10.1016/S0142-9612(96)00067-1

[57] U. Loher, R. Tognini, J. Mayer, B. Koch, E. Wintermantel, Development of a cortical bone screw made with endless carbon fiber reinforced

Polyetheretherketone (CF-PEEK) by extrusion. A new method,in: The Institute of Materials,7th international conference on polymers in medicine and surgery, Amsterdam, (1993) pp 88-97.

[58] R. Giardino, S. Glannini, M. Fini, G. Giavaresi, L. Martini, L. Orienti, Experimental in vivo model to evaluate resorable implants into bone,in: H.L. Kim, L.P. Chung (Eds.), Biodegradable Implants in Fracture Fixation, (1993) pp. 143-51.

[59] P. Tormala, J. Vasenius, S. Vainionpaa, J. Laiho, T. Pohjonen, P. Rokkanen, Ultra-high-strength absorbable self-reinforced polyglycolide (SR-PGA) composite rods for internal fixation of bone fractures: In vitro and in vivo study. J. Biomed. Mater. Res. 25(1) (1991) 1-22.
http://dx.doi.org/10.1002/jbm.820250102

[60] J. Choueka, J.L. Charvet, H. Alexander, Y.H. Oh, G. Joseph, N.C. Blumenthal, W.C. LaCourse, Effect of annealing temperature on the degradation of reinforcing fibers for absorbable implants, J. Biomed. Mater. Res. 29(11) (1995) 1309-1315.
http://dx.doi.org/10.1002/jbm.820291102

[61] C. Scholz, D. Huimacher, W. Bahr, L. Cales, The development and testing of three biodegradable screw plate-systems for maxillofacial surgery,in: H.L. Kim, L.P. Chung (Eds.), Biodegradable Implants in Fracture Fixation, (1993) pp. 67-73.

[62] M. Dauner, L. Caramaro, Y. Missirlis, E. Panagiotopoulos, Resorable continuous-fiber reinforced polymers for osteo-synthesis. J. Mater. Sci.: Mater. in Med. 9(3) (1998) 173-179.
http://dx.doi.org/10.1023/A:1008823804460

[63] A. Nazre, S. Lin, Theoretical strength comparison of bioabsorbable (PLLA) plates and conventional stainless steel and titanium plates used in internal fracture fixation, in: J.P. Harvey Jr., F. Games (Eds.), Clinical and Laboratory Performance of Bone Plates, ASTM, Philadelphia, (1994) pp. 53-64.
http://dx.doi.org/10.1520/STP12221S

[64] J.R. Parsons, A. Alexander, A.B. Weiss, Absorbable polymer-filamentous carbon composites: a new concept in orthopedic biomaterials. In: M. Szycher (Eds.) Biocompatible Polymers, Metals, and Composites. Lancaster: Technomic Publishing Co. Inc, (1983) pp. 873-905.

[65] P. Tormala, S. Vainionpaa, J. Kilpikari, P. Rokkanen, The effects of fiber reinforcement and gold plating on the flexural and tensile strength of PGA/PLA copolymer materials in vitro. Biomaterials. 8(1) (1987) 42-45.

http://dx.doi.org/10.1016/0142-9612(87)90027-5

[66] M. Zimmerman, J.R. Parsons, H. Alexander, The design and analysis of a laminated partially degradable composite bone plate for fracture fixation. J. Biomed. Mater. Res. 21(43) (1987) 345-361.

[67] J.S. Roman, P.G. Garcia, Partially biodegradable polyacrylic-polyester composites for internal bone fracture fixation. Biomaterials. 12 (1991) 236-241. *http://dx.doi.org/10.1016/0142-9612(91)90206-P*

[68] L.R. Rubin Biomaterials in maxillo-facial surgery. In: Szycher M, editor. Biocompatible Polymers, Metals, and Composites. Lancaster, USA: Technomic Publishing, (1983) pp. 941-951.

[69] T.W. Lin, A.A. Corvelli, C.G. Frondoza, J.C. Roberts, D.S. Hungerford, Glass peek composite promotes proliferation and osteo-calcin production of human osteoblastic cells. J. Biomed. Mater. Res. 36(2) (1997) 37-144. *http://dx.doi.org/10.1002/(SICI)1097-4636(199708)36:2<137::AID-JBM1>3.0.CO;2-L*

[70] J. Kettunen, A. Makela, H. Miettinen, T. Nevalainen, M. Heikkila, P. Tormala, P. Rokkanen, Fixation of femoral shaft osteotomy with an intramedullary composite rod: An experimental study on dogs with a two-year follow-up, J. Biomat. Sci. Polym. Ed. 10 (1999) 33-45.

[71] A. Ignatius, K. Unterricker, K. Wenger, M. Richter, L. Claes, A new composite made of polyurethane and glass ceramic in a loaded implant model: a biomechanical and histological analysis. J Mater Sci Mater Med. 8(12) (1997) 753-756. *http://dx.doi.org/10.1023/A:1018508511787*

[72] L. Claes, M. Schultheiss, S. Wolf, H.J. Wilke, M. Arand, L. Kinzl, A new radiolucent system for vertebral body replacement: its stability in comparison to other systems. J Biomed Mater Res. 48(1) (1999) 82-89. *http://dx.doi.org/10.1002/(SICI)1097-4636(1999)48:1<82::AID-JBM14>3.0.CO;2-E*

[73] M. Marcolongo, P. Ducheyne, J. Garino, E. Schepers, Bioactive glass fiber/polymeric composites bond to bone tissue, J. Biomed. Mater. Res. 39(1) (1998) 161-170. *http://dx.doi.org/10.1002/(SICI)1097-4636(199801)39:1<161::AID-JBM18>3.0.CO;2-I*

[74] A.D.C. Valdevit, N. Inoue, B.A. MacWilliams, L.L. Anderson, Methods for mechanical testing of spinal constructs. Spine. 10(2) (1996) 231-248.

[75] E. Wintermantel, A. Bruinink, K. Eckert, K. Ruffiex, M. Petitmermet, J. Mayer, Tissue engineering supported with structured biocompatible materials: goals and achievements,in: M.O. Speidel MO, P.J. Uggowitzer (Eds.), Materials in Medicine, vdfHochschulverlag AG an der ETH, Zurich, 1998, pp. 1-136.

[76] J.W. Brantigan, A.D. Steffee, J.M. Geiger, A carbon fiber implant to aid interbody lumbar fusion Mechanical testing, Spine (Phila Pa 1976). 16(6 Suppl) (1991) S277-82.
http://dx.doi.org/10.1097/00007632-199106001-00020

[77] P. Ciappetta, S. Boriani, G.P. Fava, A carbon fiber reinforced polymer cage for vertebral body replacement: a technical note. Neurosurgery. 41(5) (1997) 1203-6.
http://dx.doi.org/10.1097/00006123-199711000-00040

[78] Q.B. Bao, G.M. McCullen, P.A. Higham, J.H. Dumbleton, H.A. Yuan, The artificial disc: theory, design and materials, Neurosurgery. 41(5) (1997) 1203-1206.

[79] K.H.G. Schmitt-Thomas, Z.G. Yang, T. Hiermer, Performace characterization of polymeric composite implant rod subjected to torsion, in: Proceedings of ICCM-11, Gold Coast, Australia, (1997) pp. V277-V286.

[80] P.S. Walker, G.W. Blunn, Biomechanical principles of total knee replacement design,in: V.C. Mow, W.C. Hayes (Eds.), Basic Orthopaedic Biomechanics, Lippincott- Raven Publishers, Philadelphia (1997) pp. 461-93.

[81] N. Inoue, Y. Hirasawa, T. Hirai, T. Katayama, Composite materials in bio-medical engineering. Materiaux and Techniques 4-5 (1994) 23-26.

[82] C. Silverton, A.O. Rosenberg, R.M. Barden, M.B. Sheinkop, J.O. Galante, The prosthesis-bone interface adjacent to tibial components inserted without cement. J. Bone Joint Surg. Am. 78(3) (1996) 340-347.

[83] M. Deng, W. Shalaby, Properties of self-reinforced ultra-high-molecular weight polyethylene composites. Biomaterials. 18(9) (1997) 645-55.
http://dx.doi.org/10.1016/S0142-9612(96)00194-9

[84] J.O. Esslinger, E.J. Rutkowski, Studies on the skeletal attachment of experimental hip prostheses in the pygmy goat and the dog. J. Biomed. Mater. Res. 7(3) (1973) 187-193.
http://dx.doi.org/10.1002/jbm.820070312

[85] E. Sclippa, K. Piekarski, Carbon fiber reinforced polyethylene for possible orthopedic uses. J. Biomed. Mater. Res. 7(1) (1973) 59-70.
http://dx.doi.org/10.1002/jbm.820070105

[86] St. John KR. Applications of Advanced Composites in Orthopedic Implants. In: M. Szycher (Eds.) Biocompatible Polymers, Metals, and Composites. Lancaster: Technomic Publishing, (1983) pp. 861-871.

[87] N.C. Blumenthal, V. Cosma, W. Jaffe, S. Stuchin, A new technique for quantitation of metal particulates and metal reaction products in tissues near implants. J Applied Biomater. 5(3) (1994) 191-193. *http://dx.doi.org/10.1002/jab.770050303*

[88] P.K. Buchert, B.K. Vaughn, T.H. Mallory, C.A. Engh, J.D. Bobyn, Excessive Metal Relase due to Loosening and Fretting of Sintered Particles on Porous-Coated Hip Prostheses. J. Bone Joint Surg. Am. 68(4) (1986) 606-609.

[89] E. Schneider, C. Kinast, J. Eulenberger, D. Wyder, G. Eskilsson, S.M. Perren, A comparative study of the initial stability of cementless hip prostheses. Clin Orthop Relat Res. 248 (1989) 200-209. *http://dx.doi.org/10.1097/00003086-198911000-00032*

[90] P.S. Walker, D. Schneeweis, S. Murphy, P. Nelson, Strains and micromotions of press-fit femoral stem prostheses. J Biomech. 20(7) (1987) 693-702. *http://dx.doi.org/10.1016/0021-9290(87)90035-2*

[91] H. Weinans, Mechanically induced bone adaptations around hip implants. PhD thesis. University of Nijmegen, Netherlands, 1991.

[92] L.A. Whiteside, The effect of stem fit on bone hypertrophy and pain relief in cementless total hip arthroplasty. Clin Orthop Relat Res. (247) (1989) 138-147. *http://dx.doi.org/10.1097/00003086-198910000-00023*

[93] J.A. Simoes, A.T. Marques, G. Jeronimidis, Design of a controlled stiffness composite proximal femoral prosthesis. Compos Sci Technol. 60(4) (2000) 559–567. *http://dx.doi.org/10.1016/S0266-3538(99)00155-4*

[94] F.K. Chang, J.L. Perez, J.A. Davidson, Stiffness and strength tailoring of a hip prostheses made of advanced composite materials, J: Biomed: Mater: Res. 24(7) (1990) 873-99. *http://dx.doi.org/10.1002/jbm.820240707*

[95] M. Akay, N. Aslan, Numerical and experimental stress analysis of a polymeric composite hip joint prostheses. J. Biomed. Mater. Res. 31(2) (1996) 167-182. *http://dx.doi.org/10.1002/(SICI)1097-4636(199606)31:2<167::AID-JBM3>3.0.CO;2-L*

[96] E. Wintermantel, J. Mayer, J. Blum, K.L. Eckert, P. Luscher, M. Mathey, Tissue engineering scaffolds using superstructures. Biomaterials. 17(2) (1996) 83-91. *http://dx.doi.org/10.1016/0142-9612(96)85753-X*

[97] J.L. Goldner, J.R. Urbaniak, The clinical experience with silicone- Dacron metacarpophalangeal and interphalangeal joint prostheses. J. Biomed. Mater. Res. 7(3) (1973) 137-163. *http://dx.doi.org/10.1002/jbm.820070310*

[98] S. Saha, S. Pal, Mechanical properties of bone cement: a review. J. Biomed. Material. Res. 18 (1984) 435-462. *http://dx.doi.org/10.1002/jbm.820180411*

[99] C.R. Bragdon, S. Biggs, W.F. Mulroy, W.F. Kawate, W.H. Harris. Defects in the cement mantle: a fatal flaw in cemented femoral stems for THR. San Francisco: Society for Biomaterials, 1995.

[100] S.P. James, M. Jasty, J. Davies, H. Piehler, W.H. Harris, A fracto-graphic investigation of PMMA bone cement focusing on the relationship between porosity reduction and increased fatigue life. J. Biomed. Mater. Res. 6(5) (1992) 651-662. *http://dx.doi.org/10.1002/jbm.820260507*

[101] W. Krause, R.S. Mathis, Fatigue properties of acrylic bone cements: Review of the literature. J. Biomed. Mater. Res. 22(A1 Suppl) (1988) 37-53.

[102] R. Feith, Side effects of acrylic cement, implanted to bone. Acta Orthop Scand Suppl. 1975;161:3-136.

[103] R.M. Pilliar, R. Blackwell, I. Macnab, H.U. Cameron, Carbon fiber-reinforced bone cement in orthopedic surgery. J Biomed. Mater. Res. 10 (1976) 893-906. *http://dx.doi.org/10.1002/jbm.820100608*

[104] B. Pourdeyhimi, H.D. Wagner, Elastic and ultimate properties of acrylic bone cement reinforced with ultra-high-molecular-weight polyethylene fibers. J Biomed. Mater. Res. 23(1) (1989) 63-80. *http://dx.doi.org/10.1002/jbm.820230106*

[105] L.D. Timmie Topoleski, P. Ducheyne, J.M. Cuckler, The fracture toughness of titanium-fiber-reinforced bone cement. J. Biomed. Mater. Res. 26(12) (1992) 1599-1617. *http://dx.doi.org/10.1002/jbm.820261206*

[106] H.D. Wagner, D. Cohn, Use of high-performance polyethylene fibers as a reinforcing phase in poly(-methylmethacrylate) bone cement. Biomaterials, 10(2) 1989 139–141.
http://dx.doi.org/10.1016/0142-9612(89)90049-5

[107] J.M. Yang, P.Y. Huang, M.C. Yang, S.K. Lo, Effect of MMA-g-UHMWPE grafted fiber on mechanical properties of acrylic bone cement. J Biomed. Mater. Res. 38(4) (1997) 361-369.
http://dx.doi.org/10.1002/(SICI)1097-4636(199724)38:4<361::AID-JBM9>3.0.CO;2-M

[108] J.L. Gilbert, D.S. Ney, E.P. Lautenschlager, Self-reinforced com-posite poly(methyl methacrylate): static and fatigue properties. Biomaterials. 16(14) (1995) 1043-1055.
http://dx.doi.org/10.1016/0142-9612(95)98900-Y

[109] J. Tamura, K. Kawanabe, T. Yamamuro, T. Nakamura, T. Kokubo, S. Yoshihara, T. Shibuya, Bioactive bone cement: The effect of amounts of glass powder and histologic changes with time. J. Biomed. Mater. Res. 29(5) (1995) 551-559.
http://dx.doi.org/10.1002/jbm.820290502

[110] R. Lakes, Composite biomaterials. In: Bronzino JD (Eds.) The Biomedical Engineering Handbook. Bota Racon, Florida, USA: CRC Press, (1995) pp. 598-610.

[111] M. Okazaki, H. Ohmae, Mechanical and biological properties of apatite composite resins. Biomaterials. 9(4) (1988) 345-348.
http://dx.doi.org/10.1016/0142-9612(88)90031-2

[112] J.B. Park, R.S. Lakes, Biomaterials An Introduction. New York: Plenum Press, (1992) pp. 169-183.
http://dx.doi.org/10.1007/978-1-4757-2156-0

[113] T.N. Gerhart, R.L. Miller, S.J. Kleshinski, W.C. Hayes, In vitro characterization and biomechanical optimization of a biodegradable particulate composite bone cement. J. Biomed. Material. Res. 22(11) (1988) 1071-1082.
http://dx.doi.org/10.1002/jbm.820221110

[114] J.C. Knowles, G.W. Hastings, H. Ohta, S. Niwa, N. Boeree, Development of a degradable composite for orthopedic use: in vivo biomechanical and histological evaluation of two bioactive degradable composites based on the polyhydroxybutyrate polymer. Biomaterials. 1992;13(8):491-6.
http://dx.doi.org/10.1016/0142-9612(92)90099-A

[115] W. Bonfield, Composites for bone replacement. J. Biomed. Eng. 10(6) (1988) 522-526.
http://dx.doi.org/10.1016/0141-5425(88)90110-0

[116] W. Bonfield, M.D. Grynpas, A.E. Tully, J. Bowman, J. Abram, Hydroxyapatite reinforced polyethylene — a mechanically compatible implant material for bone replacement. Biomaterials. 2(3) (1981) 185-186.
http://dx.doi.org/10.1016/0142-9612(81)90050-8

[117] S. Deb, M. Wang, K.E. Tanner, W. Bonfield, Hydroxyapatite- polyethylene composites: effect of grafting and surface treatment of hydroxyapatite. J. Mater. Sci. Material in Med. 7(4) (1996) 191-193.
http://dx.doi.org/10.1007/BF00119729

[118] L.L. Hench, Bioceramics: from concept to clinic. J. Am. Ceram. Soc. 74(7) (1991) 1487-1510.
http://dx.doi.org/10.1111/j.1151-2916.1991.tb07132.x

[119] J. Suwanprateeb, K.E. Tanner, S. Turner, W. Bonfield, Influence of Ringer's solution on creep resistance of hydroxyapatite reinforced polyethylene composites. J. Mater. Sci. Mater. Med. 8(8) (1997) 469-472.
http://dx.doi.org/10.1023/A:1018522025474

[120] D. Williams, Concise Encyclopedia of Medical and Dental Materials. Oxford, UK: Pergamon Press, 1990.

[121] Q. Liu, J.R. de Wijn, C.A. van Blitterwijk, Composite biomaterials with chemical bonding between hydroxyapatite filler particles and PEG/PBT copolymer matrix. J. Biomed. Mater. Res. 40(3) (1998) 490-497.
http://dx.doi.org/10.1002/(SICI)1097-4636(19980605)40:3<490::AID-JBM20>3.0.CO;2-M

[122] S. Hidaka, R.B. Gustilo, Refracture of bones of the forearm after plate removal. J. Bone Joint Surg. Am. 66(8) (1984) 1241-1243.

[123] L.M. Rodriguez-Lorenzo, A.J. Salinas, M. Vallet-Regi, J. San Roman, Composite biomaterials based on ceramic polymers. I. Reinforced systems based on Al2O3/PMMA/PLLA. J. Biomed. Mater. Res. 30(4) (1996) 515-522.
http://dx.doi.org/10.1002/(SICI)1097-4636(199604)30:4<515::AID-JBM10>3.0.CO;2-G

[124] C.C.P.M. Verheyen, J.R. de Wijin, C.A. van Blitterwijk, K. de Groot, Evaluation of hydroxyapatite/ poly(L-lactide) composites: mechanical behavior. J. Biomed. Mater. Res. 26(10) (1992) 1277-1296.
http://dx.doi.org/10.1002/jbm.820261003

[125] C.C.P.M. Verheyen, J.R. de Wijn, C.A. van Blitterwijk, K. de Groot, P.M. Rozing, Hydroxyapatite/poly(L-lactide) composites: an animal study on push-out strengths and interface histology. J. Biomed. Mater. Res. 27(4) (1993) 433-444. *http://dx.doi.org/10.1002/jbm.820270404*

[126] Knowles JC, Hastings GW. In vitro and in vivo investigation of a range of phosphate glass-reinforced polyhydroxybutyrate- based degradable composites. J Materials Science: Materials in Medicine 1993;4:102-6. *http://dx.doi.org/10.1007/BF00120377*

[127] C.P.A.T. Klein, H.B.M. van der Lubbe, K. de Groot, A plastic composite of alginate with calcium phosphate granules as implant material: an in vivo study. Biomaterials. 8(4) (1987) 308-310.

[128] K.C. Kennedy, T. Chen, R.P. Kusy, Behavior of photopolymerized silicate-glass-fiber-reinforced dimethacrylate composites subjected to hydrothermal ageing. J. Mater. Sci-Mater. M. 9(11) (1998) 651-660.

[129] W.R. Krause, S.H. Park, R.A. Straup, Mechanical properties of BIS-GMA resin short glass fiber composites. J. Biomed. Mater. Res. 23(10) (1989) 1195-1211. *http://dx.doi.org/10.1002/jbm.820231008*

[130] E.P. Hoag, T.G. Dwyer, A comparative evaluation of three post and core techniques. J. Prosthet. Dent. 47(2) (1982) 177-181. *http://dx.doi.org/10.1016/0022-3913(82)90183-4*

[131] P.E. Lovdahl, J.I. Nicholls, Pin-retained amalgam cores Vs cast- gold dowel-cores. J. Prosthet. Dent. 38(5) (1977) 507-514. *http://dx.doi.org/10.1016/0022-3913(77)90025-7*

[132] K.C. Trabert, J.P. Cooney, The endodontically treated tooth, restorative concepts and techniques. Dent. Clin. North. Am. 28(4) (1984) 923-951.

[133] F. Issidor, P. Odman, K. Brondum, Intermitttent loading of teeth restored using prefabricated carbon fiber posts. Int. J. Prosthodont. 9(2) (1996) 131-136.

[134] A. Torbjorner, A. Karlsson, M. Syverud, A. Hensten-Pattersen, Carbon fiber reinforced root canal posts, mechanical properties and cytotoxic properties. Eur. J. Oral Sci. 104(5-6) (1996) 605-611. *http://dx.doi.org/10.1111/j.1600-0722.1996.tb00149.x*

[135] D.G. Purton, J.A. Payne, Comparison of carbon fiber and stainless steel root canal posts. Quintessence. Int. 27(2) (1996) 93-97.

[136] V.K. Ganesh, S. Ramakrishna, H.J. Leck, Fiber reinforced composite based functionally gradient materials. Adv. Composites Lett. 7 (1998) 111-115.

[137] S. Ramakrishna, V.K. Ganesh, S.H. Teoh, P.L. Loh, C.L. Chew, Fiber reinforced composite product with graded stiffness. Singapore Patent Application No. 9800874-1, 1998.

[138] P.L. Loh, K. Ravi, U.K. Ganesh, S. Ramakrishna, C.L. Chew, Moisture absorption of carbon fiber reinforced posts. J. Dental Res. 79(5) (2000) 1317.

[139] D. Adams, D.F. Williams, J. Hill, Carbon fiber-reinforced carbon as a potential implant material. J. Biomed. Mater. Res. 12(1) (1978) 35-42.
http://dx.doi.org/10.1002/jbm.820120104

[140] H. Bruckmann, K.J. Huttinger, Carbon, a promising material in endoprosthetics. Part 1: the carbon materials and their mechanical properties. Biomaterials. 1(2) (1980) 67-72.
http://dx.doi.org/10.1016/0142-9612(80)90002-2

[141] H. Bruckmann, G. Keuscher, K.J. Huttinger, Carbon, a promising material in endoprosthetics. Part 2: tribological properties. Biomaterials. 1(2) (1980) 73-81.
http://dx.doi.org/10.1016/0142-9612(80)90003-4

[142] J.P. Louis, M. Dabadie, Fibrous carbon implants for the maintenance of bone volume after tooth avulsion: first clinical results. Biomaterials. 11(7) (1990) 525-528.
http://dx.doi.org/10.1016/0142-9612(90)90071-W

[143] N. More, C. Baquey, X. Barthe, F. Rouais, J. Rivel, M. Trinquecoste, A. Marchand, Biocompatibility of carbon-carbon materials: in vivo study of their erosion using 14carbon labeled samples. Biomaterials. 9(4) (1988) 332-334.
http://dx.doi.org/10.1016/0142-9612(88)90028-2

[144] J.D. Henderson, R.H. Mullarky, D.E. Ryan, Tissue biocompatibility of kevlar aramid fibers and polymethylmethacrylate, composites in rabbits. J. Biomed. Mater. Res. 21(1) (1987) 59-64.
http://dx.doi.org/10.1002/jbm.820210110

[145] V.M. Miettinen, P.K. Vallittu, Release of residual methyl metha-crylate into water from glass fiber-poly (methyl methacrylate) composite used in dentures. Biomaterials. 18(2) (1997) 181-5.
http://dx.doi.org/10.1016/S0142-9612(96)00123-8

[146] N. Bjork, K. Ekstrand, I.E. Ruyter, Implant-fixed, dental bridges from carbon/graphite fiber reinforced poly(methyl methacrylate). Biomaterials. 7(1) (1986) 73-75.
http://dx.doi.org/10.1016/0142-9612(86)90093-1

[147] K.W.M. Davy, S. Parker, M. Braden, I.M. Ward, H. Ladizesky, Reinforcement of polymers of 2,2 bis-4(2-hydroxy-3-methacry- loyloxy propoxy) phenyl propane by ultra-high modulus polyethylene fibers. Biomaterials. 13(1) (1998) 17-19.
http://dx.doi.org/10.1016/0142-9612(92)90088-6

[148] M.A. Freilich, J.C. Meiers, J.P. Duncan, A. Jon Goldberg, Fiber Reinforced Composites in Clinical Dentistry. IL: Qintessence Publishing, 2000.

[149] J. Jancar, A.T. Dibenedetto, Fiber reinforced thermoplastic composites for dentistry, Part I hydrolytic stability of the interface. J. Mater. Sci. - Mater. Med. 4(6) (1993) 555-561.
http://dx.doi.org/10.1007/BF00125593

[150] J. Jancar, A.T. Dibenedetto, A.J. Goldberg, Thermoplastic fiber- reinforced composites for dentistry Part II Effect of moisture on flexural properties of unidirectional composites. J. Mater. Sci. - Mater. Med. 4(6) (1993) 562-568.
http://dx.doi.org/10.1007/BF00125594

[151] T. Imai, F. Watari, S. Yamagata, M. Kobayashi, K. Nagayama, S. Nakamura, Effects of water immersion on mechanical properties of new esthetic orthodontic wire. Am. J. Orthod. Dentofacial. Orthop. 116(5) (1999) 533-538.
http://dx.doi.org/10.1016/S0889-5406(99)70185-X

[152] S.W. Zufall, K.C. Kennedy, R.P. Kusy, Frictional characteristics of composite orthodontic arch wires against stainless steel and ceramic brackets in the passive and active configurations. J. Mater. Sci. - Mater. Med. 9(11) (1998) 611-620.
http://dx.doi.org/10.1023/A:1008977106648

[153] L.L. Hench, E.C. Ethridge Biomaterials An Interfacial Approach. New York: Academic Press, 1982.

[154] S. Kocvara, C.H. Kliment, J. Kubat, M. Stol, Z. Ott, J. Dvorak, Gel-fabric prostheses of the ureter. J. Biomed. Mater. Res. 1(3) (1967) 325-336.
http://dx.doi.org/10.1002/jbm.820010304

[155] A.G. Andreopoulos, M. Evangelatou, P.A. Tarantili, Properties of maxillofacial silicone elastomers reinforced with silica powder. J. Biomater. Appl. 13(1) (1998) 66-73.

[156] S.P. Arnoczky, P.A. Torzilli, R.F. Warren, A.A. Allen, Biologic fixation of ligament prostheses and augmentations. An evaluation of bone in growth in the dog. Am. J. Sports Med. 16(2) (1988) 106-112.
http://dx.doi.org/10.1177/036354658801600204

[157] J.L. Berry, W.S. Berg, T.M. Stahurski, J.M. Moran, E.M. Morgan, A.S. Greenwald, Evaluation of Dacron-covered and plain bovine xenografts as replacements for the anterior cruciate ligament, Clin. Orthop. Relat. Res. 236 (1988) 270-278.
http://dx.doi.org/10.1097/00003086-198811000-00037

[158] L. Durselen, L. Claes, A. Ignatius, S. Rubenacker, Comparative animal study of three ligament prostheses for the replacement of the anterior cruciate and medial collateral ligament. Biomaterials. 17(10) (1996) 977-982.
http://dx.doi.org/10.1016/0142-9612(96)84671-0

[159] I.W. Forster, Z.A. Ralis, D.H. Jenkins, Filamentous carbon fiber induction of new tendon: tissue reactions and environmental conditions, in: G.D Winter, J.L.Leray, K. de Groot (Eds.), Evaluation of Biomaterials, John Wiley & Sons Ltd, New York, 1980. pp. 367-371.

[160] D.H.R. Jenkins, Experimental and clinical application of carbon fiber as an implant in orthopedics. J. Bone Joint Surg. 59-B(4) (1977) 501.

[161] M.W. King, N. Poddevin, R. Guidoin, Y. Marois, B. Cronier, A.Y. Belanger, Y.P. Delagoutte, Designing textile structures to repair and replace knee ligaments. Canadian Textile Journal.113(3) (1996) 53-57.

[162] Y. Marois, R. Roy R, T. Vidovszky, M.W. King, A.Y. Bélanger, C. Chaput, R. Guidoin, Histopathological and immunological investigations of synthetic fibres and structures used in three prosthetic anterior cruciate ligaments: in vivo study in the rat. Biomaterials. 14(4) (1993) 255-262.
http://dx.doi.org/10.1016/0142-9612(93)90115-I

[163] D.G. Mendes, M.Iusim, D. Angel, A.Rotem,D: Mordehovich, M. Roffman, S.Lieberson, J. Boss, Ligament and tendon substitution with composite carbon fiber strands. J. Biomed. Mater. Res. 20(6) (1986) 699-708.
http://dx.doi.org/10.1002/jbm.820200604

[164] K.Y. Pedro, D.L. Christiansen, R.A. Hahn, S.J. Shieh, J.D. Goldstein, F.H. Silver, Mechanical properties of collagen fibers: a comparison of reconstituted and rat tail tendon fibers. Biomaterials. 10(1) (1989) 38–42.
http://dx.doi.org/10.1016/0142-9612(89)90007-0

[165] C.J. Peterson, J.H. Donachy, A. Kalenak, A segmented poly-urethane composite prosthetic anterior cruciate ligament in vivo study. J. Biomed. Mater. Res. 19(5) (1985) 589-594. *http://dx.doi.org/10.1002/jbm.820190510*

[166] J.V. Wening, H. Marquardt, A. Katzer, K.H. Jungbluth, Cytotoxicity and mutagenicity of Kelvar®: an in vitro evaluation. Biomaterials. 16(4) (1995) 337-340. *http://dx.doi.org/10.1016/0142-9612(95)93262-C*

[167] M.N. Pradas, R.D. Calleja, Reproduction in a polymer composite of some mechanical features of tendons and ligaments,in: M. Szycher (Ed.), High Performance Biomaterials:A Comprehensive Guide to Medical and Pharmaceutical Applications. Technomic Publishing, Lancaster, (1991) pp. 519-523.

[168] S. Iannace, G. Sabatini, L. Ambrosio, L. Nicolais, Mechanical behavior of composite artificial tendons and ligaments. Biomaterials. 16(9) (1995) 675-680. *http://dx.doi.org/10.1016/0142-9612(95)99693-G*

[169] L. Ambrosio, P.A. Netti, S. Iannace, S.J. Huang, L. Nicolais, Composite hydrogels for intervertebral disc prostheses. J. Mater. Sci: Mater. M. 7(5) (1996) 251-254. *http://dx.doi.org/10.1007/BF00058561*

[170] L. Ambrosio, R. De Santis, S. Iannace, P.A. Netti, L. Nicolais, Viscoelastic behavior of composite ligament prostheses. J. Biomed. Mater. Res. 42(1) (1998) 6-12. *http://dx.doi.org/10.1002/(SICI)1097-4636(199810)42:1<6::AID-JBM2>3.0.CO;2-U*

[171] B. Gershon, D. Cohn, G. Marom, Compliance and ultimate strength of composite arterial prostheses. Biomaterials. 13(1) (1992) 38-43. *http://dx.doi.org/10.1016/0142-9612(92)90093-4*

[172] B. Gershon, D. Cohn, G. Marom, Utilization of composite laminate theory in the design of synthetic soft tissues for biomedical prostheses. Biomaterials. 11(8) (1990) 548-552. *http://dx.doi.org/10.1016/0142-9612(90)90076-3*

[173] N. Klein, M.L. Carciente, D. Cohn, G. Marom, G. Uretzky, H. Peleg, Filament-wound composite soft tissue prostheses: controlling compliance and strength by water absorption and degradation. J. Mater. Sci-Mater M. 4(3) (1993) 285-291. *http://dx.doi.org/10.1007/BF00122282*

[174] G.C. Robin, Below-knee drop-foot braces: stresses during use and evaluation of design, in: D.N. Ghista (Eds.), Biomechanics of Medical Devices, Marcel Dekker Inc., New York, (1981) pp. 535-567.

[175] A.G.A. Coombes, C.D. Greenwood, J.J. Shorter, Plastic materials for external prostheses and orthoses, in: D.L. Wise, D.J. Trantolo, D.E. Altobelli, M.J. Yaszemski, J.D. Gresser (Eds.), Human Bio-materials Applications, Humana PressTotowa, New York, (1996) pp. 215-255.
http://dx.doi.org/10.1007/978-1-4757-2487-5_11

[176] M.A. Tallent, C.W. Cordova, D.S Cordova, D.S. Donnelly, Ther-moplastic fibers for composite reinforcement, in: S.M. Lee (Eds.), International Encyclopedia of Composites, VCH Publishers, New York, (1990) pp. 466-480.

[177] Z.M. Huang, S. Ramakrishna, Development of knitted fabric reinforced composite material for prosthetic application. Adv. Compos. Lett. 8(6) (1999)289-294.

[178] V.K. Ganesh, S. Ramakrishna, Prothesis: another concept. The European periodical for technical textile users. Quarterly Magazine, 29 (1998) 56-60.

[179] S.l. Toh, J.C.H. Goh, P.H. Tan, T.E. Tay, Fatigue testing of energy storing prosthetic feet. Prosth. Oth. Int. 17 (1993) 180-188.

[180] J.C.H. Goh, P.H. Tan, S.L.Toh, T.E. Tay, Gait analysis study of an energy-storing prosthetic foot-a preliminary study. Gait & Posture. 2(2) (1994) 95–101.
http://dx.doi.org/10.1016/0966-6362(94)90098-1

[181] R. Ward, R.J. Minns, Woven carbon fiber mesh patch versus Dacron mesh in the repair of experimental defects in the lumbar fascia of rabbits. Biomaterials. 10(6) (1989) 425-428.
http://dx.doi.org/10.1016/0142-9612(89)90135-X

[182] J.A. Werkmeister, G.A. Edwards, F. Casagranda, J.F. White, J.A.M. Ramshaw, Evaluation of a collagen-based biosynthetic material for the repair of abdominal wall defects, J. Biomed. Mater. Res. 39 (199) 429-436.
http://dx.doi.org/10.1002/(sici)1097-4636(19980305)39:3<429::aid-jbm12>3.0.co;2-5

[183] S. Ramakrishna, S. Ramaswamy, S.H. Teoh, C.T. Tan, Develop-ment of a knitted fabric reinforced elastomeric Composite intervertebral disc prosthesis. Proc ICCM-11, Conard Jupiters- Golad oast, Australia, 1 (1997) 458-466.

[184] S.H. Teoh, Z.G. Tang, S. Ramakrishna, Development of thin elastomeric composite membranes for biomedical applications, J. Mater. Sci. Mater. Med. 10(6) (1999) 343-352. *http://dx.doi.org/10.1023/A:1026421606939*

[185] B.M. Cherian, A.L. Leão, S.F. de Souza, L.M.M. Costa, G.M. de Olyveira, M. Kottaisamy, E.R. Nagarajan, S. Thomas, Cellulose nanocomposites with nanofibres isolated from pineapple leaf fibers for medical applications. Carb. Polym. 86(4) (2011) 1790-1798. *http://dx.doi.org/10.1016/j.carbpol.2011.07.009*

CHAPTER 3

Evaluation of the Performance Characteristics of Polymeric Geomembranes

Parvathi Krishnan[1], Ramesh Babu Padamati[2], Priya Dasan K[2]
[1]School of Civil and Chemical Engineering, VIT University, TN, India-14
[2]CRANN, Trinity College Dublin, Ireland -2
Address for correspondence: priyajeenetd@gmail.com

Abstract

The increasing consumerism and improved lifestyle has resulted in huge waste generation. The proper disposal of this waste is a real challenge and lots of research are going on around the world to develop new techniques or to improvise the existing techniques to face this challenge. Land filling has been in practice for a long time and different improvisation methodologies have been adopted from time to time. Landfill liners or geomebranes are one of the most important components of landfilling which ensure the safe disposal of wastes. This article summarizes the various physical, chemical, thermal and mechanical factors that affect the performance of a geo-membrane. The modes and mechanisms of failure of these geo-membranes are briefly looked upon. The effect of operating conditions on the projected life of geo-membrane is discussed in detail. Also some of the factors that must be considered before choosing a specific geo-membrane material and the pros and cons of various commonly used geo-membrane materials are also included in this chapter. This chapter looks into providing information that would help in the selection of proper geo-membrane material with respect to exposure conditions, expected life and available resources

Keywords

Geomembrane, Landfill, Polymer Composites, Nanoclay

Contents

1. Introduction

Management of solid waste is a problem that plagued mankind from the dawn of civilizations. Solid waste has been identified as the most visible environment problem, especially in urban areas. Increasing population, urbanization, changing consumption patterns, and industrialization have resulted in an increased solid waste generation and diversification of the type of solid waste generated. The piling up of solid waste has significant adverse impact on human health conditions, environment, society, and thereby affecting the growth, development and the economy of a nation as a whole [1]. Some of the effects of environmental degradation caused due to improper and inadequate waste disposal are surface and groundwater contamination through leachate, air pollution due to burning of wastes, soil contamination, uncontrolled release of methane gas, spreading of diseases by different vectors like birds, rodents, insects, etc.

A landfill can be defined as a site for the disposal of waste materials by burial. Landfill is the oldest form of waste treatment and has been the most common method of organized waste disposal. Landfills include internal waste disposal sites, where a producer of waste carries out his own waste disposal at the place of production, as well as sites used by many producers. Other than as disposal sites they also function as temporary storage, consolidation and transfer, or processing of waste material, i.e. sorting, treatment, or recycling. Landfills are designed in such a way that the environment including air, water and land remains protected. They have been successful in greatly eliminating the risks that the disposal of wastes may pose to the public health and environmental quality. Modern landfills are engineered and operated, in large part and consists of: (a) liner or a geomembrane (b) garbage fill (c) a cap (d) a leachate collection system and (e) a gas recovery system. Fig. 1 represents a general landfill site. Modern landfills are considered to be engineered and operated, in a massive manner. The daily waste is spread fully and well compacted to reduce the volume. Later a cover is placed to spread to prevent infiltration of rainwater, prevent littering due to blowing of wind, reduce odours and keep out the pest. Once the landfill reaches up to its full capacity, an impermeable cap, typically composed of clay soil is placed over it. In general a sanitary landfill site comprises of a liner (synthetic or natural), compacted clay buffer, geocomposite drainage layer, garbage fill, a cap (landfill cover), a landfill gas recovery system, leak detection system, and a leachate collection system. The site for landfill is found to be suitable for the purpose of constructing a landfill only if the features of land behave as natural buffers between the landfill and the environment. For example, the area may have clayey soils, surface water bodies are absent and the water table in that area may be very low, this way water contamination is prevented.

One of the major environmental issues related to landfills involves the leaching of fluids from the landfill into the surrounding soils and groundwater. Thus, there is a need to provide a landfill liner which significantly addresses these issues. A clean landfill utilizes a dirt liner to disengage the garbage from the earth. Disposal of hazardous waste material into a landfill makes it crucial for the landfill liners to be installed. The liner prevents toxic leachate permeating down into the underlying aquifers or nearby rivers. These liners need to provide a well-designed liner system.

Several studies have been conducted to determine and finalize on the type of liner framework to be adopted for different types of landfill. From the studies it has been revealed that the liner selection could be done concerning the impacts posed by the waste on the inhabitants and the surrounding environment. The inhabitants and the local area surrounding the landfill site are the worst sufferers of the potential risk possessed by the dumped waste. Today's state-of-the-art landfill liner ought to satisfy a series of demands; some of which includes preventing leachate flow into groundwater, capacity to hold wastes without failure, minimizing wrinkles, long term durability and providing stability on steep side slopes.

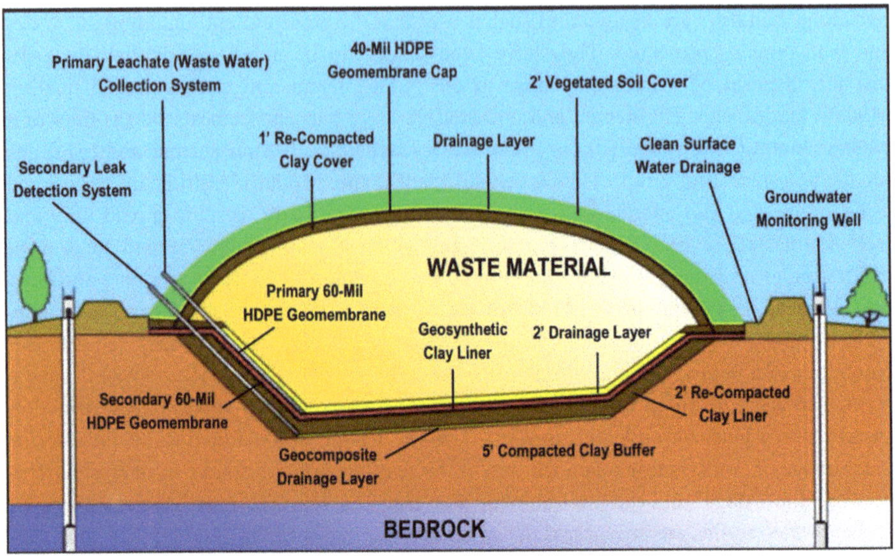

Fig. 1. Pictorial representation of a general sanitary landfill site

From the time of inception of landfills, clay has been used as landfill liners. Landfill liners which are made up of clay have high resistance to permeation of solvents. This is because clay materials have small pore sizes, and due to their complex porous structures their permeability is very low, high plasticity characteristics, good swelling characteristics, and high ability to adsorb cations and allows low rate of diffusive transport of chemicals through it [2]. Natural liners can be made up of either compacted native clay or bentonite clay. For the design of compacted native clay liners, it is crucial for compatibility between waste and natural clay to be attained. The volume change and permeability characteristics of the compact have to be ascertained. As a result of differential settlement of foundation materials, there is a possibility of bulk transport of liquids through clay liners. While in the case of bentonite clay liner, the bentonite clay is often mixed with the existing subgrade material so as to form a low permeability liner. But due to swelling, shrinkage, fracture and cracks of the liner material, the hydraulic conductivity of clay liners can be significantly reduced. Also it has been observed that the transportation and the handling costs of clay liners have been high. Then with the advancement of technology in the field of solid waste management, several artificial materials have come to replace the natural clay liners, the most common being polymer.

Fig 2 shows the classification of geosynthetic liners. This is done based on their manufacturing method.

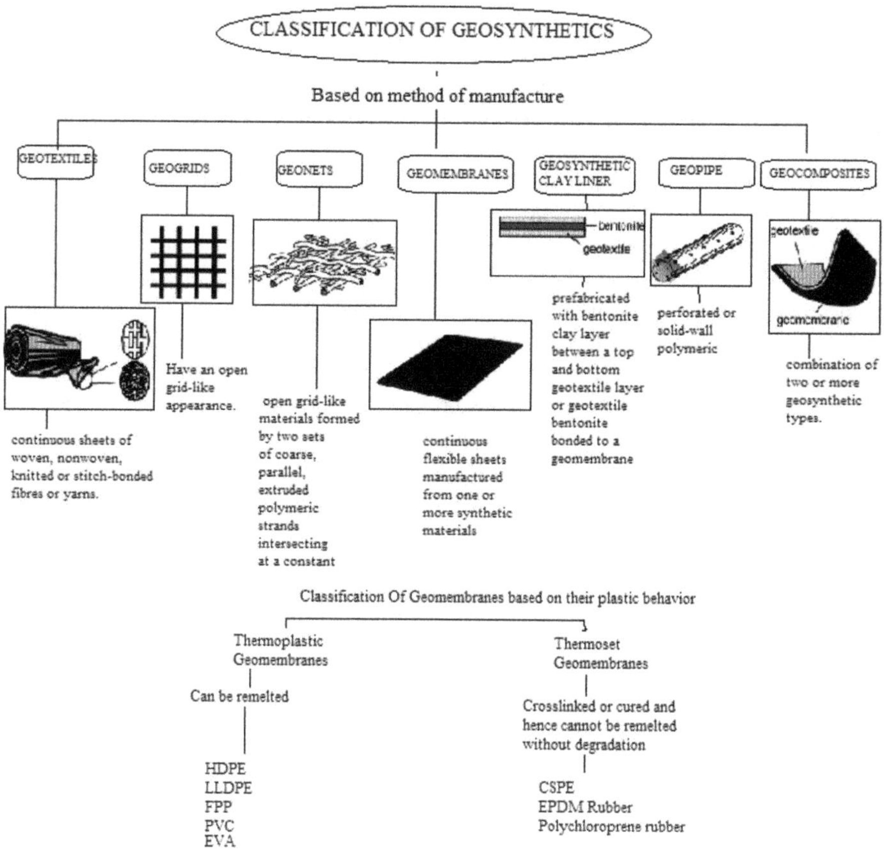

Fig. 2. Classification of geosynthetics based on the method of manufacture.

Municipal solid waste (MSW) landfill uses a synthetic (polymer) liner to isolate the waste from the environment. It is, therefore, particularly important to use robust geosynthetic systems to reliably contain these leachates. The drainage and storage or treatment of the leachate become an important challenge to be solved by geosynthetic systems as clay type solutions are not capable provide effective mechanisms to safely and effectively address this element. These liners have been successfully implemented due to its ability to prevent the permeation of biodegraded leachate components to the soil and ground water below it.

Besides functioning as a liner in landfills, geomembranes have wide range of applications including waste management, water collection ponds, municipal solid waste and hazardous waste landfills, floating covers, water protection systems and other containment facilities. They are generally impervious sheets of polymeric material primarily used to reduce the hydraulic conductivity of *in-situ* soils. They are specifically designed to enhance the building properties of natural soils. In particular, ever since their inception in the early 80's, high density polyethylene (HDPE) geomembranes are the most common choice. Geomembranes made of HDPE and PVC represent the safest, most efficient, reliable and economical from the very beginning of application of geomembranes. Geomembranes made of HDPE are much more popular as they have high tensile strength; high inherent flexibility achieved without additives, high chemical resistance. However, they are not environmental friendly since they do not permit the natural degradation of organic constituents present in the waste. Landfill liners which are synthetic in nature are usually classified into three types depending on the nature of plastic material that is used: thermoplastic geomembrane, elastomeric geomembrane and thermoset resins. Thermoset geomembranes exhibit better efficiency and long term durability since they cannot be reheated without degradation. While the thermoplastic geomembranes such as HDPE, EPDM (ethylene propylene diene mixture), EVA, PVC (Poly-vinyl chloride), CSM (Chlorosulphonated polyethylene), LLDPE (linear low density polyethylene), etc. can be easily reheated without any degradation of the polymer. Elastomers such as polyisobutylene-polyisoprene, polychloroprene etc. are geomembrane materials that under the influence of minor stresses tends to show large elastic deformations [3]. The properties, advantages and disadvantages of the most frequently used types of artificial geomembranes are discussed in Table 1.[4]

Table 1. Comparison of various properties of geomembranes.

Sl. No	TYPE	PROPERTIES	ADVANTAGES	DISADVANTAGES
1.	HDPE	Semi crystalline thermoplastic, most chemical resistant liner material, low permeability, UV resistant	Broad chemical resistance, good weld strength, good low temperature properties, relatively inexpensive	Potential for stress cracking, high degree of thermal expansion, poor puncture resistance, poor multiaxial strain properties
2	LDPE	More flexible than HDPE , withstands tension, high elongation capacity, puncture and stress crack resistant, good chemical resistant, low permeability	Better flexibility than HDPE, better lay flat than HDPE, good multi axial strain properties	Inferior UV resistance to HDPE, inferior chemical resistance to HDPE
3.	FPP	Usually scrim reinforced for high tensile strength, used for floating covers on surface impoundments, other high stress applications, good chemical, UV resistance	Factory fabricated and folded	Limited resistance to hydrocarbons and chlorinated water

4.	PVC	Good puncture resistance, very good chemical resistance, poor UV resistance, excellent flexibility	Good workability and layflat behavior, easy to seam, and be folded so fewer field fabricated seams	Poor resistance to UV and ozoneunless specially formulated,poor resistance to weathering, poor performance at high and lowtemperatures
5	EPDM	Good UV résistance, excellent temperature performance	High strength, good UV résistance	Low resistance to oils, hydrocarbons and solvents, and poor seam quality
6	CSPE	Very good chemical resistance, excellent UV and temperature resistant, good tear and puncture resistant	Outstanding resistance to UV, chemicals, acids , oils and ozone, good performance at low temperatures	Cannot be thermally welded after ageing

The factors which decide on the type of geomembrane depend upon the type of polymer used, mechanical properties, liner-waste stream association, sufficient contact, kind of fabric fortification, representing all the burdens forced, chemical resistance and product life expectancy [4]. It also depends upon the purpose and the application for which it is being used [5]. It is likewise crucial to pick the right blend of added substances, for example, stabilizers, carbon dark, fungicides and anti-oxidants in the formulation of geomembranes [4, 6].

The present chapter focuses on the various properties of polymeric geomembranes and their importance in the selection of a particular type of material during the manufacture of geomembranes. There are certain aspects to be considered before the selection of geomembrane materials. Some of which include choice of polymers, mechanical properties, liner-waste stream interaction, adequate friction, type of fabric reinforcement, accounting for all the stresses imposed, chemical resistance and product life expectancy.

It is also essential to choose the correct combination of additives such as stabilizers, carbon black, fungicides and anti-oxidants.

2. Factors effecting performance of geomembranes

The main factors affecting the performance of geomembranes are density of membrane, its mechanical properties, thermal properties, permeability, and environmental stress. As a result of degradations the material undergoes several changes in its performance thereby reducing its service life. The changes in properties of these materials are often termed as ageing. The effects of these aging factors on the performance of geomembranes are discussed below.

2.1. Oxidative degradation

Oxidative degradations are considered to be the most harmful to geomembranes when exposed to thermal or radioactive energy [4, 6, 7]. Changes in the molecular structure and morphology of the geomemberanes occur due to chemical reactions between polymer chains and oxygen. This type of degradation often takes place in three stages as shown in Figure 3:

Depletion of antioxidants or stabilizers (A): Antioxidants which are usually added during the manufacture of geomembranes undergoes depletion at several stages of ageing of the geomembrane. This has been found to be mainly due to direct exposure to sunlight and UV radiation. The process of depletion of antioxidants has been found to follow a typical kinetics of first order reaction of decay.

Oxidative induction time (OIT) without any degradation of polymers (B): During this induction period, the antioxidant depletion just begins and then the oxidation of polymers becomes adequate enough to cause a variation in the engineering properties of the geomembranes.

Complete degradation of polymers resulting in ultimate failure (C): This stage of degradation begins when the engineering properties of geomembranes have changed considerably, finally leading to the ultimate failure of polymers

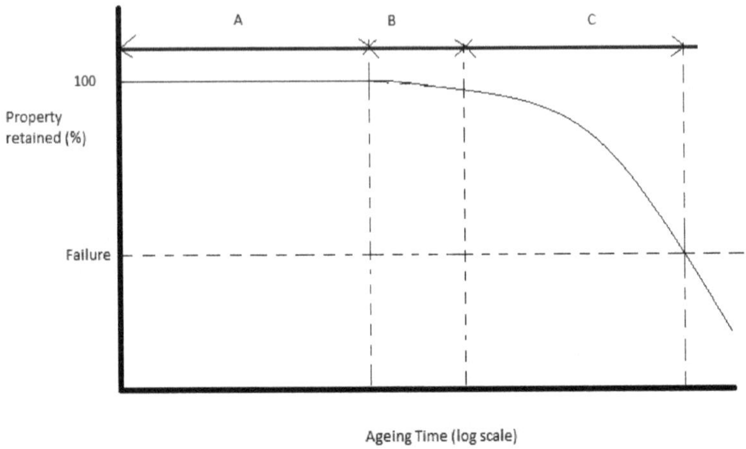

Fig. 3. Stages of degradation of polymers

The antioxidant depletion stage involves the depletion of stabilizers in high ranges of temperature without any loss in the environmental properties of the material. Usually antioxidants used as processing stabilizers have their effective temperature at ambient temperatures (e.g. Phosphites) and at elevated temperatures (e.g. HALS). Even long term stabilizers are used in order to provide protection during a low temperature service life [8]. Depletion of antioxidants may take place not under normal conditions but by natural processes such as migration, dissolution, extraction, evaporation, etc. For HDPE geomembranes, the antioxidant depletion depends on the type of antioxidants and not on the initial oxidative induction time (OIT) values since antioxidant depletion depends largely on the diffusion potential of antioxidants. The geomemberanes which contain hindered amine light stabilizers (HALS) will have longer antioxidant depletion time than those without HALS [9]. The key constituent of leachate affecting antioxidant depletion is surfactant but when the concentration of surfactant reached 5 mg/L its effect diminishes. Rowe et al. [10] identified the critical characteristics of municipal solid waste (MSW) leachate with respect to the antioxidant depletion. A 1.5mm thick HDPE geomembrane was incubated in four different leachates which possessed different combinations of main chemical combinations found in MSW leachate. There was no significant difference in the antioxidant depletion rate in all four different leachates. However, the leachate composition effects on the OIT were later investigated by Rowe et al [11]. Two series of accelerated ageing tests were conducted on a 1.5mm thick HDPE geomembrane immersed in different synthetic leachates. One test series was conducted at five different temperatures (22–85 °C) using four different synthetic leachates. The results suggested that the leachate should be replaced every two weeks in this type of test. The OIT mechanism was found to be similar in all the four leachates since they exhibited a very small difference in activation energies of the corresponding leachates [12]. Studies conducted by Gulec et al. [13] involved a 1.5 mm HDPE geomembrane incubated in synthetic acid mine drainage (AMD), acidic water with pH=2, and deionized water at three different temperatures (60, 40 and 20 °C). Their results showed a faster antioxidant depletion rate in synthetic AMD than in acidic or deionized water.

The HDPE geomembranes are usually subjected to oxidative degradation by virtue of which their long term durability is affected. As oxidation proceeds, the physical and mechanical properties of the polymer start to change, eventually leading to failure of the geomembrane as an effective hydraulic barrier. The anti-oxidant depletion rates of a HDPE geomembrane was analysed by oven aging followed by immersion in acidic and alkaline buffer solutions. The rate of antioxidant depletion increased with increase in temperature and was greater with exposure to alkaline solutions. The values of antioxidant depletion time were found to be longest on exposure to acidic condition and shortest on exposure to alkaline conditions as per the estimates by Arrhenius modelling (Han-Yong Jeon et al [14]. Werner Mueller [15] investigated the oxidative resistance of HDPE geomembranes and estimated that the service life of HDPE geomembrane is mainly determined by a slow loss of stabilizers due to migration. There was no change in the mechanical properties of the geomembrane samples as long as there was a considerable value of OIT at temperatures greater than the melting point. It was also estimated that the oxidative degradation started when the OIT value had reduced to very low values or zero. For HDPE geomembranes, the antioxidant depletion depends on the type of antioxidants and not on the initial OIT values since antioxidant depletion depends largely on the diffusion potential of antioxidants. The geomembranes which contain hindered amine light stabilizers (HALS) will have longer antioxidant depletion time than those without HALS [16]. The effect of UV light on the thermo-oxidative stability of low density polyethylene (LDPE) films using isothermal differential scanning calorimetry (DSC) was investigated and observed that films which contained only antioxidant were sensitive to gamma radiation. On the other hand, films cross-linked by γ-radiation and containing an antirad agent along with the antioxidants and UV stabilizers attained improved thermo oxidative stability [17].

2.2 Chemical degradation

During the installation of geomembranes, a number of stresses act on it. In addition to the internal stresses, fluid heads can occur causing adequate load on the material. Chemical attacks along with physical attacks occur as a result of which the polymer will swell and soften on absorption of waste. James Steven Hodge et al [18] investigated the chemical resistance of geomembrane liners. Cyclic loading, a combination of physical testing and chemical exposure was adopted for this purpose. The test samples under an aggressive chemical environment were subjected to a cyclic force. This resulted in failure at the end of 3 cycles. It was found that the results had an implication on the design of a land disposal facility. The medium density polyethylene (MDPE) geomembranes are found to have excellent chemical resistance as they do not undergo degradation themselves. Few chemical compounds such as aromatics and halocarbons swell the polymer structure and increase the permeability [19].

2.3 Permeability of geomembranes

Contaminant migration through intact geomembranes is a molecule activated process which involves three key steps; (i) partition of the contaminant between inner surface of the geomembrane and the medium containing contaminant (adsorption); (ii) diffusion of the permeant through the geomembrane and, (iii) partition between the outer surface of the geomembrane and the outer medium (desorption). These methods were evaluated on performing sorption and desorption tests on a 2mm thick HDPE geomembrane by using three chlorinated and four aromatic hydrocarbons. In the diffusion test, apparatus consisted of a closed system with source and receptor reservoirs separated by geomembrane to be examined. It was observed that the partitioning parameter controlled the decrease in the source concentration while the permeation coefficient dominated the increase in receptor concentration. The variations in the source and receptor were different for each contaminant indicating that the migration of the organic contaminants through HDPE geomembrane is contaminant dependent [10].

2.4 Density of geomembranes

Data on density of the plastic film are useful because the material properties vary greatly with the density of the polymer [20]. The density is also an indicator of the crystallinity of the material. Since the crystalline parts of the LDPE are well ordered and the material included in them is well packed, it covers less volume than the equal quantity of amorphous material. Thus, films with more crystallinity will exhibit higher density, than films which are less crystalline. Furthermore, the incorporation of additives to LDPE films, such as antioxidants, pigments etc. will also affect the density of the material.

2.5 Mechanical properties of geomembranes

The changes in tensile properties are useful to assess the durability of the geomembrane. Polymer degradation due to ageing leads to decrease in strength and strain at break but increase in tensile modulus and yield stress [9]. The tensile strength analysis of 2.5mm thick HDPE geomembrane samples showed that there was a small but statistically significant difference between the yield strength of virgin and exhumed geomembrane exposed for 7 years in the field. A small increase in tensile strain at yield from 20.4 to 21.8% was noted. However it was later examined that there was no significant change in the tensile properties of these samples at break [21]. A video controlled tensile set up was adopted to study the mechanism of tensile deformation of HDPE/LDPE blends on the basis of stress-strain measurements. It was observed that HDPE has a smaller critical strain due to its rigid amorphous structure compared with that of LDPE. The tensile deformation process involved two steps, with HDPE crystallites being broken down initially at a strain of 0.4 and then followed by LDPE crystallites being broken at 0.6. The critical stress was first reached for HDPE and then for LDPE [22]. Antonio F. Avila, et al [23] investigated the mechanical properties of polyethylene terepthalate (PET)/HDPE composites. Several PET/HDPE blends exhibited different values of stress and strain. Due to larger concentration of PET, the stiffness of the material was found to increase with the increase in density of the PE. Tensile strength at yield and stiffness increased while elongation and flexibility were decreased. This is because crystalline regions significantly promote the increase of the modulus of elasticity and the ability of the plastic to support loads at elevated temperatures. LDPE has a modulus of elasticity which varies between 96.5 and 262 MPa and a relatively low tensile strength, in the range of 4.1–15.9 MPa. Despite the large variation in the tensile values, LDPE is still one of the weakest films (in terms of tensile strength) used as covering of greenhouses. Farizah Hamid et al. [24] investigated the mechanical effects of adding HDPE-g-MAH (maliec anhydrate) on various blends of PA6/HDPE in increasing proportion of 2% wt. of HDPE-g-MAH (compatabilizer) upto 8 % wt. It was observed that the addition of excess amount of compatabilizer could cause a drag in the structure thereby weakening the blends. This ultimately resulted in a decrease in the tensile modulus and tensile strength values of the resultant blends which contained HDPE-g-MAH exceeding to 2% wt. It was later estimated that the addition of 2% wt. HDPE-g-MAH improved the interfacial adhesion of HDPE and polyamide 6 probably due to the formation PA6-g-HDPE copolymer during blending. This thereby caused a rapid decrease in the interfacial tension. Abdullah M. Alhamdan, et al. [25] investigated the influence of arid conditions on the mechanical properties of polyethylene greenhouse covers at different exposure times. Mechanical resistance of polyethylene films was reduced when exposed to harsh environmental conditions. The data revealed that the elongation at break considerably reduced over a period of 12 months and further reduced to as low as 150% after 28 months of exposure. Ioannis Arvanitoyannisa, et al. [26] investigated the effects of adding starch into the LDPE/Rice starch blends. It was observed that with the increase in starch content in the blends, the tensile strength as well as the modulus of elasticity were reduced beyond 20% wt. of starch content.

2.6 Environmental stress crack resistance

The various types of general stresses/ degradation that the geomembranes are prone to are shown in Fig. 3. The degradation of polymers is often considered to be a very complex mechanism The main types of degradation are biological degradation, chemical degradation, ultraviolet (UV) degradation, oxidative degradation, mechanical and thermal degradation[4].

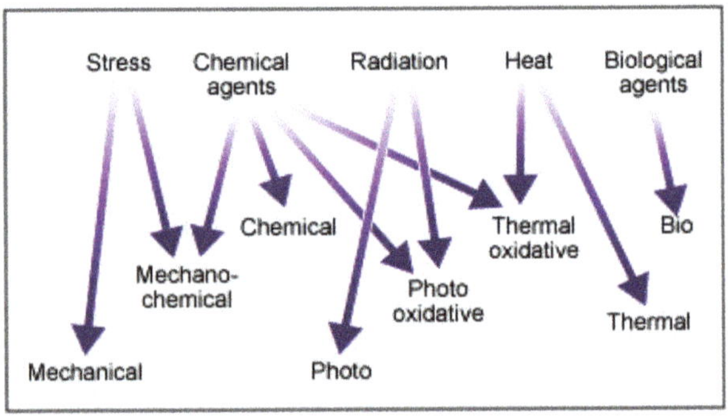

Figure 3 types of degradation of polymers

In addition to the internal stresses due to fluid heads causing adequate load on the material, chemical attacks occur resulting in polymer swelling and softening [18]. It has been observed that the defects in geomembrane liner have prompted the transport of leachate from the waste to the soil and groundwater. Molecular diffusion and leakage through holes and defects has contributed greatly to the movement of contaminant through geomembranes [27]. The three main ways in which the molecule activated processes of leachate transport through geomembranes include adsorption, diffusion and desorption.

The stress crack resistance of polyethylene depends largely on its density. It also depends on molecular weight, molecular weight distribution, comonomer type and concentration. The stress crack resistance of medium density polyethylene (MDPE) varies significantly and does not depend on its index properties alone. The very low density polyethylene (VLDPE) exhibited greater stress crack resistance than MDPE because of its low crystallinity and high branching facility [28].

2.7 Thermal properties of geomembranes

Thermal properties of geomembranes play a major role as the geomembranes are subjected to variation in temperature during its service. R.N Jana et al. [29] investigated the thermal stability of LDPE and poly-dimethyl siloxane (PDMS) rubber by non-isothermal thermogravimetry. It was observed that with increasing proportion of PDMS rubber to the blends, the degradation temperature increased marginally. This was mainly due to incompatibility of the components of blends which led to two-phase morphology in the blends. The blend obtained on the addition of ethylene methyl acrylate (EMA) compatabilizer showed high stability at an optimum proportion of EMA copolymer i.e. the blends containing 2 wt. % ethylene acrylic acid showed maximum initiation temperature. Whereas the blends containing EMA and ethylene vinyl acetate (EVA) showed a decrease in initiation temperature. Santra et al. [30] studied the thermal stability of miscible blends of EMA copolymer and poly (dimethyl siloxane) (PDMS) rubber. They reported that the degradation of the blends and the pure components followed first-order reaction kinetics. Moreover, the blend had exhibited thermal stability intermediate between the blend constituents. DSC instruments are widely used for the thermal characterization of polymers. It was later on confirmed that the melting temperature could be used as an essential criteria differentiating for HDPE pipe samples of different diameter from same source and pipe samples from different sources [31]. Farizah Hamid [32] investigated the thermal properties of various blends of (Polyamide 6) PA6/HDPE and HDPE-g-MAH. The DSC and TGA analyses showed that the thermal stability of the PA6/HDPE improved after the inclusion of 2-4% of HDPE-g-MAH. 98% PA6/HDPE with 2% HDPE-g-MAH showed the highest value of mechanical properties.

2.8 Ultraviolet (UV) degradation of geomembranes

Geomembranes are prone to several attacks by UV radiation due to which the polymeric materials crack or disintegrate. Infrared spectroscopy is the most commonly used technique to detect the crack caused by this mode of degradation. The effects of UV degradation on materials that require long service life can be measured by means of accelerated exposure tests. Ahmed A. Basfar [33] investigated the UV stability of LDPE and LLDPE for greenhouse applications. It was found that the UV stability of those samples which contained HALS and other antioxidants improved considerably. This was confirmed through accelerated weathering tests i.e. Xenon and by natural weathering tests, where the PE films were protected against UV degradation.

3. Prospects of future research and conclusions

Geomembranes are used extensively for many fundamentally different applications. It is hence unfair from the part of the engineer to expect all the materials to operate similarly under varying conditions. Geomembranes vary extensively in their physical, and chemical properties. Therefore there is no one geomembrane material that fits well for all scenarios. Hence it is necessary to consider all the required operating parameters before the selection of a geomembrane material. Resistance to chemical changes, flexibility, susceptibility to stress cracks, response to UV radiation, ease of seaming, mechanical strength and sensitivity to temperature are some of the factors that must be considered. To conclude, the selection of suitable material for the geomembrane must be made after careful consideration of its application, cost incurred and operating conditions. Further investigations are needed in order to develop an environmental friendly geomembrane with improved properties and desirable ageing life.

4. Performance evaluation of a few geomembranes

HDPE still remains the most popular choice when it comes to various applications as geomembrane. The main reason being its low cost, availability and high durability. Attempts are going on to improve the performance of HDPE by blending it with different fillers or by reinforcement. These modifications are generally carried out for specifc applications. The performance of a few geomembranes made of HDPE; nanofiller reinforced HDPE and EVA (polyethylene co vinyl acetate) modified EVAwere studied and compared with pure HDPE. Polymer nanocomposites of HDPE were prepared by melt mixing of various polymers with pre-exfoliated (MA2HT) and multiwall carbon nanotube modified clays (MWCNT) using a Brabender mixing head. The melt mixed samples were pressed into 0.5-3 mm sheets for mechanical testing and thermogravimetric analysis (TGA). Meanwhile samples of HDPE/EVA blends were prepared using the Brabneder mixer. The compression molding technique was employed to cast the melted samples into thin plastic sheets. The formulas of the samples are given in Table 2.

Table 2 Composition of various blends of EVA, nanoclay and HDPE.

SAMPLE	EVA (Wt %)	DCP (Wt %)	HDPE (Wt %)
EH1 (EVA + HDPE)	50	3	50
EH2 (EVA + HDPE)	75	3	25
EH3 (EVA + HDPE)	25	3	75

Figure 4 shows the modulus of HDPE nanocomposites prepared with various modified nanoclays. Composites prepared with organoclays modified with CNT shows 20-25% higher modulus and also 30-37% improvement in the maximum load. However HDPE composites prepared with only organoclay only did not show any improvement in their mechanical properties.

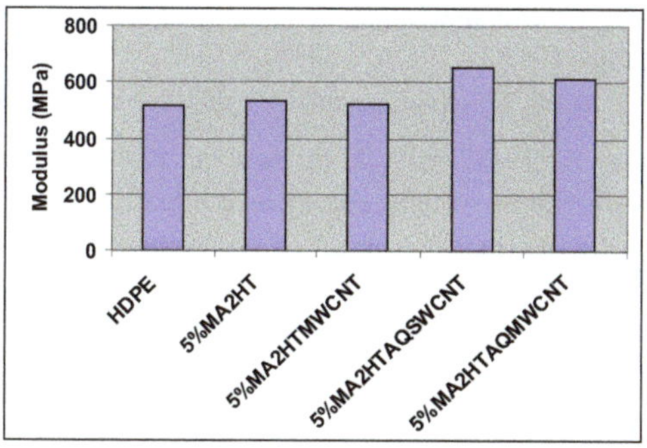

Figure 3: Tensile Modulus of HDPE Nanocomposites.

Table 2 shows the tensile properties of EVA/HDPE blends. The tensile properties included modulus of elasticity, tensile strength at break, tensile strength at yield, and elongation at break. The maximum value of young's modulus of 75.68 N/mm^2 was obtained by the EVA/HDPE blend with 25% EVA and 75%HDPE by weight. The tensile strength results clearly indicated that the sample EH3 possess the maximum tensile strength of 10.780 N/mm^2.

Table 2 Tensile strength of various blends of sample.

Specimen	Modulus Of Elasticity (N/mm^2)	Tensile Strength at break (N/mm^2)	Tensile strength At yield(N/mm^2)	Elongation at break (mm)
EH1	50.37	9.250	6.077	42.83
EH2	28.05	8.267	5.833	50.90
EH3	75.68	10.780	6.668	28.44

References

[1] MufeedSharholy, Kafeel Ahmad, R.C. Vaishya, R.D. Gupta: Municipal solid waste characteristics and management in Allahabad, India, Waste Management. 27(2007) 490–496.
http://dx.doi.org/10.1016/j.wasman.2006.03.001

[2] Cuevas, J., Ruiz, A. I, de Soto, I. S, Sevilla, T, Procopio, J. R, Da Silva, P. Leguey, The performance of natural clay as a barrier to the diffusion of municipal solid waste landfill leachates. Journal of environmental management, 95 (2012) 175-181.
http://dx.doi.org/10.1016/j.jenvman.2011.02.014

[3] Gamski, K.Geomembranes: classification, use and performance. Geotextiles and Geomembranes, 1, 2 (1984), 85-117.
http://dx.doi.org/10.1016/0266-1144(84)90009-8

[4] Scheirs, John. A guide to polymeric geomembranes: a practical approach. John Wiley & Sons, 2009.
http://dx.doi.org/10.1002/9780470748213

[5] Peggs, I.D, Thiel, R. Selecting a geomembrane material. Proceedings of the Sixth International Conference on Geosynthetics, Vol.1, Atlanta, March. Industrial Fabrics Association International, 1998, St. Paul, MN, 381–388

[6] R. K Rowe, Effect of ageing on the stress crack resistance of an HDPE geomembrane, Polymer Degradation and Stability, 109 (2014) 194-208.
http://dx.doi.org/10.1016/j.polymdegradstab.2014.06.013

[7] Hsuan, Y. G., and R. M. Koerner. "Long term durability of HDPE geomembrane: Part I—Depletion of antioxidant." GRI Report 16 (1995) 35

[8] Mueller, Werner, Ines Jakob, Oxidative resistance of high-density polyethylene geomembranes. Polymer Degradation and Stability, 79, 1 (2003) 161-172. *http://dx.doi.org/10.1016/S0141-3910(02)00269-0*

[9] Rowe R. K, Sangam H. P,Durability of HDPE geomembranes. Geotextiles and Geomembranes, 20, 2 (2002) 77-95. *http://dx.doi.org/10.1016/S0266-1144(02)00005-5*

[10] R.K. Rowe, Migration of dilute aqueous organic pollutants through HDPE geomembranes, Geotextiles and Geomembranes, 19, 6 (2001) 329-357. *http://dx.doi.org/10.1016/S0266-1144(01)00013-9*

[11] R. K. Rowe, M. Z Islam, Y. G. Hsuan. Leachate chemical composition effects on OIT depletion in HDPE geomembranes, Geosynthetics International, 15,2 (2008) 136-151 *http://dx.doi.org/10.1680/gein.2008.15.2.136*

[12] R. K. Rowe, M. Z. Islam, Y. G. Hsuan. Leachate chemical composition effects on OIT depletion in HDPE geomembranes." Geosynthetics International, 15 (2), (2008), 136-151 *http://dx.doi.org/10.1680/gein.2008.15.2.136*

[13] S.B Gulec, T.B Edil, C.H Benson, Geosynthetics International 11,2 (2004), 60-72 *http://dx.doi.org/10.1680/gein.2004.11.2.60*

[14] Han-Yong Jeon, Nanotechnology Formulations and Modeling of Hydraulic Permeability Improvement for Nonwoven Geotextiles. *http://dx.doi.org/10.5772/61997*

[15] Mueller, Werner, Ines Jakob, Oxidative resistance of high-density polyethylene geomembranes, Polymer Degradation and Stability, 79, 1 (2003) 161-172. *http://dx.doi.org/10.1016/S0141-3910(02)00269-0*

[16] A.M.R Ewais, R.K Rowe, Geotextiles and Geomembranes 42, 2, (2014) 111–126 ·

[17] M. Sen, A. A. Basfar, The effect of UV light on the thermooxidative stability of linear low density polyethylene films crosslinked by ionizing radiation, Radiation Physics and Chemistry, 52,1 (1998) 247-250 Stessel, Richard Ian; Hodge, James Steven, Chemical resistance testing of geomembrane liners. Journal Of Hazardous Materials. 42,3 (1995) 265-287

[18] A. Nicholas, J. I. Apse, W. A. Hoffman, W. M. Song. "MDPE/VLDPE materials development." Geotextiles and Geomembranes, 10,5 (1991) 613-620
http://dx.doi.org/10.1016/0266-1144(91)90051-W

[19] P. A. Dilara, D. Briassoulis, Standard Testing Methods for Mechanical Properties and Degradation of Low Density Polyethylene (LDPE) Films Used as greenhouse covering materials: a critical evaluation, Polymer Testing, 17 (1998) 549-585
http://dx.doi.org/10.1016/S0142-9418(97)00074-3

[20] R. Kerry Rowe, Durability of fluorinated high density polyethylene geomembrane in the Arctic Geotextiles and Geomembranes, 28 (2010) 100-107

[21] Q. Fu, et al., Understanding of the tensile deformation in HDPE/LDPE blends based on their crystal structure and phase morphology, Polymer, 44 (2003) 1927-1933
http://dx.doi.org/10.1016/S0032-3861(02)00940-0

[22] F. A Antonio vila, A mechanical analysis on recycled PET/HDPE composites, Polymer Degradation and Stability 80 (2003) 373–382
http://dx.doi.org/10.1016/S0141-3910(03)00025-9

[23] Farizah Hamid, Mechanical and thermal properties of polyamide 6/HDPE-g MAH/HDPE, Procedia Engineering, 68 (2013) 418-424.
http://dx.doi.org/10.1016/j.proeng.2013.12.201

[24] Abdullah M. Alhamdan, Mechanical deterioration of polyethylene greenhouses covering under arid conditions, journal of materials processing technology, 209 (2009) 63-69

[25] Arvanitoyannis, Ioannis, Costas G. Biliaderis, Hiromasa Ogawa, and Norioki Kawasaki. "Biodegradable films made from low-density polyethylene (LDPE), rice starch and potato starch for food packaging applications: Part Mechanical properties of covering materials for greenhouses, part II: quality assessment1. Carbohydrate Polymers. 36 2, (1998) 89-104

[26] T. Katsumi, Performance based design of landfill liners, Engineering geology 60 (2001) 139-148
http://dx.doi.org/10.1016/S0013-7952(00)00096-X

[27] A. Nicholas, J. I. Apse, W. A. Hoffman, W. M. Song, MDPE/VLDPE materials development, Geotextiles and Geomembranes10,5, (1991) 613-620

[28] R. N. Jana, G. B. Nando, Thermogravimetric analysis of blends of low-density polyethylene and poly (dimethyl siloxane) rubber: The effects of compatibilizers, Journal of applied polymer science 90,3 (2003) 635-642. *http://dx.doi.org/10.1002/app.12599*

[29] Santra, N.Robin, P. G. Mukunda, G. B. Nando, and T. K. Chaki. Thermogravimetric studies on miscible blends of ethylene-methyl acrylate copolymer (EMA) and polydimethylsiloxane rubber (PDMS), Thermochimicaacta, 219 (1993) 283-292 *http://dx.doi.org/10.1016/0040-6031(93)80505-5*

[30] Sajwan, Madhuri, Saroj Aggarwal, R. B. Singh, Forensic characterization of HDPE pipes by DSC, Forensic science international 175,2 (2008) 130-133 *http://dx.doi.org/10.1016/j.forsciint.2007.05.020*

[31] Farizah Hamid, Mechanical and thermal properties of polyamide 6/HDPE-g MAH/HDPE, Procedia Engineering, 68 (2013) 418-424. *http://dx.doi.org/10.1016/j.proeng.2013.12.201*

[32] Ahmed A. Basfar, UV stability and radiation crosslinking of linear low density polyethylene and low density polyethylene for greenhouse applications, Polymer Degradation and Stability, 82 (2003) 229–234. *http://dx.doi.org/10.1016/S0141-3910(03)00216-7*

CHAPTER 4

A Concise Overview of Biofuel Cells

Sufia ul Haque and Inamuddin
Department of Applied Chemistry, Aligarh Muslim University, Aligarh, India
*Corresponding author E-mail address: inamuddin@rediffmail.com

Abstract

Nowadays, the amount of implantable devices with lots of applications has increased the need for supplying the required energy to sense, monitor, vibrate, pump and report the collected data of such devices. This tendency has concerned the scientists and engineers around the world. Such a power source should have a long life time and be of ecofriendly nature. It should be capable of producing energy directly from the living organisms' fluids or organic matter. These required properties, light up the idea of using enzymatic biofuel cells (EBFCs) to generate the power. Unfortunately, EBFCs are limited by their short life time and poor electron transfer from the enzyme active site to the electrode surface. This issue can be overcome by using certain biocompatible mediators and some enzyme immobilization techniques such as: physical adsorption, electrostatic attraction, covalent coupling and entrapment. These techniques enhance the power density obtained from EBFCs.

Keywords

Biofuel Cell, Enzyme Catalysis, Enzyme Electrodes, Glucose

Contents

1. Fuel cells

The utilization of the electrochemical reactions by the device known as fuel cells for the conversion of the chemical energy of a fuel into electricity has emerged as one of the interesting research areas [1]. The term fuel cell is often peculiarly applied to hydrogen-oxygen fuel cells. For the generation of electricity, electrons are stripped from hydrogen gas molecules to form protons (H^+). The protons, so formed, then pass through a proton exchange membrane (PEM) and combine with oxygen (O_2) to form water. The fuel cells are usually much more efficient than other types of energy converters as the energy conversion efficiencies as high as 60-80% have been achieved [1]. In addition to the high levels of efficiency accessible, another benefit of fuel cells is the fact that they are able of providing electrical energy over a much longer period of time. In contrast to a battery, which contains only a restricted amount of fuel material and oxidant, fuel cells are continuously supplied with fuel and air (or O_2) from an external source. Fuel cells differ from internal-combustion devices, where a fuel is burned and gas is expanded to do work, but fuel cell converts chemical energy directly into electrical energy. Fuel cells offer distinct benefits over other types of power sources as they provide a high efficiency, do not emit pollutants and provide a truly sustainable power output if hydrogen is produced from a renewable energy source [2]. The technologies involved in fuel cells are that they generate electricity through electrochemical processes, rather than combustion. There are many types of fuel cells, but the most useful include the direct methanol fuel cell (DMFC), proton exchange membrane (PEMFC) fuel cell, alkaline fuel cell (AFC), phosphoric acid fuel cell (PAFC), molten carbonate fuel cell (MCFC), solid oxide fuel cell (SOFC) and biofuel cell (BFC). Most of these are commercially available today.

Each type of fuel cell has its own especial chemistry, such as different operating temperatures, electrolytes and catalysts. A fuel cell's operating condition help in defining its application - for example, for stationary power generation, higher temperature MCFC and PAFC fuel cells are used while lower temperature PEM and DMFC fuel cells are used to power passenger vehicles and forklifts. Research work is continued for the improvement of fuel cell technologies, examining different electrolytes and catalysts in order to get better performance and reduce costs. New fuel cell technologies, such as biofuel cells, are also being examined in the laboratory. Types of fuel cells with their applications are listed in **Table 1.**

2. Biofuel cell

The term biofuel cell is employed to explain any fuel cell in which the fuel stock's energy is obtained from biological carbon fixation [3]. Fuel cells that utilize proteins, carbohydrates, ethanol, amino acids or lipids can all be reported as biofuel cells if the path for fuel making included the use of biological matter whether it is an animal, plant or microbial based. The method for efficiently extracting chemical energy from the fuel stock varies with the complexity of the biofuel used, as well. In conventional fuel cells ruthenium and platinum based catalysts were utilized to extract energy from fuels such as ethanol or methanol. However, microbial based fuel cells required complex molecules, such as proteins, glucose, etc. [4,5]. Biofuel cells can be classified as: abiotic biofuel cells (ABFC) (which generate power from biofuels without the use of any biological organism); microbial biofuel cells (MBFC) (that produce power from biofuels using living micro-organisms); and enzymatic biofuel cells (EBFC) (that generate power from biofuels with the help of enzymes) [1–6].

The initial user of ABFCs can be traced back to the birth of implantable biofuel cells in the 1970s in the flank of a dog and delivered stable power output for a short time, but then stability was lacking as the power rapidly decreased within an hour of operation. These devices were typically very inept, since the conversion of glucose they could achieve was very low. In present scenario with the increasing pressure to find alternative energy sources to fossil fuels, researchers are trying to develop a source of energy from available and natural resources which are plentiful.

Table 1: Classification and comparison of various types of fuel cells.

Type	Electrolyte	Fuel at anode	Fuel at cathode	OperatingTemperature(°C)	Application
Direct methanol fuel cell (DMFC)	Proton conductive electrolyte	Methanol (CH_3OH)	Oxygen (O_2)	90-120	Stationary and transportation
Polymer electrolyte membrane fuel cell (PEMFC)	Proton conductive electrolyte	Hydrogen (H_2)	Oxygen (O_2)	90-120	Stationary and transportation
Alkaline fuel cell (AFC)	KOH	Hydrogen (H_2)	Oxygen (O_2)	150-200	Space craft and submarine
Phosphoric acid fuel cell (PAFC)	Phosphoric acid	Hydrogen (H_2)	Oxygen (O_2)	150-200	Stationary and transportation
Molten carbonate fuel cell (MCFC)	Molten alkaline carbonate	Hydrogen (H_2)	Oxygen (O_2)	650	Stationary and transportation
Solid oxide fuel cell (SOFC)		Hydrogen (H_2)	Oxygen (O_2)	800-1000	Stationary and transportation
Biofuel cell	Fuel in aqueous medium	Organic matter (eg glucose, alcohol)	Oxygen (O_2)	Body temperature	Portable Devices

The schematic view of general fuel cells is shown in **Figure1**.The fuel at the anodic compartment is reacted (usually with the help of a catalyst, either in solution, or on the electrode surface), resulting in the rescue of an electron and a proton. The electron is conducted via the anode to the cathodic compartment, while the proton remains in solution. The oxidant has to be separated since most of the catalysts lack specificity for the fuel and thus the proton exchange membrane is used (as a separator) for the diffusion of protons to the cathodic compartment [7]. The drop in pH at the cathodic half-cell then facilitates the electrochemical reaction between an oxidant (typically oxygen) and an electron from the cathode. This is the simplest case, usually adopted in ABFCs with metal catalysts on the anode facilitating the oxidation of biofuels.

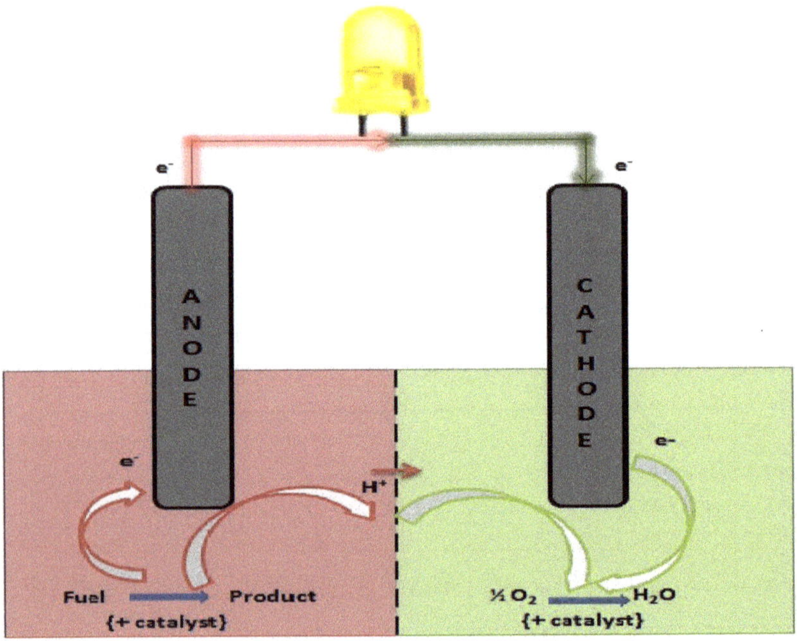

Figure 1: Schematic representation of general fuel cell.

In 1911, the initial work on developing a fully characterized MBFC (arguably the oldest fuel cell technology) was reported. Here the main objective was not to induce power but to evolve a live sensor for microbial activity by determining their electrical potential [8]. Since then there has always been considerable interest in fuel cells for research, it had a huge boost in interest with the birth of the space age in the 1960s and 1970s. In time, a number of projects were launched worldwide leaning to determine the possibility of employing biofilms of bacteria such as C. butyricum and E. coli at the anode [9]. An extensive military backed research was conducted by the United States Department of Defense[10].

In MBFCs, microbes were used as a catalyst for the production of protons from fuels as it was done in ABFCs [11]. Afterwards, the transfer of electrons from the microbes occurs, either aided with the use of electron carriers (mediators) or directly through conducting biofilms. Typically a dual chamber system is used with a proton exchange membrane in the middle[12–14].

It is hard to determine the exact origin of EBFCs since many techniques involving the use of microbes in biofuel cells could potentially have relied on the lyses of microbes used. For example, in 1966 Kenneth Lewis explained a method of metabolic pathways both inside and outside of microbial cell membranes in order to fully oxidize glucose, which could only be achieved by lyses of the cells [2]. Hence, this technique could be considered as an enzyme based method for generating energy since no living microbes are required, but instead freshly lysed ones.

The first deliberate use of enzymes in a biofuel cell happened in 1962, when glucose oxidase used in solution of glucose dependent biofuel cells was tested [11]. Since then, many researchers have started to focus on enzymatic biofuel cells as the enzymes used are more specific and can achieve higher levels of conversion than metal catalysts. EBFCs can also work at low temperatures utilizing traditional fuels such as ethanol, and carbohydrates such as glucose which were previously not considered a commercial fuel source [1].

In comparison, EBFCs can be much simpler in their design depending on the method adopted. In order to separate the oxidizing enzymes from the reducing enzymes when enzymes are used in solution, the exact same general design that is used for ABFCs has to be adopted [1]. In contrast, there is no need for a proton exchange membrane if the enzymes are immobilized onto the surfaces of the anode and cathode [15]. Consequently the specificity of enzymes on each bio electrode ensures that there is no interference from the oxidant at the anode and *vice versa.* Thus, EBFCs **(Figure2)** can be used with mixed fuels and oxidants with much smaller size than that of both ABFCs and MBFCs.

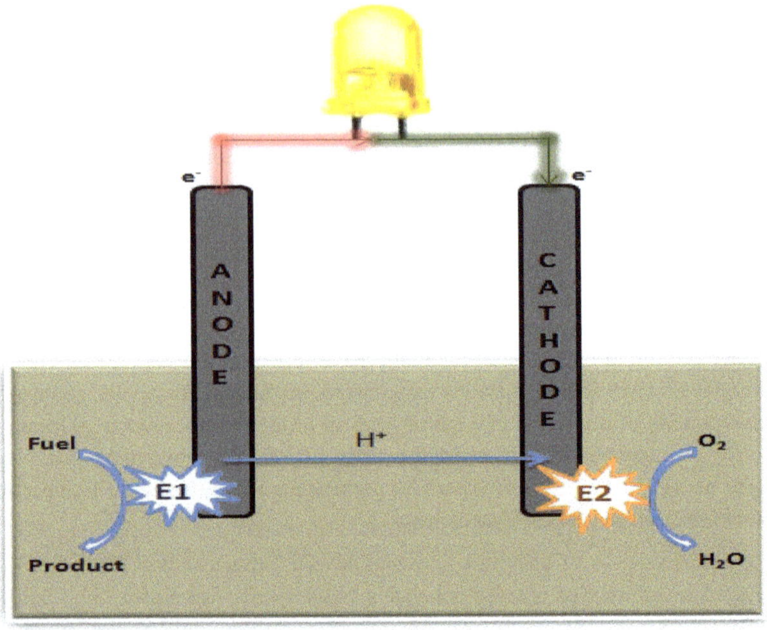

Figure 1: Schematic representation of an EBFC utilizing immobilized enzymes (E1 andE2)

Unfortunately, from such biofuel cells the power density obtained is typically much lower than obtained in non-biological fuel cells. This is due to the physical size of the chemical mechanisms utilized here. For example, in a hydrogen fuel cell the whole electrode surface can be agile for oxidation as the metal catalyst used is conductive and is itself works as the electrode. On the electrode surface, the oxidation of hydrogen occurs in bulk and the electrons generated are immediately conducted away.

On the other hand, in EBFCs each oxidation site (the enzymes active site) requires a relatively large surface area of electrode as the catalyst is much larger. Furthermore, the enzymes are also typically not conductive or only weakly conductive and hence, more time and energy is spent in transferring the electrons from the enzymes to the conductive surface of the electrodes. **Figure 2,** shows a simplified mechanism of reaction for a glucose oxidase (GOx) electrode. Once a molecule of glucose (which is relatively large in size) has successfully reached, an immobilised GOx (which is sparsely distributed), diffused to the active site (which is deep seated inside the enzyme) and oxidized, the released electrons then have to be transferred from the enzyme to the electrode surface which typically occurs through secondary reactions. Consequently, though EBFCs do not require proton exchange membranes, they do require extremely large surface area electrodes to generate useful range of power. As far as the choice for oxidants is concerned, there is only one clear choice and that is dissolved dioxygen. This is because most other oxidants present in the human body vary greatly in concentration and are alkaline. Thus, a fuel cell that uses oxidants other than oxygen would have a variable power output and would risk changing the pH of the blood and would result in serious health issues.

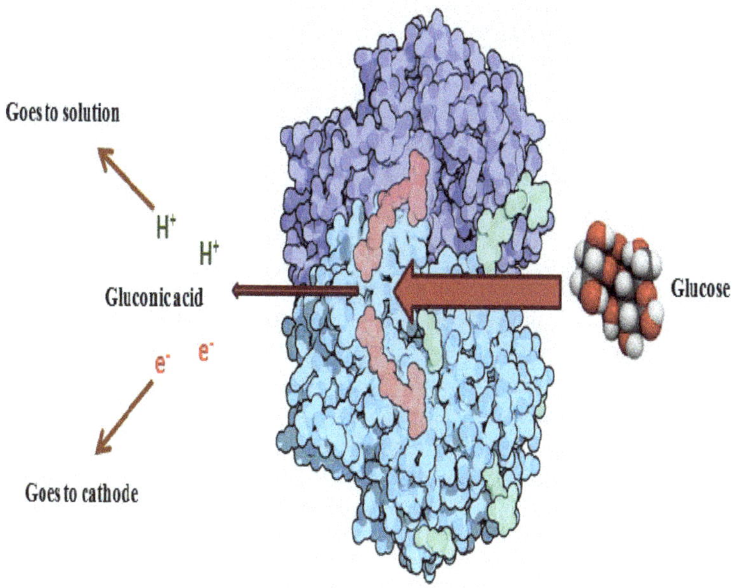

Figure 2: Simplified schematic representation of reaction scheme involved with a GOx bioelectrode

In glucose-oxygen fuel cell, following reactions are involved:

$$\underbrace{C_6H_{12}O_6}_{glucose} \rightarrow \underbrace{C_6H_{10}O_6}_{D-glucono-1,5-lactone} + 2e^- + 2H^+ \quad (1)$$

$$C_6H_{10}O_6 + H_2O \rightarrow \underbrace{C_6H_{12}O_7}_{GluconicAcid} \quad (2)$$

$$1/2O_2 + 2e^- + 2H^+ \rightarrow H_2O \quad (3)$$

First glucose is converted to D-glucono-1,5-lactone in the anodic half-cell (Equation 1), which is successively hydrolysed to form gluconic acid (Equation 2). The electrons released pass through the circuit to generate electric current, the hydrogen ions travel to the cathodic half-cell where they react with oxygen and electrons from the cathode to form water (Equation 3).

Each type of biofuel cells discussed is efficient to work on these reaction schemes. But there are two routes of utilizing catalysts: First the use of enzymes inside the human body would not be commonly supported, and since catalysts in solution would wash away or potentially stimulate an immune response. The second approach is to develop an EBFC utilizing immobilized enzymes as electrode, since enzymes illustrate a very high level of specificity, are generally active at low temperatures and could easily be used with fuel and oxidant mixtures **(Figure 3)**. However, the attention should be given to the types of materials used to ensure biocompatibility, biostability and to prevent long term toxicity of the device.

The obvious choice for a fuel stock for this purpose is glucose, since it is found in relatively high concentrations in blood (approximately 1.0 g L^{-1} under normal conditions) and also in the interstitial fluid of the human body (at concentration of approximately 0.5 g L^{-1}) [16]. Glucose is also very dense in energy providing up to 16 kJ g^{-1} by generating 12 electrons per molecule during complete oxidation [17]. The waste metabolites produced from the incomplete oxidation of glucose (such as gluconic acid) are also naturally present in the human body and as such can easily be disposed of without the risk of waste toxicity. Other possible fuel stocks include lipids and amino acids, however these are much harder to metabolize electrochemically and as such are not well suitable for use in fuel cells.

The first implantation of a cardiac pacemaker took place in 1960 **(Figure 4)**[18], many attempts have been made to develop miniature implantable power devices, which would be able to work continuously for a long time without the need for replacement. Many different technologies have been considered such as thermoelectric [19] and mechanical [20] power generators which utilize the excess energy created by the human body. Technologies such as inductive power transfer *via* radiofrequencies have also been checked for the continuous supply of power such as in cochlear implants[21]. Each technology challenged the researchers with its own limitations. For example, mechanical power generators only give power when a person is in motion and thus would not produce power when someone is sleeping. Thermoelectric power supply relies on the presence on a significant temperature gradient between the body and the surrounding air, and so would fail to produce power on a hot day. Inductive power transfer techniques on the other hand require the use of an external power source which may not always be available. By definition inductive power transfer techniques also uses strong magnetic and electrical fields which could cause interference with the standard operation of medical implants such as pacemakers [22].

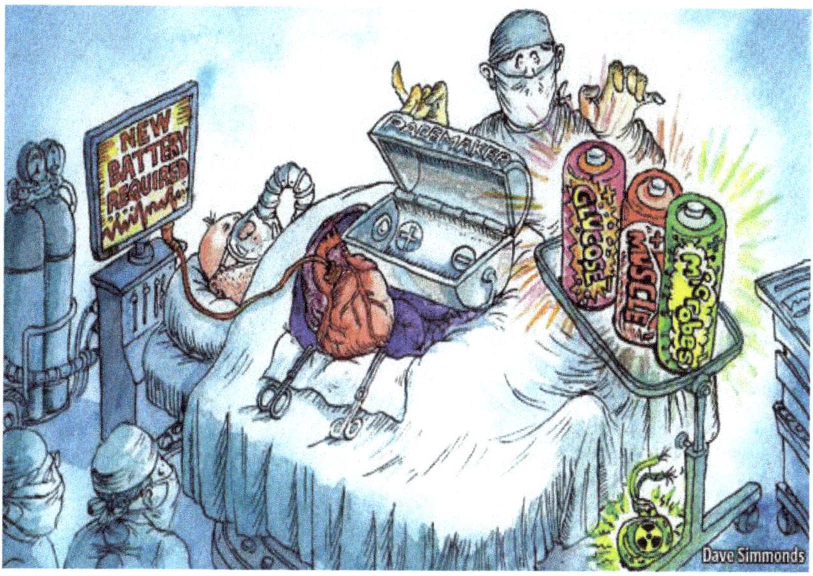

Figure 3: "A sweet idea" [20].

The use of nuclear powered devices were also very common initially as these devices were able to produce the required power output and could last for decades without the need of any maintenance [23]. Ultimately, this type of device was abandoned in the 1980s in favour of lithium based power cells. The main reason for this was that the associated cost of refining the radioactive material required to power the devices was extremely high, with a single nuclear powered pacemaker costing approximately £1,000 in 1974 (equivalent to £9,250 in 2014) [24]. In this context, biofuel cells which use naturally occurring metabolites as fuel (typically glucose) are a promising alternative source for power production.

2.1 Enzymatic biofuel cell

Appreciable attempts in the wiring of enzymes (methods for improving the electrical communication from enzymes to electrode surfaces) have recently given rise to several promising alternatives to abiotic systems for *in vivo* applications which in turn have led to the development of a wide range of prototype devices. These devices have emerged as significant means of power generation when incorporated in living organisms (Figure 5).

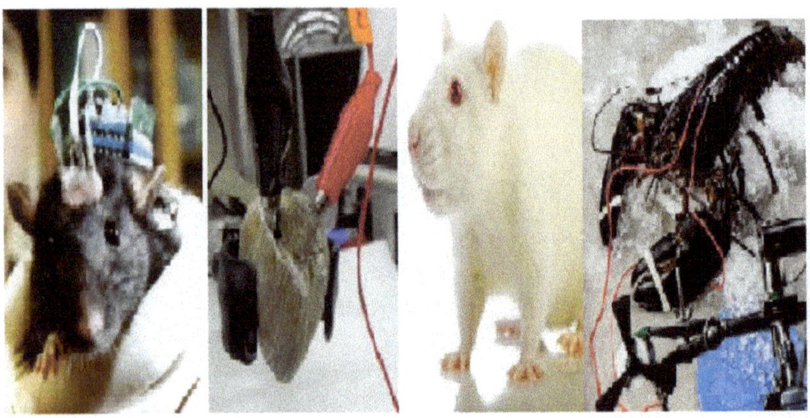

Rat - 2010	*Clam - 2012*	*Rat - 2013*	*Lobusters - 2013*
0.25 V, 6 μW	**0.8 V, 5.2 μW**	**0.5 V, 40 μW**	**0.5 V, 150 μW**

Figure 4: EBFCs implanted into live animals [25–32]

In particular, the improvements in nanostructured materials have enhanced enzyme wiring by lowering the over potentials of glucose oxidation and oxygen reduction. The anode typically comprised of GOx can be used to induce the catalytic oxidation of glucose at potentials between -0.4 V and -0.1 V against a saturated calomel electrode (SCE). Cathodes (comprised of laccases (LAC) or bilirubin oxidase (BOD)) can achieve oxygen reduction at potentials between 0.5 V and 0.6 V vs SCE[33]. In addition to improving enzyme wiring, the development of nanostructured electrodes also greatly improves the surface area and thus the number of wired enzymes per surface or volume [34]. For the first time an EBFC implantation in an animal was performed in 2010 by a research group from the University Joseph Fourier in France when it was surgically implanted in a rat[29]. They considered that covalently bound enzymes have greatly reduced functionality, so they opted to encapsulate the enzymes used.

This strategy also permitted them to use a combination of different enzymes and electron mediators in the same electrodes. Their design comprised of an enzyme and graphite mixture compressed into pellets, which they encapsulated in dialysis tubing. This shielded the enzymes from metabolites in the bodies of the rats and, in theory, would protect the rats from any potential carcinogenic effects of the implanted graphite[35]. They computed an initial open circuit potential of 0.275V with a power density of 24.4 $\mu W\ mL^{-1}$ during an *in vivo* practical with a rat, which then lowered to a steady value of 0.22 V with a power density of 7.5 $\mu W\ mL^{-1}$ for several hours more [32]. Long term power production was not explained and ultimately the long stability of such a device is questionable. However, no signs of inflammation or rejection were observed over a 3 month period when the fuel cell was not in constant use.

With an intention to improve the stability of implanted EBFCs, many researchers turned their focus to the use of carbon nanotubes (CNTs) and their composites. Due to their shape and size, CNTs are able to intertwine with the enzymes. Therefore, allowing for direct wiring from the enzyme's active site to the electrode surface without affecting the activity of the enzyme **(Figure 6)**[36]. As a result, direct electron transfer (DET) between the enzyme and electrode surface can be achieved. By using bucky paper (a CNT aggregate), a research group from Clarkson University in the United States of America has created several different prototype EBFCs which they implanted in invertebrates [26,29,31]. Though, these implantable EBFCs did not create any complications that could appear in mammals (such as biocompatibility, toxicity, sterilization and immune responses), they made appreciable approaches in enhancing the levels of power production possible. In one case, they even showed that two fuel cells implanted in lobsters could provide power to a digital watch for approximately 1 hour **(Figure 7)**, and five EBFCs running in simulated physiological conditions could generate enough power for the normal operation of a pacemaker for 5 hours [31]. To better understand the conditions within the human body many researchers have started to implant EBFCs in lab animals such as rabbits and rats.

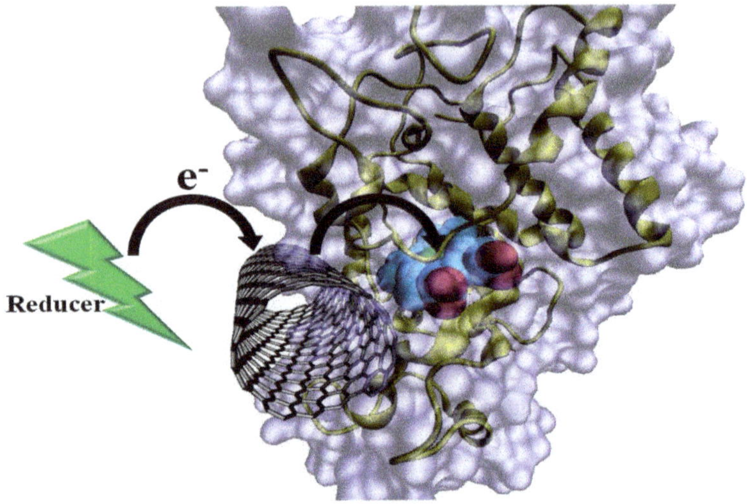

Figure 5: Illustration of protein and CNT intertwinement [36]

A research group from Tohoku University in Japan developed a 'semi-implantable' EBFC. The device comprised of a needle tip enzymatic anode with an external air breathing cathode utilizing carbon nanoparticle composites [30]. This way the problems related with low oxygen levels in the living organisms could easily be controlled, and thus power production was only limited to the concentration of glucose. This means that their device could potentially be a feasible candidate for live glucose sensing with minimal invasive procedures. When this device was tested in the vein of a rabbit (Figure 8), a power output as high as 0.42 μW was generated. This is a notable achievement as the anode fits inside a needle with a cross section of just 0.057 cm2. This kind of technology may be well suited for small subcutaneous or sensing purposes.

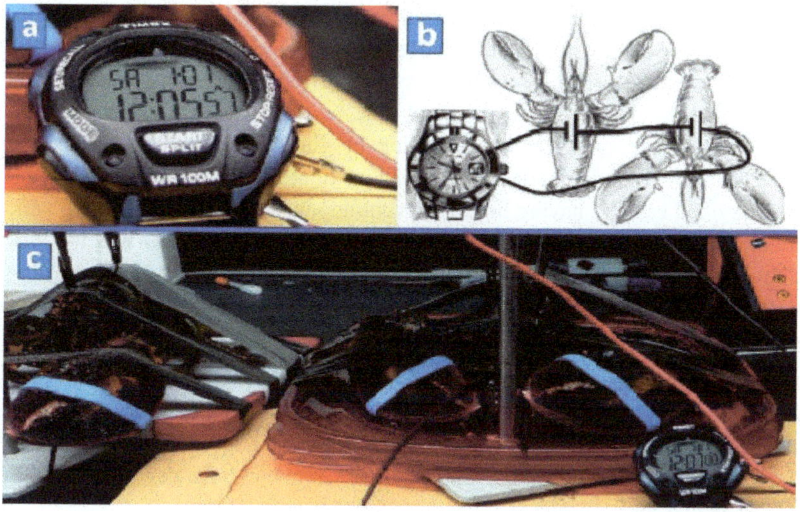

Figure 6: Two lobsters implanted with EBFCs connected in series to power a digital watch [31]

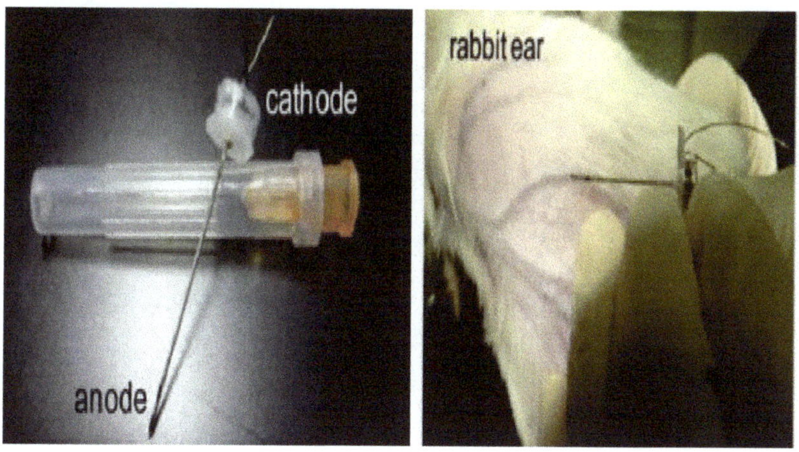

Figure 7: Needle tip anode and external cathode EBFC implanted in a rabbit's ear [30]

One of the two approaches must be kept in mind in order to develop fully implantable EBFCs that can be used with medical devices such as pacemakers. Devices can either be implanted in the soft tissue, or in blood vessels. Each of these ways have their own advantages and disadvantages [32,33,37]. In the case of devices to be implanted in the soft tissue, the issue of lower concentration of glucose and oxygen exists. Devices designed for use in blood vessels exhibited the continuous flow through operation. These devices provided ease by having higher concentrations of glucose and oxygen which are continuously supplied by the flow of blood. But the continuous flow was the reason of extra strains on the system, since the flow could result in the removal of enzymes from the electrodes and interfere with the electron transfer between enzymes and the electrode surface[33].

Several groups of researchers have focused to implant EBFC in the soft tissue of animals. In an expansion to their pioneering research utilizing compressed enzyme and graphite disks implanted in a rat. The research group from the University Joseph Fourier (in collaboration with a group from the University de Grenoble) have developed new EBFCs using CNT pellets and compressed enzyme [38] **(Figure 9)**.

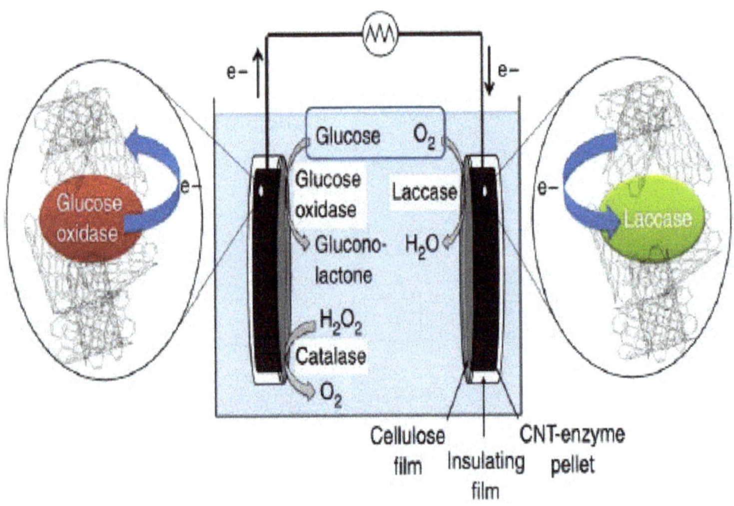

Figure 8: EBFC with compressed enzyme and CNT pellets as electrodes [38]

In their work, they only partially depended on the DET between GOx and the CNT electrodes. Instead of selecting a simple DET or mediated electron transfer (MET) mechanism, they relied on the production of H_2O_2 from GOx which then oxidized using a secondary enzyme (catalase). Therefore, this device relied on a two stage DET mechanism with H_2O_2 as an intermediate electron mediator. Special measures had to be taken to ensure biocompatibility when implanted in a rat. The pellet fuel cell was first wrapped and sealed in a dialysis bag and then sealed in a biocompatible Dacron bag (conventionally used for implants) [25]. When the EBFC was implanted in the retroperitoneal space of a rat, it delivered a maximum power output of 40 μW and had an OCP of 0.57 V. This was sufficient to charge a capacitor and (with the aid of a boost converter) intermittently power an LED and an electric thermometer **(Figure 10)**.

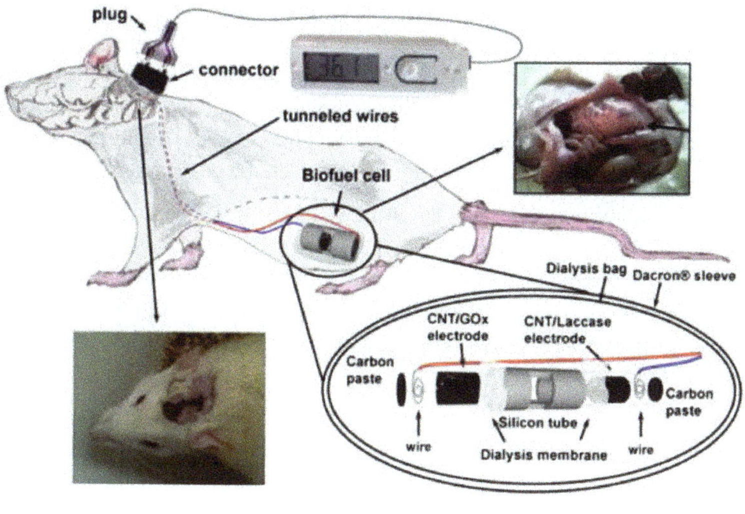

Figure 9: EBFC implanted in the abdomen of a rat powering an electric thermometer [25]

Succeeding on from this work with invertebrates, the research group from Clarkson University also took an attempt to implant EBFCs in rats. Now they used the Bucky paper based enzymatic electrodes, but this time placed the electrodes directly in contact with partially excised Cremaster muscle whilst the rat was under anesthesia[39]. As compared to the results with invertebrates, the OCP in this case was only 0.14 V and a maximum power output of only 0.35 μW was achieved.

One more group of researchers has advanced their EBFC research to the point of implantation. By utilizing a GOx and modified carbon fibre anode and a platinum and carbon fiber cathode, the group (from Brazil) developed a hybrid enzymatic/abiotic fuel cell [40]. *In vitro* test of this fuel cell yielded 24 nW of power after implantation into the jugular vein of a rat the fuel cell produced a 9.5 nW of power over a 24 hour period of operation. This is mainly due to the small size of the fuel cell with a reported electrode area of just 1×10^{-4} cm^2. Since the use of a catheter bag around the fuel cell greatly increased its overall size so the use of a small size electrode was necessary.

As reported so far, the EBFCs used carbon based electrode materials such as CNTs, carbon nanoparticle composites or carbon fibers. These electrodes showed problems with their long term biocompatibility, toxicity and stability (owing to the fact that many are in fact biodegradable) [41,42]. However, the EBFCs that rely on the use of more biocompatible and stable electrode materials such as gold have caught much attention. In order to develop the devices that can be equivalent to the performance of carbon based EBFCs, it is important to construct gold surfaces that imitate the properties of carbon based electrodes which would be well suited for immobilization and enzyme wiring. In this context, the use of nanostructured gold is considered a promising alternative.

A research collaboration with members from Sweden, Russia, and Austria made a major development in this new emerging field. They developed an EBFC comprising of gold microwire electrodes modified with gold nanoparticles [43]. The resulting nanostructures are alike in morphology to nanoporous gold (nPG) films. They implanted their EBFC in the brain of a rat where it generated 2 $\mu W\ cm^{-2}$ of power from the glucose found in cerebrospinal fluid.

From Kyoto University in Japan, a research group also shared their notable contribution to this research. They developed enzymatic electrodes of functionalized nPG films by immobilizing enzymes on it [44]. In slightly acidic conditions by using a GOx anode and a LAC cathode in an EBFC, which were able to generate a maximum power output of 4.7 μW, though no evidence is given on how long they were able to maintain this level of power.

The collaboration between Irish and Austrian researchers has also made advances using nPG surfaces in EBFC development. They used osmium based polymers co-immobilized with enzymes on a nPG support **(Figure 11)**. Since osmium based compounds are very famous in MET mechanisms so the use of an osmium polymer allowed to increase the efficiency of electron transfer drastically [45]. The maximum power output of 7.3 μW was achieved from this prototype EBFC under physiological conditions. However, within the first 100 minutes of operation, this power output had a rapid decay of 50% and no further data is given on the long term stability of this EBFC.

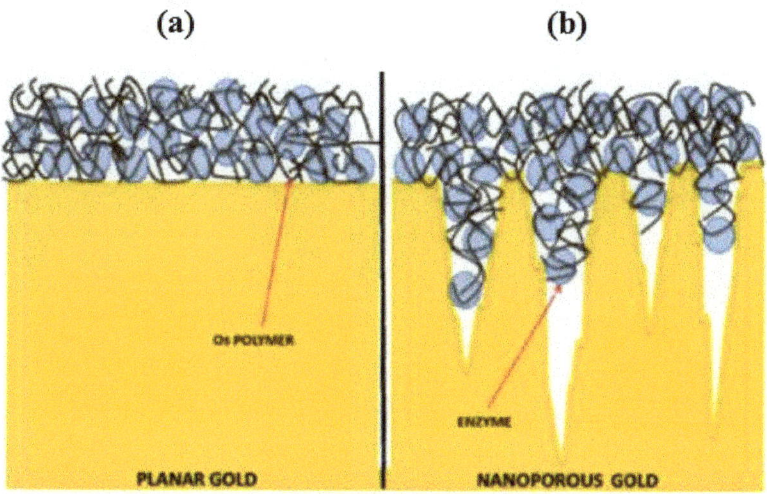

Figure 10: nPG-based electrodes used in osmium mediated EBFC. (a), (b) Schematic of osmium polymer and enzyme immobilisation on polished gold and nPG, respectively [46]

2.1.1 General enzyme classification

Enzymes are proteinaceous substances that catalyze specific chemical reactions. They are generally categorized according to the type of reaction they catalyze. For example, the kinases transfer phosphate group from adenosine triphospate (ATP) to particular substrates. For the ease to understand these things, enzymes are also classified into the following three groups based on the location of their redox center (Figure 12).

Class I enzymes are those in which the redox center is located near to, or on the periphery of the protein's shell **(Figure 12 (a))**. The enzyme correctly orientated on the electrode surface allows the redox center to communicate directly with an electrode. These types of enzymes are relatively rare. For example, Cytochrome c, Azurin and Horseradish peroxidase.

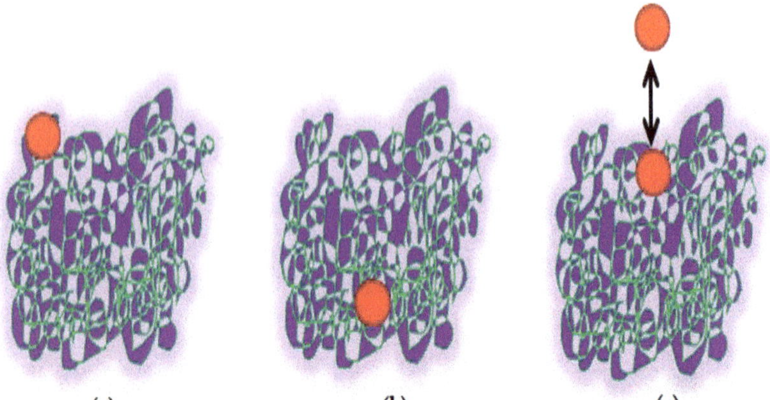

Figure 12: Sorted according to the location of the enzyme's redox centre, schematic showing the three main classes of enzymes. In each case the purple potion shows the insulating bulk of the enzyme and the red portion depicts its redox centre. (a) Class I enzyme with a peripheral active site, (b) Class II enzyme with a redox site buried within the protein shell and (c) Class III enzyme with loosely bound active site.

Class II enzymes are those in which the active site has concealed inward (deep seated) within its insulating protein or glycoprotein shell **(Figure 12 (b)).** Such a settlement hinders direct electrical communication between the redox centre and an electrode, and generally needs the utilization of a redox mediator to shuttle out the electrons from active site to the electrode. Well known example of this type is Glucose oxidase.

Class III enzymes are those in which the redox center is present in a cofactor that is loosely bound to the enzyme. The cofactor effectively serves as a redox mediator and therefore able to diffuse away to an electrode surface **(Figure 12 (c)).** The most common cofactors are nicotinamide adenine dinucleotide (NADH) and nicotinamide adenine dinucleotide phosphate (NADPH).

2.1.2 Direct and mediated electron transfer

Up to date the large numbers of enzymes that have been used in biofuel cells have not been able to provide the efficient transfer of electrons between the substrate and the electrode surface during oxidation and reduction. Therefore, polymers and small redox active particles have been used as electron carriers in MET mechanisms **(Figure13)**. The typical mediators include organic dyes, ferrocene and its derivatives, modified vitamin complexes and conducting salts [47].

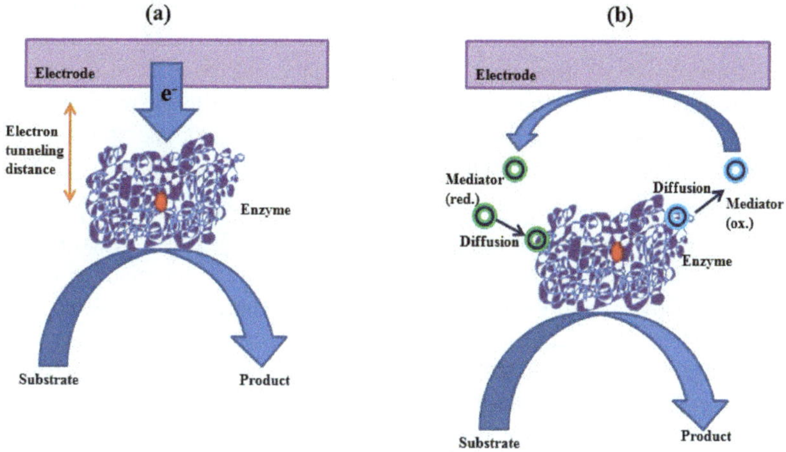

Figure 13: Electron transfer mechanisms utilized in biofuel cell technology. a) Direct electron transfer, b) Mediated electron transfer.

These mediators are often utilized with flavin adenine dinucleotide (FAD) and nicotinamide adenine dinucleotide (NAD^+) dependent enzymes, such as GOx, as they can react directly with either the enzymes themselves or their cofactors to become charged. If the mediator is used in solution then its diffusion to the electrode surface allows for a more rapid electron transfer compared to the direct transfer from the enzyme itself [48]. Alternatively, the mediators can also be polymerized directly onto the electrode surface or co-immobilized onto the electrode surface with the reacting enzymes to further enhance the rate of electron transfer [49,50].

The disadvantage related with the use of redox active electron carriers as mediators is their short life spans, as they are not biocompatible or can leach away from the electrode surface. As a result, the utilization of enzymes able of DET became the primary focus of many researchers in the 1980s. Transition metal rich enzymes such as LAC, which is capable of catalyzing the reduction of oxygen to water through the direct catalytic transfer of electrons between the substrates and electrode surface, provided major advancements [51].

In most EBFCs developed so far, DET has been achieved at the anode by using electrodes modified by CNTs. This amazing property such as hydrophobic outer walls which permit the strong adsorption of enzymes for enzyme wiring, which facilitate electron transfer between enzyme active site and electrode [37]. DET has also been possible when using pyrroloquinoline quinone (PQQ) dependent enzymes. Haem group associated with the PQQ dependent enzymes, which is able to exist in some different redox states, function as electron acceptors in a similar manner to the redox active metal centers in enzymes such as LAC. In particular pyrroloquinoline quinone glucose dehydrogenase (PQQ-GDH) has been used with great success at the anode of EBFCs[52].

The power outputs secured from EBFCs using MET mechanisms is typically much higher than those achieved when using DET mechanisms [53]. The issue associated with DET is the direct correlation between the electron transfer rate and improper orientation of the enzyme active site to the electrode surface [54]. Thus, by utilizing fast diffusing mediators, the rate of electron transfer and overall power output can be increased. However, DET can be improved by the specific immobilization of enzymes in a favorable orientation which controls the proximity of the enzyme's active site to the electrode surface [51]. Another method for improving electron transfer is *via* the immobilization of mediator molecules onto the surface of the electrode [43].

2.1.2.1 Physical adsorption

The simplest method for enzyme immobilization is physical adsorption. This method depends on a physical interaction between the enzyme and the surface of the electrode (by salt linkages, hydrogen bonds, and Van der Waal's forces), usually that does not require reagents and only a minimum number of activation steps [55]. Thus, adsorption is easily carried out, cost effective, and tends to be less disruptive to the enzyme than chemical means of attachment **(Figure 14)**[56].

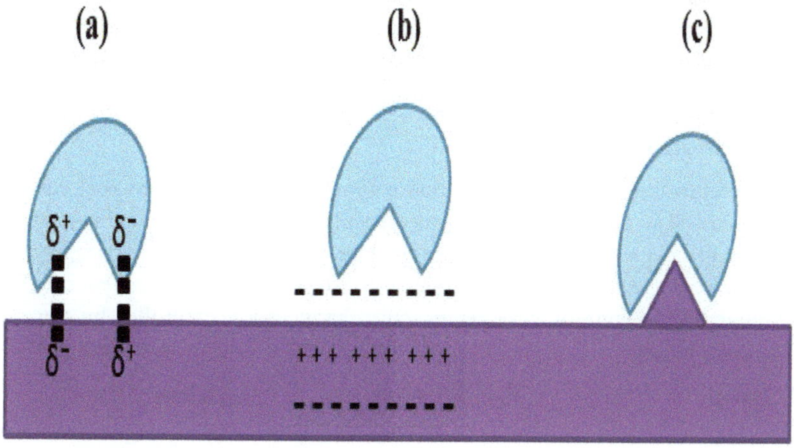

Figure 114: Popular physical adsorption mechanisms; (a) Van der Waals forces, (b) ionic bonding and (c) affinity binding

The use of hydrophobic interactions was another approach. Due to the size and the hydrophobic nature of CNTs, they have been shown capable to intertwine with and spontaneously adsorb enzymes. Thus, much success has been reported utilizing hydrophobic physical adsorption with CNTs and GOx, even achieving DET since the CNTs are able to come in very close proximity to the FAD center in flavoenzymes such as GOx [57,58].

For possible use as a substrate for adsorbed enzymes, porous gold has also been investigated with a specific focus to creating LAC electrodes. One study reported that, highly comparable results obtained with a physical adsorption method (which involved simply incubating enzymes in freshly prepared nPG), and the covalent coupling method with regards to amount of enzymes immobilized, specific activity, and enzyme leakage. They found that with smaller nPG structures (close to those of gold nanoparticles) the adsorption affinity could be greatly improved and the enzyme loading increased while decreasing leakage, without using any reagents to facilitate binding [59]. This suggests that the adsorption here is driven by affinity binding (improved Van der Waal's or hydrogen bonding) assisted by the complimentary morphologies of enzyme and substrate.

However, adsorption methods are typically reversible due to the involvement of weak bonds. Changes in temperature, pH, ionic strength or even the presence of substrate can result in desorption of enzymes[60,61]. Therefore, adsorption is not the most widely used method for developing reusable enzymatic electrodes or electrodes intended for constant use. It is, however, often used as a first step prior to the covalent coupling or entrapment methods described later.

2.1.2.2 Electrostatic attraction

Electrostatic attraction is also an adsorption method but, unlike the physical adsorption methods discussed so far, it is an active process and not a passive one. In this method the basic idea is to take advantage of the ionic nature of proteins. Much like amino acids, enzymes can exist as cations, zwitterions or anions depending on the ambient pH (Figure 15). Moreover the strength of an enzymes ionic charge can be controlled by controlling the pH of the solution.

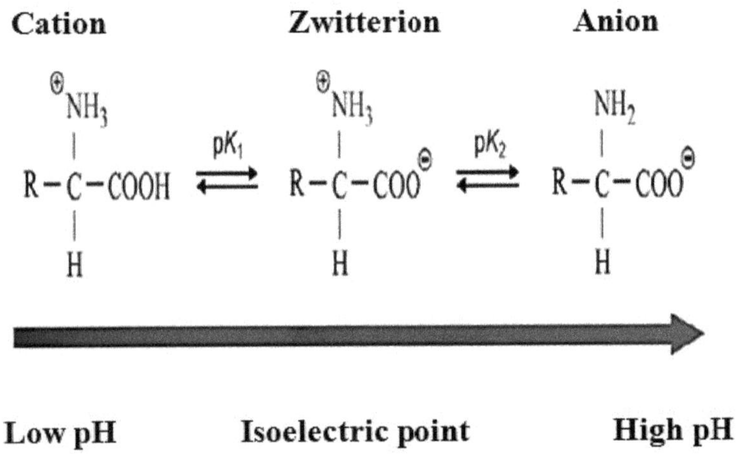

Figure 15: Ionic nature of amino acids [62]

Immobilization by electrostatic attraction takes advantage of this charge on enzymes by applying a potential to an electrode of opposite charge to the enzyme [60]. Thus enzymes can attach to the electrode by ionic bonding. Furthermore, applying a charge to the electrode induces a local electrical field which can actually draw enzymes from the solution to the electrode surface, greatly increasing the enzyme loading [61].

This method is also reversible. As the enzyme can be detached and even repel from the electrode surface by changing the charge on the electrode after immobilization. Thus, it is important to use enzymes with charges which are complimentary to the electrodes for the motive of using electrostatic attraction immobilization in fuel cells. Particularly, a positively charged enzyme should be used at the cathode and a negatively charged enzyme should be used at the anode. But this method cannot always be implemented since the pH of a fuel cell system is often fixed, and the choice of enzymes is limited, finding an enzyme with the right complimentary charge at the given pH is often not possible.

2.1.2.3 Covalent coupling

By the formation of covalent bonds the immobilization of enzymes is among the most widely used method **(Figure 16)**. Here the immobilization procedure is considered irreversible which is the greatest advantage of such method and thus the enzyme will not detach upon use [57]. This is achieved in the ideal situation by forming a peptide bond with either the carboxylic acid or amine terminal groups of the enzyme. This is not always possible since the terminal groups of a protein are often folded into unreachable positions. Thus the side chains of the following amino acids are frequently used to form peptide or sulphide bonds: aspartic and glutamic acids (carboxylic group), lysine (amino group), and cysteine (thiol group). However, it is important to ensure that amino acids essential for catalytic activities are not involved in the covalent coupling process so that the activity of the enzymes is not affected [63].

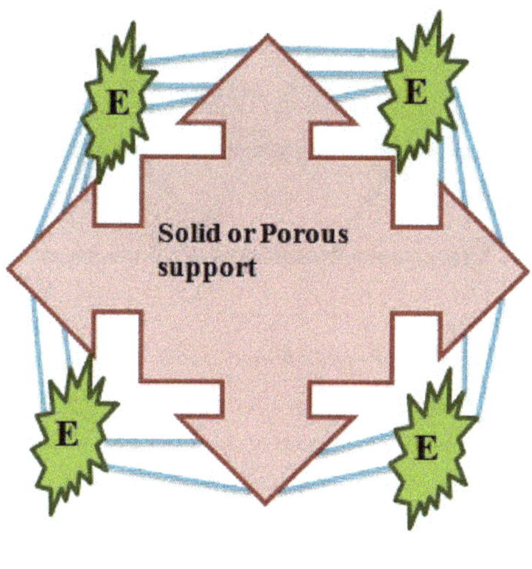

Figure 126: Illustration of covalent binding of enzyme with solid or porous support.

2.1.2.4 Entrapment

This method is based on the entrapment of an enzyme within a polymer which allows the conversion of substrate into the product but retains the enzyme **(Figure 17)**[64]. Entrapment is carried out by mixing the biocatalyst into a monomer solution, followed by polymerization initiated by a change in temperature or by a chemical reaction. Different approaches are there for the entrapment of enzymes such as gel encapsulation, micro-encapsulation and sandwich entrapment. In practical the application of these methods is typically restricted by mass transfer limitations through membranes or gels.

Metal or semi-metal oxide matrices are formed in gel encapsulation technique around enzymes. SiO_2 is the most common material for the encapsulation of enzymes [65]. Silica is commonly used due to the fact that the resultant gels can be tailored to a large range of porous textures, surface functionalities and processing conditions, and can thus be adapted to a particular enzyme [50].

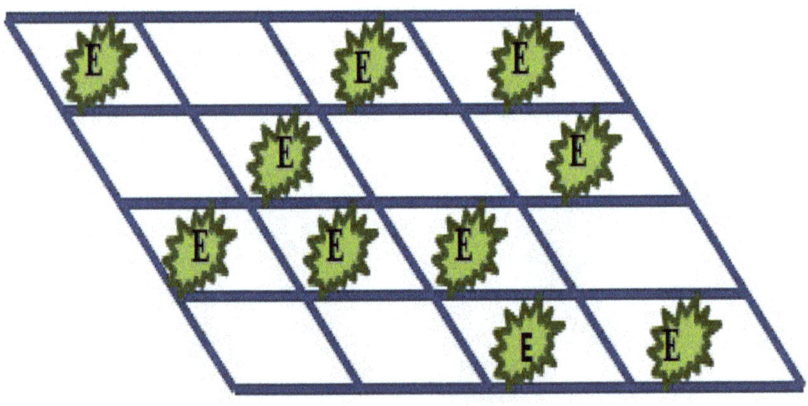

Figure 17: Entrapment of enzyme (E) in polymer matrix

In the construction of a gel matrix, a porous wall is built around the enzyme (which acts as template) by the chemical condensation of a silica gel network. Consequently the enzyme is immobilized within the gel, while the electron mediators and substrates can freely move around, permitting for immobilization without inhibiting enzyme activity (**Figure 16**). The electrodes that employ gel encapsulation are generally intended for single use, so this technique is not ideal for the construction of reusable bio-electrodes [41].

Sandwich entrapment techniques involve the layering of enzyme followed by coating of another material (typically a sol-gel matrix or a polymer). The limiting mass transport of substrates to the active site of the enzyme is the effect of this coating. Further, nafion polymer has also been used to coat modified electrodes, but it reduces the enzyme activity due to the acidic side chains of nafion resulting in an unfriendly environment for the enzymes [66]. Some attempts have been made to produce electrodes which could function perpetually in a biofuel cell like, the utilization particles of graphite mixed with GOx and ferrocene successively compressed into a pellet before encapsulation in a dialysis membrane [40]. This method allows the high level of proximity of the enzymes to conductive surfaces of the composite electrode.

The physical entrapment of an enzyme in matrices or the pores of a membrane at the electrode surface is known as microencapsulation. However, it follows the same principle, works on either gel encapsulation or sandwich entrapment, but with a much greater degree of control (through electrochemical assembly for example) in order to produce very thin functional matrices around enzymes or enzymes embedded within a thin membrane coating.

Success has been reported with the controlled assembly of simple CNT and chitosan matrices [67], as well as with more complicated CNT, poly-cation polyethylenimine and nafion matrices[68] and dithiol-gold nanoparticle-cystamine matrices constructed by sequential self-assembly [69], with all achieving a DET response from GOx. This method of encapsulation was also shown to protect the encapsulated enzymes from rise in temperature and external changes in pH [70].

3. Comparative analysis of immobilization methods

On the basis of their complexity, relative stability and the types of electron transfer techniques achievable in each case, a comparison of the different immobilization techniques regarding to construct devices with long term stability with irreversible processes has been presented in **Table 2**.

Table 2: Qualitative comparison of different enzyme immobilization techniques [61,66,70–72]

Immobilization Method	**Stability**	**Complexity**	**Electron transfer**
Van der waals Adsorption	Very low	Low	DET/MET
Ionic bonding	Low	Low	DET/MET
Affinity binding	Variable (sensitive to pH)	Variable (controlling surface morphology can be complex)	DET/MET
Electrostatic attraction	Variable (sensitive to pH and charge)	Low	DET/MET
Covalent coupling	Irreversible	High	DET/MET
Gel encapsulation	Irreversible (but subject to mediator leaching)	High	MET
Sandwich entrapment	Irreversible	High	DET/MET
Micro encapsulation	Irreversible (improved pH resistance)	High	DET/MET

So far we have discussed the major developments in enzymatic fuel cells for power generation applications over the past five years and their large scale applications [70], enzyme immobilization and electrode materials [72–75]. The focus here is on the application of BFCs for electrical power production. The two major problems in enzyme-based systems are the short lifetime of the enzyme caused by a reduction in its stability when functioning in a foreign environment, and the low power densities resulting from a slow electron transfer rate from the enzyme active site of the electrode [72].

The bulk of the research in enzymatic fuel cells has been directed to enzyme/electrode integration methods that mitigate these problems. Enzymes have short life time which is an inherent characteristic of enzymes even in their natural environment, but the lifetime may be increased to a few days by immobilization [73]. The researchers have focused on polymers available that help in the fabrication of the efficient electrodes for the utilization in EBFCs. Here are some procedures to fabricate the electrodes.

4. Enzyme electrodes with layered structures

The most irreversible and stable immobilization technique is covalent immobilization. Enzyme electrodes with layered structure based on covalent bindings have been developed. Enzymes immobilized on the electrode surface with self – assembled monolayer (SAM) or layer-by-layer structure binding act as mediator to transfer electrons from the site of fuel oxidation at the enzyme to the electrode surface.

Katz and Willner [76] developed a method to establish direct electron transfer through a defined structured path by reconstitution of the enzyme with nitrospiropyran-modified and 2-aminoethyl modified flavin adenine dinucleotide, FAD cofactor between the active center of glucose oxidase and the electrode surface [74,76–79]. They produced a fuel cell using enzymes on both anode and cathode. The electrode substrate was gold. The anodic reactions (defined as below) were glucose oxidation using reconstituted glucose oxidase connecting with a monolayer of pyrroloquinoline quinone (PQQ) as the mediator, and the cathodic reaction was the reduction of hydrogen peroxide by microperoxidase-11(MP-11) [79]. The maximum power density was around 160 $\mu W\ cm^{-2}$.

$$\text{Electrode–PQQ–FAD–GOx} + \text{Glucose} \rightarrow \text{Electrode–PQQ–FADH2–GOx} + \text{Gluconic acid} \quad (1)$$

$$\text{Electrode–PQQ–FADH2–GOx} \rightarrow \text{Electrode–PQQH2–FAD–GOx} \quad (2)$$

$$\text{Electrode–PQQH2–FAD–GOx} \rightarrow \text{Electrode–PQQ–FAD–GOx} + 2H^+ + 2e^- \quad (3)$$

5. Enzyme electrodes with polymer matrix

Although enzyme electrodes with layered structures have shown efficient electron transfer in various applications, but there are some limitations. First, the amount of enzymes immobilized on the electrode is limited by the electrode surface due to a monolayer covalent binding nature. Second, if more molecular layers immobilized on the electrode surface, the more electric resistance would be introduced to the electrode, which in turn could affect the electronic response of the electrode.

Also, the electrode activity will be influenced by the orientation of the enzymes and mediator molecules. Conducting redox polymers can be a solution to overcome these limitations. Conducting polymers, such as polypyrrole and polyaniline, are very common to immobilize enzymes and fabricating enzyme electrodes. Polypyrrole (Ppy) is one of the most extensively used conducting polymers in the design of bio analytical sensors. It has unique properties of prevention of some undesirable electrochemical interactions and facilitation of electron transfer from some redox enzymes [80,81]. Enzyme electrodes with polypyrrole are fabricated by electropolymerisation and enzymes are entrapped in the polymer as a dopant during the polymerization process [81–87]. Polypyrrole can also be functionalized by adding cationic pendant groups, such as tris(bipyridyl)ruthenium(II) complex to the polymer films to introduce an electron relay [88,89]. A two-step method consists of the adsorption of an aqueous amphiphilic pyrrole monomer-enzyme mixture on an electrode surface followed by the electropolymerization of the adsorbed monomers was developed by Cosnire [89]. Foulds *et al.*[90], and Umana *et al.*[91]have described the entrapment of glucose oxidase within the growing network of conducting polypyrroles, as the entrapment of enzymes occurs without a chemical reaction that could affect their activity and polymer films can be prepared easily in a rapid, one-step procedure. Inamuddin *et al.* [92] fabricated an enzyme anode by using a conducting polymer polypyrrole (Ppy) as electron transfer enhancer and ferritin (Frt) protein as electron transfer mediator. This anode provides a current density of 1.2 $mAcm^{-2}$ in a 45 mM glucose solution and offered a good possibility for application in biofuel cells. Inzelt *et al.*[93], MacDiarmid *et al.*[94] Wallace *et al.*[95] developed the interest in research of polyaniline which is connected to discovery of its conductivity in the form of emeraldine salt and existence of different oxidation forms. Stejskal *et al.*[96] reported the conducting protonated emeraldine in the form of green emeraldine salt, obtained as a product of electrochemical polymerization of aniline in acidic electrolytes. Both reduction of emeraldine salt to leucoemeraldine and oxidation to perningraniline states are followed by decrease in conductivity. Karyakin *et al.*[97], Kumar [98], Mattoso & Bulhoes[99], Zhang [100] described the electrochemical polymerization and co-polymerization of numerous substituted aniline derivates, resulted in polymer materials with properties different from the parent polymer. Ryang Eun Kim *et al.*[101] showed the maximum power density output of 37.4μWcm^{-2} by employing three different approaches to immobilize GOx on polyaniline nanofibers (PANFs): enzyme adsorption (EA), enzyme adsorption and crosslinking (EAC), enzyme adsorption, precipitation and crosslinking (EAPC) approaches. EAPC-GOx showed higher activity and stability than EA-GOx and EAC-GOx. For the biofuel cell application, EAPC-GOx and EAPC-LAC were applied over the carbon papers to form enzyme anode and cathode, respectively. Polyaniline is another extensively used polymer for enzyme immobilization. An enzyme-mediator-conducting polymer model using benzoquinone (Q)-polyaniline (Pani) system was established by Cooper and Hall. An enhanced current can be obtained on Pani [102]. Ratiman *et al.* [103] integrated polyaniline/poly (acrylic acid) films and redox enzymes for the study of the bioelectrocatalyzed oxidation of glucose or lactate. Improved selectivity and stability of glucose biosensor was obtained based on *in-situ* electropolymerized polyaniline-polyacrylonitrile composite film [104]. A novel method was developed by the Willner's group to generate an integrated electrically contacted glucose dehydrogenase electrode by the surface reconstitution of the apo-enzyme on a pyrroloquinoline quinone (PQQ)-modified polyaniline [105].

6. Enzyme electrode with mediator immobilization

Immobilization of the enzyme has several advantages, including isolation of the enzyme for reaction, increased selectivity, improved mass transfer and long-term stability [106]. It also acts to separate the enzyme from the mixture containing the substrate, allowing for more modular cell designs. On the other hand, it can affect the stability and/or activity of the enzyme, introduce additional mass-transfer limitations on the substrate and involves additional costs. Stability is clearly a key consideration. The stability of the immobilized enzyme will depend on the nature and strength of the bond to the support material, the conditions required for immobilization, the degree of confinement and the conditions under which the enzyme reactions occur in a functioning electrode. The method of immobilization must be selected carefully to avoid denaturing of the enzymes and loss of structural freedom required for their activity [102,107]. Ideally, enzymatic electrodes for power generation should maximize enzyme loading, activity and stability, while also minimizing inefficiencies due to substrate mass transport limitations and slow electron transfer from the enzyme to the electrode. While increasing bond strength (physical adsorption/entrapment, ionic bonding, and covalent bonding) improves enzyme retention, deactivation is also more probable[53]. Simple physical binding or entrapment in polymers is sometimes complemented by ionic binding, but this imposes an operating pH depending on the type of enzyme used and the ionomer charge. Addition of a cross-linking agent is a simple and widely used method to improve the stability and increase both the concentration of the enzyme and its activity [55]. In this thesis we have used glutaraldehyde as a cross-linking agent and stability enhancer of immobilized enzyme as well.

The random enzyme orientation, when physically adsorbed or covalently bonded, is unfavorable for efficient electron communication with the active site. The reconstruction of apo-enzymes around the active site bound to a redox relay provides excellent electrical communication. Monolayers of these electrically contacted enzymes have a high turnover number and sensitivity, although the enzyme loading is limited by the electrode surface area. This limitation can be relieved by a similar enzyme reconstitution but on an electrically active thin polymer film [75]. Immobilization of mediators and enzymes on polymer films is a simple and practical method for biofuel cell electrodes, however, the effective diffusion coefficient of the mediator (and hence the electron transport rate) is several orders of magnitude lower than that of a typical diffusional mediator [79]. The simplest method of enzyme immobilization is physical adsorption or entrapment. Enzymes can be adsorbed, for example, onto conductive particles such as carbon black or graphite powder [108]. The methods are straight forward and cost-effective. If the binding forces (primarily electrostatic) between the enzyme and the support are too weak, however, the enzymes can desorb and contaminate the solution. If they are too strong, denaturation can occur during the immobilization process. Entrapment involves the confinement of the enzyme within a polymer matrix [109], a redox hydrogel [110] or behind a semi permeable membrane. The structure must permit a sufficient degree of enzyme movement, while simultaneously preventing any leaching of the enzyme and/or mediator. Isolated enzymes can be covalently bonded to supports (e.g. porous glass, cellulose, ceramics, and metallic oxides) via different functional groups on the support and enzyme, often in the presence of enzyme inhibitors. Reagents are used to activate the functional groups on the support. The functional groups on the enzyme, which include amino, carboxylic and hydroxyl groups, should not be essential for catalytic activity. The conditions for this type of immobilization are important since they determine the level of enzyme activity retention. Cross-linking consists of joining enzymes to form three-dimensional aggregates via covalent bonding between active groups within the enzymes. The aggregates exhibit low mechanical stability and the retained enzymatic activity can be low using this method [108].

7. Review of Literature

The work published on EBFCs during the last eight years has been presented briefly in **Table 3.**

Table 3 Literature review on EBFC during the last eight years

No	Title	Remarks	Ref.
1.	A conducting polymer/ferritin anode for biofuel cell applications	An enzyme anode for use in biofuel cells (BFCs) was constructed using an electrically connected bilayer based on a glassy carbon (GC) electrode immobilized with the conducting polymer polypyrrole (Ppy) as electron transfer enhancer, and with horse spleen ferritin protein (Frt) as electron transfer mediator. This anode provides a current density of 1.2 mA cm^{-2} in a 45 mM glucose solution and offers a good possibility for application in biofuel cells.	[92]
2.	Membraneless enzymatic biofuel cells based on graphene nanosheets.	Graphene based biofuel cell exhibited a maximum power density of about 24.3 μW cm^{-2}, which is nearly two times greater than that of the SWCNTs biofuel cell, and the performance of the graphene biofuel cell lasted for 7 days.	[111]
3.	A glucose/O_2 biofuel cell base on nanographene platelet-modified electrodes.	This study demonstrated a novel nanographene platelets (NGPs)-based glucose/O_2 biofuel cell (BFC) with the glucose oxidase (GOx) as the anodic biocatalysts. The maximum power density was ca. 57.8 μW cm^{-2} for the assembled glucose/O_2 NGPs-based BFC.	[112]
4.	Surface engineering of graphene-enzyme nanocomposites for miniaturized biofuel cell	The biofuel cell employs the gold plate electrodes modified by specific graphene-enzyme conjugations,which are immobilized by electrochemical deposition of the conducting polypyrrole polymer. The power output and the biostability of the integrated biofuel cell were improved.	[113]
5.	A comparison of redox polymer and enzyme co-immobilization on carbon electrodes to provide membrane-less glucose/O_2 enzymatic fuel cells with improved power output and stability.	Glassy carbon and graphite electrodes modified with films of enzyme and osmium redox polymer, cross linked with poly (ethylene glycol) diglycidyl ether, were used for elaboration of a glucose/O_2 enzymatic fuel cell.	[114]

6.	An overview of graphene in energy production and storage applications	Graphene has incurred intense interest since its freestanding form was isolated in 2004, and with the vast array of unique and highly desirable electrochemical properties it offers.	[115]
7.	Fabrication of polymeric ionic liquid/graphene nanocomposite for glucose oxidase immobilization and direct electrochemistry	A novel polymeric ionic liquid functionalized graphene, poly(1-vinyl-3-butylimidazolium bromide)-graphene (denoted as poly(ViBuIm+Br−)-G), was synthesized which exhibited a strong positive charge. Based on self-assembly, the negatively charged glucose oxidase (GOD) was immobilized onto the poly(ViBuIm+Br−)-G to form a GOx/poly(ViBuIm+Br−)-G/glassy carbon (GC) electrode under mild conditions. Which can provide a favorable and conductive microenvironment for the immobilized GOx and thus promote their direct electron transfer at the GC electrode.	[116]
8.	An enzymatic biofuel cell based on electrochemically reduced graphene oxide and multiwalled carbon nanotubes/zinc oxide modified electrode	GOx was co immobilized at electrochemically reduced graphene oxide (ERGO) by simple electrochemical approach. The fabricated biofuel cell delivers a power density up to 54 nW cm $^{-2}$.	[117]
9.	Synthesis of sulfonated graphene/polyaniline composites with improved electroactivity	Composites of polyaniline (PANI) and sulfonated graphene (SG) are prepared by the polymerization of aniline in the presence of the SG. These materials were electrochemically active at pHs close to physiological pH. Furthermore, good conductivity values were obtained.	[118]
10.	Graphene/carbon cloth anode for high-performance mediatorless microbial fuel cells.	Graphene was electrochemically deposited on carbon cloth to fabricate an anode for a *Pseudomonas aeruginosa* mediator less microbial fuel cell (MFC). The graphene modification improved power density and energy conversion efficiency by 2.7 and 3 times, respectively.	[119]
11.	Hierarchical composites of sulfonated graphene-supported vertically aligned polyaniline nanorods for high-	Hierarchical composites of sulfonated graphene-supported-vertically aligned polyaniline nanorods (sGNS/PANI) were successfully synthesized via interfacial polymerization of aniline monomers in the presence of sulfonated graphene nanosheets (sGNS).	[120]

	performance supercapacitors.	The hierarchical composite based on the two-electrode cell showed higher specific capacitance.	
12.	Electrochemical actuator based on polypyrrole/sulfonatedgraphene/graphene tri-layer film	A tri-layer electrochemical actuator was fabricated by the electrodeposition of polypyrrole (PPy) onto a sulfonated graphene (SG)/reduced graphene oxide (RGO) bi-layer film. In this actuator, PPy and RGO were acted as actuation and conductive inert layers, respectively. The SG layer was used to enhance the interfacial interactions.	[121]
13.	Layer-by-layer assembled sulfonated graphene/polyaniline nanocomposite films: enhanced electrical and ionic conductivities, and electrochromic properties	The facile synthesis of sulfonic acid-grafted reduced graphene oxide (S-rGO) using a one-pot method under mild conditions and layer-by-layer (LbL) assembly and electrochromic properties of SrGO/polyaniline (S-rGO/PANI) nanocomposite thin films were reported. This can be attributed to the drastically increased electrical and ionic conductivities of the S-rGO/PANI films.	[122]
14.	Direct electrochemistry and electrocatalysis of glucose oxidase on three-dimensional interpenetrating, porous graphene modified electrode.	Direct electrochemistry of glucose oxidase (GOx) on three-dimensional (3D) interpenetrating porous graphene electrodes has been reported, which have been fabricated by one-step electrochemical reduction of graphene oxide (GO) from its aqueous suspension. The electrochemically reduced GO (ERGO) modified electrodes exhibited excellent electron transfer properties for GOx and enhanced the enzymeactivity and stability.	[123]
15.	Three-dimensional graphene-carbon nanotube hybrid for high-performance enzymatic biofuel cells.	Enzyme-functionalized 3D graphene-single walled carbon nanotubes (SWCNTs) hybrid electrodes using the naturally abundant glucose as the fuel were developed. Such bioanode, has high stability and power density (2.27 mW cm^{-2}).	[124]
16.	Comparison of performances of bioanodes modified with graphene oxide and graphene–platinum hybrid nanoparticles	This was the first study that constituted these nanomaterials in BFC systems. Power and current densities of these systems were calculated as 2.40 μW cm^{-2} and 211.90 μA cm^{-2} for graphene oxide (GO) based BFC and 4.88 μW cm^{-2} and 246.82 μA cm^{-2}, for Gr-Pt hybrid NPs based BFC. As a result, the effective performances of different combinations	[125]

		of graphene with Pt were realized.	
17.	Optimization of glassy carbon electrode base graphene/ferritin/glucose oxidase bioanode for biofuel cell applications	A glassy carbon (GC)/graphene/ferritin/glucose oxidase (GOx) anode was developed by using graphene/ferritin biocomposite as an electron transfer enhancer and mediator, respectively. All electrochemical measurements were carried out by cyclic voltammetry (CV) and linear sweep voltammetry (LSV). A saturation current density of 66.5 ± 2 mA cm^{-2} .	[126]
18.	New trends in enzyme immobilization at nanostructured interfacesfor efficient electrocatalysis in biofuel cells.	This review focused on the materials developed and involved during the last few years to meet certain demands via nanostructuration of electrode interfaces.	[127]
19.	Is graphene worth using in biofuel cells ?	A two-dimensional carbon nanomaterial, exhibiting excellent conductivity, good mechanical and optical properties with an affordable cost was developed. It was also found that it can be integrated quite effectively with biocatalysts for fabrication of graphene-based biofuel cells (BFCs), where the biocatalysts were used for turning a chemical energy of substrates/biofuels into electricity. Like other nanomaterials, graphene can be applied for preparation of highly structured electrode interfaces, where high amount of biocatalysts can be loaded and thus the power output of a BFC can be increased.	[128]
20.	Graphene-based nanobiocatalytic systems: recent advances and future prospects	Graphene-based nanomaterials are particularly useful nanostructured materials that show great promise in biotechnology and biomedicine. Owing to their unique structural features, exceptional chemical, electrical, and mechanical properties, and their ability to affect the microenvironment of biomolecules, graphene-based nano- materials are suitable for use in various applications, such as immobilization of enzymes.	[129]
21.	Glucose oxidase directly immobilized onto highly porous gold electrodes for sensing and fuel cell	This study reports, a simple and rapid methodology for the direct immobilisation of glucose oxidase into highly porous gold electrodes. This glucose oxidase electrode was tested as an anode in a glucose/O_2	[63]

	applications.	enzymatic biofuel cell, leading to a peak power density of 6 μW cm^{-2}.	
22.	Wired enzymes in mesoporous materials: A benchmark for fabricating biofuel cells	Mesostructured platforms with high surface area and stability have been developed to enhance mass transport, charge transfer from biocatalysts to electrodes and enzyme stability, leading to biofuel cells with improved power density (up to 602 μW cm^{-2} at physiological conditions) and overall performance (high stability after 30 h of continuous operation and after 10 days of storage). This review discussed the developments using mesoporous materials as novel platforms for effective electronic charge transfer in the context of current and emerging technologies in enzymatic fuel cell research, emphasizing their practical implications and potential improvements leading to a major impact on medical science and portable electronics.	[130]
23.	An overviewof biofuel policies and industrialization in the major biofuelproducing countries	Biofuels are important substitute for fossil fuel with the advantages such as susatainablility, environmental friendly and good adaptability. In the last 10 years, biofuels development has been driven by the government policies. By the support of the government, some biofuel projects have been commercialized.	[131]
24.	Membrane/mediator-free rechargeable enzymatic biofuel cell utilizing graphene/single-wall carbon nanotube cogel electrodes.	Novel electrodes of graphene and single-wall carbon nanotube cogel were prepared. These cogel electrodes had large surface area that enabled high enzyme loading, large porosity for unhindered glucose transport and moderate electrical conductivity for efficient charge collection.	[132]
25.	Microbial fuel cell based on Ni-tetra sulfonated phthalocyanine cathode and graphene modified bioanode	A microbial fuel cell (MFC) has been constructed using stainless steel modified with reduced graphene oxide (rGO) by layer by layer (LbL) original method as the bioanode. The rGO permitted to obtain stable power density over a period of 40 days 24.8 mWm^{-2}	[133]
26.	Graphene based enzymatic bioelectrodes and biofuel cells	The excellent electrical conductivity and ease of functionalization make graphene a promising material for use in enzymatic bioelectrodes and biofuel cells. Specific examples of enzyme immobilization methods, enzyme loading, stability	[134]

		and function on graphene, functionalized graphene and graphene based nanocomposite materials were discussed along with their advantages and limitations.	
27.	Design of an enzymatic biofuel cell with large power output	Enzymes were bound to the hydrophilic, carboxyl group functionalized graphene–gold nanoparticle hybrid, and the hybrid as electrode material was able to increase the rate of electron transfer in the EBFC. The open-circuit voltage (Eocv cell) of this designed EBFC reached 1.16 ± 0.02 V, and the maximal power density (Pmax) was as high as 1.96 ± 0.13 mW cm^{-2}.	[135]
28.	Study of direct electron transfer and enzyme activity of glucose oxidase on graphene surface	The DET property and enzyme activity of GOx on graphene surface without and with mediators were examined. Experimental results suggested that DET property and enzyme catalytic activity could not occur on the same GOx simultaneously. Therefore, keeping enzyme activity and DET of GOx at the same time is still a major challenge for biosensor and biofuel cell researches.	[136]

Acknowledgments

The authors are thankful to the Department of Applied Chemistry, Aligarh Muslim University for providing facilities and Council of Scientific and Industrial Research, India, for financial support vide project no. 01 (2702)/12EMR-II.

Reference

[1] S. Kerzenmacher, U. Kräling, T. Metz, R. Zengerle, F. von Stetten, A potentially implantable glucose fuel cell with Raney-platinum film electrodes for improved hydrolytic and oxidative stability, J. Power Sources. 196 (2011) 1264–1272. *http://dx.doi.org/10.1016/j.jpowsour.2010.08.019*

[2] D. Bhatnagar, S. Xu, C. Fischer, R.L. Arechederra, S.D. Minteer, Mitochondrial biofuel cells: expanding fuel diversity to amino acids, Phys. Chem. Chem. Phys. 13 (2011) 86–92. *http://dx.doi.org/10.1039/C0CP01362E*

[3] http://www.search.eb.com/eb/article?idxref=51245.

[4] B. Cook, Introduction to fuel cells and hydrogen technology, Eng. Sci. Educ. J. 11 (2002) 205–216.
http://dx.doi.org/10.1049/esej:20020601

[5] A. Demirbas, Political, economic and environmental impacts of biofuels: A review, Appl. Energy. 86 (2009) S108–S117.
http://dx.doi.org/10.1016/j.apenergy.2009.04.036

[6] A.T. Yahiro, S.M. Lee, D.O. Kimble, Bioelectrochemistry, Biochim. Biophys. Acta - Spec. Sect. Biophys. Subj. 88 (1964) 375–383.

[7] S. Cheng, H. Liu, B.E. Logan, Increased performance of single-chamber microbial fuel cells using an improved cathode structure, Electrochem. Commun. 8 (2006) 489–494.
http://dx.doi.org/10.1016/j.elecom.2006.01.010

[8] J. Giner, Electrochemical Glucose Oxidation on a Platinized Platinum Electrode in Krebs-Ringer Solution, J. Electrochem. Soc. 128 (1981) 2106.
http://dx.doi.org/10.1149/1.2127199

[9] H. Liu, S. Grot, B.E. Logan, Electrochemically Assisted Microbial Production of Hydrogen from Acetate, Environ. Sci. Technol. 39 (2005) 4317–4320.
http://dx.doi.org/10.1021/es050244p

[10] M.C. Potter, Electrical Effects Accompanying the Decomposition of Organic Compounds, Proc. R. Soc. B Biol. Sci. 84 (1911) 260–276.
http://dx.doi.org/10.1098/rspb.1911.0073

[11] I. Karube, T. Matsunaga, S. Tsuru, S. Suzuki, Biochemical fuel cell utilizing immobilized cells of clostridium butyricum, Biotechnol. Bioeng. 19 (1977) 1727–1733.
http://dx.doi.org/10.1002/bit.260191112

[12] H. Liu, S. Cheng, B.E. Logan, Production of Electricity from Acetate or Butyrate Using a Single-Chamber Microbial Fuel Cell, Environ. Sci. Technol. 39 (2005) 658–662.
http://dx.doi.org/10.1021/es048927c

[13] B.C.H. Steele, A. Heinzel, Review article Materials for fuel-cell technologies, Nature. 414 (2001) 345–352.
http://dx.doi.org/10.1038/35104620

[14] J.B. Davis, H.F. Yarbrough, Preliminary Experiments on a Microbial Fuel Cell, Science (80-.). 137 (1962) 615–616.
http://dx.doi.org/10.1126/science.137.3530.615

[15] H. Liu, B.E. Logan, Electricity Generation Using an Air-Cathode Single Chamber Microbial Fuel Cell in the Presence and Absence of a Proton Exchange Membrane, Environ. Sci. Technol. 38 (2004) 4040–4046. *http://dx.doi.org/10.1021/es0499344*

[16] P. Clauwaert, D. van der Ha, N. Boon, K. Verbeken, M. Verhaege, K. Rabaey, W. Verstraete, Open Air Biocathode Enables Effective Electricity Generation with Microbial Fuel Cells, Environ. Sci. Technol. 41 (2007) 7564–7569. *http://dx.doi.org/10.1021/es0709831*

[17] N. Kakehi, T. Yamazaki, W. Tsugawa, K. Sode, A novel wireless glucose sensor employing direct electron transfer principle based enzyme fuel cell, Biosens. Bioelectron. 22 (2007) 2250–2255. *http://dx.doi.org/10.1016/j.bios.2006.11.004*

[18] S. Cosnier, A. Le Goff, M. Holzinger, Towards glucose biofuel cells implanted in human body for powering artificial organs: Review, Electrochem. Commun. 38 (2014) 19–23. *http://dx.doi.org/10.1016/j.elecom.2013.09.021*

[19] D.L. HAYES, S. FURMAN, Cardiac Pacing:. How It Started, Where We Are, Where We Are Going, Pacing Clin. Electrophysiol. 27 (2004) 693–704. *http://dx.doi.org/10.1111/j.1540-8159.2004.00515.x*

[20] J.D. Salamone, M. Correa, The Mysterious Motivational Functions of Mesolimbic Dopamine, Neuron. 76 (2012) 470–485. *http://dx.doi.org/10.1016/j.neuron.2012.10.021*

[21] P.D. Mitcheson, E.M. Yeatman, G.K. Rao, A.S. Holmes, T.C. Green, Energy Harvesting From Human and Machine Motion for Wireless Electronic Devices, Proc. IEEE. 96 (2008) 1457–1486. *http://dx.doi.org/10.1109/JPROC.2008.927494*

[22] V. Leonov, T. Torfs, P. Fiorini, C. Van Hoof, Thermoelectric Converters of Human Warmth for Self-Powered Wireless Sensor Nodes, IEEE Sens. J. 7 (2007) 650–657. *http://dx.doi.org/10.1109/JSEN.2007.894917*

[23] M.W. Baker, R. Sarpeshkar, Feedback Analysis and Design of RF Power Links for Low-Power Bionic Systems, IEEE Trans. Biomed. Circuits Syst. 1 (2007) 28–38. *http://dx.doi.org/10.1109/TBCAS.2007.893180*

[24] G.A. Covic, J.T. Boys, Inductive Power Transfer, Proc. IEEE. 101 (2013) 1276–1289.

http://dx.doi.org/10.1109/JPROC.2013.2244536

[25] Prutchi. Nuclear Pacemakers, [Accessed: 2/3/2016] Available from: http://home.comcast.net/~dprutchi/nuclear_pacemakers.pdf.

[26] S. Kerzenmacher, J. Ducrée, R. Zengerle, F. von Stetten, Energy harvesting by implantable abiotically catalyzed glucose fuel cells, J. Power Sources. 182 (2008) 1–17.
http://dx.doi.org/10.1016/j.jpowsour.2008.03.031

[27] A. Zebda, S. Cosnier, J.-P. Alcaraz, M. Holzinger, A. Le Goff, C. Gondran, F. Boucher, F. Giroud, K. Gorgy, H. Lamraoui, P. Cinquin, Single Glucose Biofuel Cells Implanted in Rats Power Electronic Devices, Sci. Rep. 3 (2013).
http://dx.doi.org/10.1038/srep01516

[28] L. Halámková, J. Halámek, V. Bocharova, A. Szczupak, L. Alfonta, E. Katz, Implanted Biofuel Cell Operating in a Living Snail, J. Am. Chem. Soc. 134 (2012) 5040–5043.
http://dx.doi.org/10.1021/ja211714w

[29] E. Katz. Evgeny Katz, [Accessed: 10/03/2016] Available from http://people.clarkson.edu/~ekatz/.D. Scherson.

[30] W.E. Farneth, M.B. D'Amore, Encapsulated laccase electrodes for fuel cell cathodes, J. Electroanal. Chem. 581 (2005) 197–205.
http://dx.doi.org/10.1016/j.jelechem.2005.03.018

[31] T. Miyake, K. Haneda, N. Nagai, Y. Yatagawa, H. Onami, S. Yoshino, T. Abe, M. Nishizawa, Enzymatic biofuel cells designed for direct power generation from biofluids in living organisms, Energy Environ. Sci. 4 (2011) 5008.
http://dx.doi.org/10.1039/c1ee02200h

[32] A. Szczupak, J. Halámek, L. Halámková, V. Bocharova, L. Alfonta, E. Katz, Living battery – biofuel cells operating in vivo in clams, Energy Environ. Sci. 5 (2012) 8891.
http://dx.doi.org/10.1039/c2ee21626d

[33] X. Wei, J. Liu, Power sources and electrical recharging strategies for implantable medical devices, Front. Energy Power Eng. China. 2 (2008) 1–13.
http://dx.doi.org/10.1007/s11708-008-0016-3

[34] P. Cinquin, C. Gondran, F. Giroud, S. Mazabrard, A. Pellissier, F. Boucher, J.-P. Alcaraz, K. Gorgy, F. Lenouvel, S. Mathé, P. Porcu, S. Cosnier, A Glucose BioFuel Cell Implanted in Rats, PLoS One. 5 (2010) e10476.
http://dx.doi.org/10.1371/journal.pone.0010476

[35] M. Southcott, K. MacVittie, J. Halámek, L. Halámková, W.D. Jemison, R. Lobel, E. Katz, A pacemaker powered by an implantable biofuel cell operating under conditions mimicking the human blood circulatory system – battery not included, Phys. Chem. Chem. Phys. 15 (2013) 6278.
http://dx.doi.org/10.1039/c3cp50929j

[36] M. Holzinger, A. Le Goff, S. Cosnier, Carbon nanotube/enzyme biofuel cells, Electrochim. Acta. 82 (2012) 179–190.
http://dx.doi.org/10.1016/j.electacta.2011.12.135

[37] S. Cosnier, D. Shan, S.-N. Ding, An easy compartment-less biofuel cell construction based on the physical co-inclusion of enzyme and mediator redox within pressed graphite discs, Electrochem. Commun. 12 (2010) 266–269.
http://dx.doi.org/10.1016/j.elecom.2009.12.011

[38] L. Ren, D. Yan, W. Zhong, Enhanced enzyme activity through electron transfer between single-walled carbon nanotubes and horseradish peroxidase, Carbon N. Y. 50 (2012) 1303–1310.
http://dx.doi.org/10.1016/j.carbon.2011.10.053

[39] S. Calabrese Barton, J. Gallaway, P. Atanassov, Enzymatic Biofuel Cells for Implantable and Microscale Devices, Chem. Rev. 104 (2004) 4867–4886.
http://dx.doi.org/10.1021/cr020719k

[40] S. Rengaraj, V. Mani, P. Kavanagh, J. Rusling, D. Leech, A membrane-less enzymatic fuel cell with layer-by-layer assembly of redox polymer and enzyme over graphite electrodes, Chem. Commun. 47 (2011) 11861.
http://dx.doi.org/10.1039/c1cc15002b

[41] J.A. Castorena-Gonzalez, C. Foote, K. MacVittie, J. Halámek, L. Halámková, L.A. Martinez-Lemus, E. Katz, Biofuel Cell Operating in Vivo in Rat, Electroanalysis. 25 (2013) 1579–1584.
http://dx.doi.org/10.1002/elan.201300136

[42] F.C.P.F. Sales, R.M. Iost, M.V.A. Martins, M.C. Almeida, F.N. Crespilho, An intravenous implantable glucose/dioxygen biofuel cell with modified flexible carbon fiber electrodes, Lab Chip. 13 (2013) 468–474.
http://dx.doi.org/10.1039/C2LC41007A

[43] M. Bottini, S. Bruckner, K. Nika, N. Bottini, S. Bellucci, A. Magrini, A. Bergamaschi, T. Mustelin, Multi-walled carbon nanotubes induce T lymphocyte apoptosis, Toxicol. Lett. 160 (2006) 121–126.
http://dx.doi.org/10.1016/j.toxlet.2005.06.020

[44] K.F. O'Driscoll, Techniques of enzyme entrapment in gels, In: Methods in Enzymology 1976: pp. 169–183.

[45] A.C. Pierre, The sol-gel encapsulation of enzymes, Biocatal. Biotransformation. 22 (2004) 145–170.
http://dx.doi.org/10.1080/10242420412331283314

[46] V. Andoralov, M. Falk, D.B. Suyatin, M. Granmo, J. Sotres, R. Ludwig, V.O. Popov, J. Schouenborg, Z. Blum, S. Shleev, Biofuel Cell Based on Microscale Nanostructured Electrodes with Inductive Coupling to Rat Brain Neurons, Sci. Rep. 3 (2013).
http://dx.doi.org/10.1038/srep03270

[47] M. Hakamada, M. Takahashi, M. Mabuchi, Enzyme electrodes stabilized by monolayer-modified nanoporous Au for biofuel cells, Gold Bull. 45 (2012) 9–15.
http://dx.doi.org/10.1007/s13404-011-0038-1

[48] U. Salaj-Kosla, M.D. Scanlon, T. Baumeister, K. Zahma, R. Ludwig, P.Ó. Conghaile, D. MacAodha, D. Leech, E. Magner, Mediated electron transfer of cellobiose dehydrogenase and glucose oxidase at osmium polymer-modified nanoporous gold electrodes, Anal. Bioanal. Chem. 405 (2013) 3823–3830.
http://dx.doi.org/10.1007/s00216-012-6657-4

[49] A. Chaubey, B.D. Malhotra, Mediated biosensors, Biosens. Bioelectron. 17 (2002) 441–456.
http://dx.doi.org/10.1016/S0956-5663(01)00313-X

[50] R.A. Bullen, T.C. Arnot, J.B. Lakeman, F.C. Walsh, Biofuel cells and their development, Biosens. Bioelectron. 21 (2006) 2015–2045.
http://dx.doi.org/10.1016/j.bios.2006.01.030

[51] Shaolin Mu, Huaiguo Xue, Bioelectrochemical characteristics of glucose oxidase immobilized in a polyaniline film, Sensors Actuators B Chem. 31 (1996) 155–160.
http://dx.doi.org/10.1016/0925-4005(96)80060-8

[52] Y. Degani, A. Heller, Direct electrical communication between chemically modified enzymes and metal electrodes. I. Electron transfer from glucose oxidase to metal electrodes via electron relays, bound covalently to the enzyme, J. Phys. Chem. 91 (1987) 1285–1289.
http://dx.doi.org/10.1021/j100290a001

[53] M.J. Moehlenbrock, S.D. Minteer, Extended lifetime biofuel cells, Chem. Soc. Rev. 37 (2008) 1188.
http://dx.doi.org/10.1039/b708013c

[54] N. Yuhashi, M. Tomiyama, J. Okuda, S. Igarashi, K. Ikebukuro, K. Sode, Development of a novel glucose enzyme fuel cell system employing protein engineered PQQ glucose dehydrogenase, Biosens. Bioelectron. 20 (2005) 2145–2150.
http://dx.doi.org/10.1016/j.bios.2004.08.017

[55] N. Mano, F. Mao, W. Shin, T. Chen, A. Heller, A miniature biofuel cell operating at 0.78 V, Chem. Commun. (2003) 518–519.
http://dx.doi.org/10.1039/b211796g

[56] J. Lim, N. Cirigliano, J. Wang, B. Dunn, Direct electron transfer in nanostructured sol–gel electrodes containing bilirubin oxidase, Phys. Chem. Chem. Phys. 9 (2007) 1809–1814.
http://dx.doi.org/10.1039/B618422G

[57] J.M. Bolivar, L. Wilson, S.A. Ferrarotti, R. Fernandez-Lafuente, J.M. Guisan, C. Mateo, Stabilization of a Formate Dehydrogenase by Covalent Immobilization on Highly Activated Glyoxyl-Agarose Supports, Biomacromolecules. 7 (2006) 669–673.
http://dx.doi.org/10.1021/bm050947z

[58] B.Z. Egan, I. Chibata, Immobilized Enzymes: Research and Development, Kodansha Scientific Books, (1979) 481.

[59] T.-W. Tsai, G. Heckert, L.F. Neves, Y. Tan, D.-Y. Kao, R.G. Harrison, D.E. Resasco, D.W. Schmidtke, Adsorption of Glucose Oxidase onto Single-Walled Carbon Nanotubes and Its Application in Layer-By-Layer Biosensors, Anal. Chem. 81 (2009) 7917–7925.
http://dx.doi.org/10.1021/ac900650r

[60] A. Guiseppi-Elie, C. Lei, R.H. Baughman, Direct electron transfer of glucose oxidase on carbon nanotubes, Nanotechnology. 13 (2002) 559–564.
http://dx.doi.org/10.1088/0957-4484/13/5/303

[61] H. Qiu, C. Xu, X. Huang, Y. Ding, Y. Qu, P. Gao, Immobilization of Laccase on Nanoporous Gold: Comparative Studies on the Immobilization Strategies and the Particle Size Effects, J. Phys. Chem. C. 113 (2009) 2521–2525.
http://dx.doi.org/10.1021/jp8090304

[62] E. Generalic. “Zwitterion.” Croatian-English Chemistry Dictionary & Glossary, [Accessed: 20/3/2016] Available from:
http://glossary.periodni.com/glossary.php?en=zwitterion.

[63] H. du Toit, M. Di Lorenzo, Glucose Oxidase Directly Immobilized onto Highly Porous Gold Electrodes for Sensing and Fuel Cell applications, Electrochim. Acta. 138 (2014) 86–92. *http://dx.doi.org/10.1016/j.electacta.2014.06.074*

[64] D. Stark, U. von Stockar, In Situ Product Removal (ISPR) in Whole Cell Biotechnology During the Last Twenty Years, Adv Biochem Eng Biotechnol 80 (2003) 149–175.

[65] J. Wang, Glucose Biosensors: 40 Years of Advances and Challenges, Electroanalysis. 13 (2001) 983–988. *http://dx.doi.org/10.1002/1521-4109(200108)13:12<983::AID-ELAN983>3.0.CO;2-#*

[66] R. Sahney, S. Anand, B.K. Puri, A.K. Srivastava, A comparative study of immobilization techniques for urease on glass-pH-electrode and its application in urea detection in blood serum, Anal. Chim. Acta. 578 (2006) 156–161. *http://dx.doi.org/10.1016/j.aca.2006.06.063*

[67] K. Zeng, H. Tachikawa, Z. Zhu, V.L. Davidson, Amperometric Detection of Histamine with a Methylamine Dehydrogenase Polypyrrole-Based Sensor, Anal. Chem. 72 (2000) 2211–2215. *http://dx.doi.org/10.1021/ac9911138*

[68] Y. Liu, M. Wang, F. Zhao, Z. Xu, S. Dong, The direct electron transfer of glucose oxidase and glucose biosensor based on carbon nanotubes/chitosan matrix, Biosens. Bioelectron. 21 (2005) 984–988. *http://dx.doi.org/10.1016/j.bios.2005.03.003*

[69] D. Ivnitski, B. Branch, P. Atanassov, C. Apblett, Glucose oxidase anode for biofuel cell based on direct electron transfer, Electrochem. Commun. 8 (2006) 1204–1210. *http://dx.doi.org/10.1016/j.elecom.2006.05.024*

[70] S. Zhang, N. Wang, Y. Niu, C. Sun, Immobilization of glucose oxidase on gold nanoparticles modified Au electrode for the construction of biosensor, Sensors Actuators B Chem. 109 (2005) 367–374. *http://dx.doi.org/10.1016/j.snb.2005.01.003*

[71] M.K. Goel. Immobilized Enzymes, [Accessed: 20/3/2016] Available from: http://www.rpi.edu/dept/chem-eng/Biotech-Environ/IMMOB/goel2nd.html.

[72] B.E. Logan, Scaling up microbial fuel cells and other bioelectrochemical systems, Appl. Microbiol. Biotechnol. 85 (2010) 1665–1671.

http://dx.doi.org/10.1007/s00253-009-2378-9

[73] C.M. Moore, N.L. Akers, A.D. Hill, Z.C. Johnson, S.D. Minteer, Improving the Environment for Immobilized Dehydrogenase Enzymes by Modifying Nafion with Tetraalkylammonium Bromides, Biomacromolecules. 5 (2004) 1241–1247. *http://dx.doi.org/10.1021/bm0345256*

[74] J. Kim, H. Jia, P. Wang, Challenges in biocatalysis for enzyme-based biofuel cells, Biotechnol. Adv. 24 (2006) 296–308. *http://dx.doi.org/10.1016/j.biotechadv.2005.11.006*

[75] Y. Xiao, "Plugging into Enzymes": Nanowiring of Redox Enzymes by a Gold Nanoparticle, Science (80-.). 299 (2003) 1877–1881.

[76] D. Schumacher, J. Vogel, U. Lerche, Construction and applications of an enzyme electrode for determination of galactose and galactose-containing saccharides, Biosens. Bioelectron. 9 (1994) 85–89. doi:10.1016/0956-5663(94)80098-7. *http://dx.doi.org/10.1016/0956-5663(94)80098-7*

[77] S. Ohkoshi, A. Fujishima, K. Hashimoto, Transparent and Colored Magnetic Thin Films: (Fe II x Cr II 1- x) 1.5 [Cr III (CN) 6], J. Am. Chem. Soc. 120 (1998) 5349–5350. *http://dx.doi.org/10.1021/ja980732f*

[78] I. Willner, G. Arad, E. Katz, A biofuel cell based on pyrroloquinoline quinone and microperoxidase-11 monolayer-functionalized electrodes, Bioelectrochemistry Bioenerg. 44 (1998) 209–214. *http://dx.doi.org/10.1016/S0302-4598(97)00091-3*

[79] I. Willner, R. Blonder, E. Katz, A. Stocker, A.F. Bückmann, Reconstitution of Apo-Glucose Oxidase with a Nitrospiropyran-Modified FAD Cofactor Yields a Photoswitchable Biocatalyst for Amperometric Transduction of Recorded Optical Signals, J. Am. Chem. Soc. 118 (1996) 5310–5311. *http://dx.doi.org/10.1021/ja960228f*

[80] E. Katz, A. Riklin, V. Heleg-Shabtai, I. Willner, A.F. Bückmann, Glucose oxidase electrodes via reconstitution of the apo-enzyme: tailoring of novel glucose biosensors, Anal. Chim. Acta. 385 (1999) 45–58. *http://dx.doi.org/10.1016/S0003-2670(98)00688-6*

[81] S. Cosnier, C.H. Gondran, Fabrication of biosensors by attachment of biological macromolecules to electropolymerized conducting films, Analusis. 27 (1999) 558–563. *http://dx.doi.org/10.1051/analusis:1999270558*

[82] I. Willner, V. Heleg-Shabtai, R. Blonder, E. Katz, G. Tao, A.F. Bückmann, A. Heller, Electrical Wiring of Glucose Oxidase by Reconstitution of FAD-Modified Monolayers Assembled onto Au-Electrodes, J. Am. Chem. Soc. 118 (1996) 10321–10322.
http://dx.doi.org/10.1021/ja9608611

[83] A. Ramanavičius, A. Ramanavičienė, A. Malinauskas, Electrochemical sensors based on conducting polymer—polypyrrole, Electrochim. Acta. 51 (2006) 6025–6037.
http://dx.doi.org/10.1016/j.electacta.2005.11.052

[84] W.J. Sung, Y.H. Bae, A Glucose Oxidase Electrode Based on Electropolymerized Conducting Polymer with Polyanion−Enzyme Conjugated Dopant, Anal. Chem. 72 (2000) 2177–2181. doi:10.1021/ac9908041.
http://dx.doi.org/10.1021/ac9908041

[85] S. Cosnier, C. Gondran, J.-C. Watelet, A Polypyrrole-Bienzyme Electrode (Salicylate Hydroxylase-Polyphenol Oxidase) for the Interference-Free Determination of Salicylate, Electroanalysis. 13 (2001) 906–910.
http://dx.doi.org/10.1002/1521-4109(200107)13:11<906::AID-ELAN906>3.0.CO;2-L

[86] M. Yasuzawa, T. Nieda, T. Hirano, A. Kunugi, Properties of glucose sensors based on the immobilization of glucose oxidase in N-substituted polypyrrole film, Sensors Actuators B Chem. 66 (2000) 77–79.
http://dx.doi.org/10.1016/S0925-4005(99)00453-0

[87] S. Cosnier, A. Senillou, M. Grätzel, P. Comte, N. Vlachopoulos, N. Jaffrezic Renault, C. Martelet, A glucose biosensor based on enzyme entrapment within polypyrrole films electrodeposited on mesoporous titanium dioxide, J. Electroanal. Chem. 469 (1999) 176–181.
http://dx.doi.org/10.1016/S0022-0728(99)00223-5

[88] M. Trojanowicz, W. Matuszewski, M. Podsiadła, Enzyme entrapped polypyrrole modified electrode for flow-injection determination of glucose, Biosens. Bioelectron. 5 (1990) 149–156.
http://dx.doi.org/10.1016/0956-5663(90)80005-X

[89] S. Cosnier, A. Deronzier, J.-F. Roland, Electrocatalytic oxidation of alcohols on carbon electrodes modified by functionalized polypyrrole–RuO2 films, J. Mol. Catal. 71 (1992) 303–315.
http://dx.doi.org/10.1016/0304-5102(92)85021-7

[90] N.C. Foulds, C.R. Lowe, Enzyme entrapment in electrically conducting polymers. Immobilisation of glucose oxidase in polypyrrole and its application in amperometric glucose sensors, J. Chem. Soc. Faraday Trans. 1 Phys. Chem. Condens. Phases. 82 (1986) 1259.

[91] M. Umana, J. Waller, Protein-modified electrodes. The glucose oxidase/polypyrrole system, Anal. Chem. 58 (1986) 2979–2983. *http://dx.doi.org/10.1021/ac00127a018*

[92] Inamuddin, K.M. Shin, S.I. Kim, I. So, S.J. Kim, A conducting polymer/ferritin anode for biofuel cell applications, Electrochim. Acta. 54 (2009) 3979–3983. *http://dx.doi.org/10.1016/j.electacta.2009.02.020*

[93] G. Inzelt, Conducting Polymers, Springer Berlin Heidelberg, Berlin, Heidelberg, 2012.

[94] A.. Epstein, J.. Ginder, F. Zuo, H.-S. Woo, D.. Tanner, A. Richter, M. Angelopoulos, W.-S. Huang, A.. MacDiarmid, Insulator-to-metal transition in polyaniline: Effect of protonation in emeraldine, Synth. Met. 21 (1987) 63–70. *http://dx.doi.org/10.1016/0379-6779(87)90067-1*

[95] G.G Wallace, G.M. Spinks, L.A. Kane-Maguire, P. R. Teasdale, Conductive Electroactive Polymers, 3rd Edition, CRC Press, 10 (2008) 22., n.d.

[96] J. Stejskal, R.G. Gilbert, Polyaniline. Preparation of a conducting polymer(IUPAC Technical Report), Pure Appl. Chem. 74 (2002). *http://dx.doi.org/10.1351/pac200274050857*

[97] A.A. Karyakin, A.K. Strakhova, A.K. Yatsimirsky, Self-doped polyanilines electrochemically active in neutral and basic aqueous solutions., J. Electroanal. Chem. 371 (1994) 259–265. *http://dx.doi.org/10.1016/0022-0728(93)03244-J*

[98] D. Kumar, Synthesis and characterization of poly(aniline-co-o-toluidine) copolymer, Synth. Met. 114 (2000) 369–372. *http://dx.doi.org/10.1016/S0379-6779(00)00270-8*

[99] L.H.C. Mattoso, L.O.S. Bulhões, Synthesis and characterization of poly(o-anisidine) films, Synth. Met. 52 (1992) 171–181. *http://dx.doi.org/10.1016/0379-6779(92)90305-3*

[100] J. Zhang, D. Shan, S. Mu, Electrochemical copolymerization of aniline with m-aminophenol and novel electrical properties of the copolymer in the wide pH range, Electrochim. Acta. 51 (2006) 4262–4270. *http://dx.doi.org/10.1016/j.electacta.2005.12.005*

[101] R.E. Kim, S.-G. Hong, S. Ha, J. Kim, Enzyme adsorption, precipitation and crosslinking of glucose oxidase and laccase on polyaniline nanofibers for highly stable enzymatic biofuel cells, Enzyme Microb. Technol. 66 (2014) 35–41. *http://dx.doi.org/10.1016/j.enzmictec.2014.08.001*

[102] J.C. Cooper, E.A.H. Hall, Catalytic reduction of benzoquinone at polyaniline and polyaniline/enzyme films, Electroanalysis. 5 (1993) 385–397. *http://dx.doi.org/10.1002/elan.1140050504*

[103] O.A. Raitman, E. Katz, A.F. Bückmann, I. Willner, Integration of Polyaniline/Poly(acrylic acid) Films and Redox Enzymes on Electrode Supports: An in Situ Electrochemical/Surface Plasmon Resonance Study of the Bioelectrocatalyzed Oxidation of Glucose or Lactate in the Integrated Bioelectrocatalytic System, J. Am. Chem. Soc. 124 (2002) 6487–6496. *http://dx.doi.org/10.1021/ja012680r*

[104] J. Kopeček, K. Ulbrich, Biodegradation of biomedical polymers, Prog. Polym. Sci. 9 (1983) 1–58. *http://dx.doi.org/10.1016/0079-6700(83)90005-9*

[105] O.A. Raitman, F. Patolsky, E. Katz, I. Willner, Electrical contacting of glucose dehydrogenase by the reconstitution of a pyrroloquinoline quinone-functionalized polyaniline film associated with an Au-electrode: an in situ electrochemical SPR study, Chem. Commun. (2002) 1936–1937. *http://dx.doi.org/10.1039/b205393d*

[106] H. Xue, Z. Shen, C. Li, Improved selectivity and stability of glucose biosensor based on in situ electropolymerized polyaniline–polyacrylonitrile composite film, Biosens. Bioelectron. 20 (2005) 2330–2334. *http://dx.doi.org/10.1016/j.bios.2004.07.018*

[107] L. Cao, Immobilised enzymes: science or art?, Curr. Opin. Chem. Biol. 9 (2005) 217–226. *http://dx.doi.org/10.1016/j.cbpa.2005.02.014*

[108] R.A. Sheldon, Enzyme Immobilization: The Quest for Optimum Performance, Adv. Synth. Catal. 349 (2007) 1289–1307. doi:10.1002/adsc.200700082. *http://dx.doi.org/10.1002/adsc.200700082*

[109] A. Pizzariello, M. Stred'ansky, S. Miertuš, A glucose/hydrogen peroxide biofuel cell that uses oxidase and peroxidase as catalysts by composite bulk-modified bioelectrodes based on a solid binding matrix, Bioelectrochemistry. 56 (2002) 99–105. *http://dx.doi.org/10.1016/S1567-5394(02)00026-9*

[110] V.B. Kandimalla, V.S. Tripathi, H. Ju, Immobilization of Biomolecules in Sol–Gels: Biological and Analytical Applications, Critical Reviews in Analytical Chemistry, 36 (2006) 73-106.

[111] C. Liu, S. Alwarappan, Z. Chen, X. Kong, C.-Z. Li, Membraneless enzymatic biofuel cells based on graphene nanosheets, Biosens. Bioelectron. 25 (2010) 1829–1833.
http://dx.doi.org/10.1016/j.bios.2009.12.012

[112] W. Zheng, H.Y. Zhao, J.X. Zhang, H.M. Zhou, X.X. Xu, Y.F. Zheng, Y.B. Wang, Y. Cheng, B.Z. Jang, A glucose/O2 biofuel cell base on nanographene platelet-modified electrodes, Electrochem. Commun. 12 (2010) 869–871.
http://dx.doi.org/10.1016/j.elecom.2010.04.006

[113] C. Liu, Z. Chen, C.-Z. Li, Surface Engineering of Graphene-Enzyme Nanocomposites for Miniaturized Biofuel Cell, IEEE Trans. Nanotechnol. 10 (2011) 59–62.
http://dx.doi.org/10.1109/TNANO.2010.2050147

[114] S. Rengaraj, P. Kavanagh, D. Leech, A comparison of redox polymer and enzyme co-immobilization on carbon electrodes to provide membrane-less glucose/O2 enzymatic fuel cells with improved power output and stability, Biosens. Bioelectron. 30 (2011) 294–299.
http://dx.doi.org/10.1016/j.bios.2011.09.032

[115] D.A.C. Brownson, D.K. Kampouris, C.E. Banks, An overview of graphene in energy production and storage applications, J. Power Sources. 196 (2011) 4873–4885.
http://dx.doi.org/10.1016/j.jpowsour.2011.02.022

[116] Q. Zhang, S. Wu, L. Zhang, J. Lu, F. Verproot, Y. Liu, Z. Xing, J. Li, X.-M. Song, Fabrication of polymeric ionic liquid/graphene nanocomposite for glucose oxidase immobilization and direct electrochemistry, Biosens. Bioelectron. 26 (2011) 2632–2637.
http://dx.doi.org/10.1016/j.bios.2010.11.024

[117] A.T.E. Vilian, S.-M. Chen, C.H. Kwak, S.-K. Hwang, Y.S. Huh, Y.-K. Han, Immobilization of hemoglobin on functionalized multi-walled carbon nanotubes-poly-l-histidine-zinc oxide nanocomposites toward the detection of bromate and H2O2, Sensors Actuators B Chem. 224 (2016) 607–617.
http://dx.doi.org/10.1016/j.snb.2015.10.099

[118] E. Coşkun, E.A. Zaragoza-Contreras, H.J. Salavagione, Synthesis of sulfonated graphene/polyaniline composites with improved electroactivity, Carbon N. Y. 50 (2012) 2235–2243. *http://dx.doi.org/10.1016/j.carbon.2012.01.041*

[119] J. Liu, Y. Qiao, C.X. Guo, S. Lim, H. Song, C.M. Li, Graphene/carbon cloth anode for high-performance mediatorless microbial fuel cells, Bioresour. Technol. 114 (2012) 275–280. *http://dx.doi.org/10.1016/j.biortech.2012.02.116*

[120] B. Ma, X. Zhou, H. Bao, X. Li, G. Wang, Hierarchical composites of sulfonated graphene-supported vertically aligned polyaniline nanorods for high-performance supercapacitors, J. Power Sources. 215 (2012) 36–42. *http://dx.doi.org/10.1016/j.jpowsour.2012.04.083*

[121] A. Liu, W. Yuan, G. Shi, Electrochemical actuator based on polypyrrole/sulfonated graphene/graphene tri-layer film, Thin Solid Films. 520 (2012) 6307–6312. *http://dx.doi.org/10.1016/j.tsf.2012.06.039*

[122] J. Lu, W. Liu, H. Ling, J. Kong, G. Ding, D. Zhou, X. Lu, Layer-by-layer assembled sulfonated-graphene/polyaniline nanocomposite films: enhanced electrical and ionic conductivities, and electrochromic properties, RSC Adv. 2 (2012) 10537. *http://dx.doi.org/10.1039/c2ra21579a*

[123] M. Cui, B. Xu, C. Hu, H.B. Shao, L. Qu, Direct electrochemistry and electrocatalysis of glucose oxidase on three-dimensional interpenetrating, porous graphene modified electrode, Electrochim. Acta. 98 (2013) 48–53. *http://dx.doi.org/10.1016/j.electacta.2013.03.040*

[124] K.P. Prasad, Y. Chen, P. Chen, Three-Dimensional Graphene-Carbon Nanotube Hybrid for High-Performance Enzymatic Biofuel Cells, ACS Appl. Mater. Interfaces. 6 (2014) 3387–3393. *http://dx.doi.org/10.1021/am405432b*

[125] Y. Tepeli, U. Anik, Comparison of performances of bioanodes modified with graphene oxide and graphene–platinum hybrid nanoparticles, Electrochem. Commun. 57 (2015) 31–34. *http://dx.doi.org/10.1016/j.elecom.2015.05.002*

[126] Inamuddin, K. Ahmad, M. Naushad, Optimization of glassy carbon electrode based graphene/ferritin/glucose oxidase bioanode for biofuel cell applications, Int. J. Hydrogen Energy. 39 (2014) 7417–7421. *http://dx.doi.org/10.1016/j.ijhydene.2014.02.171*

[127] A. de Poulpiquet, A. Ciaccafava, E. Lojou, New trends in enzyme immobilization at nanostructured interfaces for efficient electrocatalysis in biofuel cells, Electrochim. Acta. 126 (2014) 104–114. *http://dx.doi.org/10.1016/j.electacta.2013.07.133*

[128] J. Filip, J. Tkac, Is graphene worth using in biofuel cells?, Electrochim. Acta. 136 (2014) 340–354. *http://dx.doi.org/10.1016/j.electacta.2014.05.119*

[129] I. V. Pavlidis, M. Patila, U.T. Bornscheuer, D. Gournis, H. Stamatis, Graphene-based nanobiocatalytic systems: recent advances and future prospects, Trends Biotechnol. 32 (2014) 312–320. *http://dx.doi.org/10.1016/j.tibtech.2014.04.004*

[130] P.N. Catalano, A. Wolosiuk, G.J.A.A. Soler-Illia, M.G. Bellino, Wired enzymes in mesoporous materials: A benchmark for fabricating biofuel cells, Bioelectrochemistry. 106 (2015) 14–21. *http://dx.doi.org/10.1016/j.bioelechem.2015.06.006*

[131] Y. Su, P. Zhang, Y. Su, An overview of biofuels policies and industrialization in the major biofuel producing countries, Renew. Sustain. Energy Rev. 50 (2015) 991–1003. *http://dx.doi.org/10.1016/j.rser.2015.04.032*

[132] A.S. Campbell, Y.J. Jeong, S.M. Geier, R.R. Koepsel, A.J. Russell, M.F. Islam, Membrane/Mediator-Free Rechargeable Enzymatic Biofuel Cell Utilizing Graphene/Single-Wall Carbon Nanotube Cogel Electrodes, ACS Appl. Mater. Interfaces. 7 (2015) 4056–4065. *http://dx.doi.org/10.1021/am507801x*

[133] J. Champavert, S. Ben Rejeb, C. Innocent, M. Pontié, Microbial fuel cell based on Ni-tetra sulfonated phthalocyanine cathode and graphene modified bioanode, J. Electroanal. Chem. 757 (2015) 270–276. *http://dx.doi.org/10.1016/j.jelechem.2015.09.012*

[134] A. Karimi, A. Othman, A. Uzunoglu, L. Stanciu, S. Andreescu, Graphene based enzymatic bioelectrodes and biofuel cells, Nanoscale. 7 (2015) 6909– *http://dx.doi.org/10.1039/C4NR07586B*

[135] M. Liu, J. Qian, Y. Zhao, D. Zhu, L. Gan, L. Chen, Core–shell ultramicroporous@microporous carbon nanospheres as advanced supercapacitor electrodes, J. Mater. Chem. A. 3 (2015) 11517–11526. *http://dx.doi.org/10.1039/C5TA02224J*

[136] B. Liang, X. Guo, L. Fang, Y. Hu, G. Yang, Q. Zhu, J. Wei, X. Ye, Study of direct electron transfer and enzyme activity of glucose oxidase on graphene surface, Electrochem. Commun. 50 (2015) 1–5. *http://dx.doi.org/10.1016/j.elecom.2014.10.016*

CHAPTER 5

Synthesis and Characterization of Sodium Dodecyl Sulphate Intercalated Zirconium Phosphate (SDS-ZrP): A New Zinc Selective Ion- exchange Material

Arshid Bashir and Altaf Hussain Pandith*

Department of Chemistry, University of Kashmir, Hazratbal, Srinagar-190006, J&K, India

Email: altafpandit23@gmail.com , Phone: +91-(194)-2424900: Fax No.: +91-(194)-2414049

Abstract

Herein, we report the synthesis of surfactant based sodium dodecyl sulphate zirconium phosphate (SDS-ZrP) intercalated cation exchange material. The material was characterized by Fourier transform infrared spectroscopy (FTIR), scanning electron microscopy(SEM), powder X-ray diffraction(XRD), thermo gravimetric analysis(TGA) and differential thermal gravimetry (DTG). The synthesized composite material showed good ion exchange capacity of 3.11 mequiv g^{-1} for strontium ion. Ion-exchange studies on the synthesized material have also been performed for heavy metal ions in neutral and acidic media .The composite exhibited high selectivity towards Zn (II) with distribution coefficient value of 4900mLg^{-1}. The presence of surfactant in the matrix of composite highly enhances its adsorption capacity towards heavy metal ions. On the basis of significant dimensionless separation factor (SF) in a particular media, some binary separations of Zn (II) ion from other heavy metal ions were performed on the column loaded with SDS-ZrP.

Keywords

Sodium dodecyl sulphate, Ion exchange material, Zirconium phosphate

Contents

1. Introduction

The growing world population, large scale urbanization and fast rising global economies have caused unprecedented levels of water pollution, which has, in recent years, assumed gigantic dimensions across the globe. This has generated a renewed interest and concern amongst the analytical and material scientists to devise new and efficient methods of purification of water and design of new materials useful for removal and uptake of pollutants and hazardous materials such as heavy metals and toxic dyes from polluted waters. Extensive efforts have been made to develop efficient materials for the remediation of waste water containing heavy metal ions and other hazardous wastes. Traditionally, the metal ions present in the waste waters have been effectively removed by using an ion exchange process [1-4]. As such, a number of inorganic ion exchange materials [5-7] as well as organic resins [8, 9] have been reported for the remediation of waste water containing heavy metal ions from industries and laboratories.

No doubt, organic resins possess good ion exchange capacity towards metal ions but these are unstable towards ionizing radiations and cannot be employed over a varied range of pH. Thus, there is a need to synthesize new ion exchange materials which improve the treatment of specific types of radioactive wastes and the removal of hazardous metal ions like Zn^{2+},Cd^{2+}, Hg^{2+}, Pb^{2+},Cu^{2+}, Ni^{2+}etc, from water bodies over a wide range of pH. For this purpose, many researchers were motivated to study inorganic-organic hybrid type ion exchangers [10-13],which could offer enhanced properties in terms of its chemical, mechanical, and thermal stabilities, with high selectivity for heavy toxic metals for environmental cleanup [14-18].In order to synthesize an efficient layered type hybrid ion exchange material, zirconium phosphate proved to be the versatile host matrix [19-21] for the incorporation of organic dopants within its layered framework, to enhance its physico-chemical properties. These organic dopants offer better functionality to the hybrid material. Organic dopants such as poly-o-toluidine[22], acrylonitrile [11], polyacrylamide [23], polyaniline [24,25], polystyrene [26], L-(+)-phosphoserine [27] polymethylacrylate [28], sulphonic acid [29], nylon-66 [10], and cellulose acetate [30] have been incorporated into zirconium phosphate. Such materials find application in separation, purification and detection of heavy metal ions. Keeping in view the intractableness and defiant nature of Pb^{2+},Cd^{2+} and Zn^{2+} ions, special emphasis has been laid on the synthesis of hybrid material that could effectively be used for the removal of aforesaid metal ions from wastewaters [31,32].

Heavy metal ions such as cadmium, lead, mercury, zinc etc. have an adverse effect on the aquatic environment. Manufacturing of brass and bronze alloys and galvanization are the main sources of zinc in the environment [33, 34].Zinc is the basic constituent of various metalloenzymes for their normal functioning and acts as an essential micronutrient [33]. The maximum acceptable concentration of zinc as recommended by WHO in drinking water is 5.0 mg/L [35], beyond this value , Zn^{2+} becomes highly toxic [36].

Surfactants, which are surface-active agents, play an important role in modifying the adsorption behavior of ion exchangers[37].In aqueous solutions, surfactants at concentrations above CMC values, spontaneously form micelles. Micelles are spherical and dynamic aggregates possessing anisotropic interface composed of polar head groups, counter ions and water between their long hydrocarbon chain and the surrounding bulk aqueous phase[38].The Surfactants are surface-active agents capable of reducing the interfacial tensions between solid and liquid phases by adsorbing themselves on the solid surfaces [39, 40], thereby enhance the selective adsorption of some of the heavy metals onto the surface of exchangers [37].

In the present study, we have synthesized a new ion- exchange material by incorporating sodium dodecyl sulphate in the matrix of zirconium phosphate and studied the ion exchange properties of the synthesized material towards heavy metal ions in presence of surfactant media. The synthesized ion exchange material has high ion exchange capacity and is observed to be thermally stable as well as highly selective towards Zn^{2+} ion. The material was characterized by Fourier transform infrared spectroscopy (FTIR), scanning electron microscopy (SEM), powder X-ray diffraction (XRD), thermo gravimetric analysis (TGA) differential thermal gravimetry (DTG) and ion exchange properties to identify various functional groups and ion-exchange sites present in the material. Distribution studies were performed on the synthesized material so as to determine the selectivity and the capacity for heavy metal up-take in different solvent systems including surfactant media.

2. Materials and methods

2.1.1 Reagents and chemicals

Zirconium oxychloride and orthophosphoric acid (H_3PO_4) were obtained from Merck, sodium dodecyl sulphate ($C_{12}H_{25}SO_4Na$) was obtained from Himedia. All the chemicals were of A.R. grade and were used as purchased. Solutions of zirconium oxychloride were prepared in 0.1M H_2SO_4, orthophosphoric acid and SDS solutions were prepared in demineralized water (DMW).

2.1.2 Material synthesis

A number of samples were prepared by mixing zirconyl oxychloride ($ZrOCl_2.8H_2O$), sodium dodecyl sulphate, and orthophosphoric solutions in different volume ratios. Solution of SDS was mixed with an aqueous zirconyl oxychloride. To this mixture, a solution of orthophosphoric acid was added drop-wise with constant stirring using a magnetic stirrer at a temperature of 50 $\pm 1^{\circ}$C to obtain a gel. The gel was aged in the mother liquor for 24 hours at room temperature and filtered using suction. Before drying in an oven at 60°C, the excess acid was removed by washing with DMW. The dried gel was then cracked into small granules by putting it into DMW. The resultant granules were converted into H^+ form by treating them with 1M HNO_3 for 24 hours. Finally, the material was washed with DMW, dried and sieved. Various samples were synthesized, varying the concentration and mixing volume ratios of reactants in order to optimize ion exchange capacity of the ion exchange material. Among all the samples, the one containing $ZrOCl_2$ (0.1M), SDS (0.01M) and H_3PO_4 (0.1M) in the volume ratio of 1:1:1 was found to possess the maximum ion exchange capacity. Finally, the material was selected for further studies on the basis of its highion exchange capacity towards alkali, alkaline and heavy metals ions. The material was repeatedly synthesized, using the same method of synthesis, and the ion exchange capacity of the samples from each lot was determined using the same procedure.

2.1.3 Instrumentation and characterization methods

The pH measurements were performed using Eutech Instrument PC 550. A shaking incubator, Yellow Line OSC. An FTIR spectrum of the sample was recorded by KBr disc method using Interspec-2020 FTIR spectrometer. X-ray powder diffraction patterns were obtained using a diffractometer (Bruker AXS D8 Advance) with Ni-filtered Cu-Kα

radiation. The XRD patterns were recorded over the 2θ range of 05–60. Exstar 6000 TGA/DTG instrument was used for thermal studies and the curves were recorded from 0°Cto 700°C under nitrogen atmosphere. Surface morphology of exchange material was characterized by Saatchi-S 3000H scanning electron microscope.

2.1.4 Ion exchange capacity

Ion exchange capacity is the measure of the number of replaceable H^+ ions per unit mass of the exchanger. For determining ion exchange capacity, 1 g of the exchanger (H^+ form) was taken in a column of internal diameter of 1cm, fitted with sintered glass at the bottom and 250 ml of 1.0 M metal nitrate solution was passed through it at the flow rate of (0.5 mL min^{-1}). The effluent was titrated against a standard 0.01M NaOH solution using phenolphthalein as indicator to determine the total H^+-ions liberated during the ion exchange process.

2.1.5 Elution behavior

The column containing one gram of the SDS-ZrRP in H^+ form was eluted out with 1M $NaNO_3$ solution (100ml) having a standard flow rate of 0.5ml min^{-1}. Several 10 ml fractions of the effluent were collected and checked for H^+ ion concentration tritrimetrically using standard alkali solution. Almost all of the H^+ ions were found to be released by the column within the first 70 ml of the effluent.

2.1.6 Distribution studies

Distribution studies were carried out by batch method, in which 300 mg of the ion exchange material was taken in each flask to which 30 mL of different metal nitrate solutions in the required medium were added. After continuous shaking for 6 hours, the flasks were kept for 24 hours in a temperature controlled incubator shaker at 25±1°C to attain equilibrium. The amounts of metal ions in the solution before and after equilibrium were determined by titrating against standard 0.001M EDTA solution. Distribution coefficients (K_d) is the measure of the fractional uptake of metal ions competing for H^+ ions from a solution by an ion exchange material. Distribution coefficients may be calculated as the ratio of the amount of specific metal ions in the exchanger phase and in the solution phase, by using the formula as given bellow;

$$K_d = \frac{C_0 - C_e}{C_0}\frac{V}{M}(\mathrm{mLg^{-1}}) \quad (1)$$

where K_d is the distribution coefficient (mLg^{-1}), C_0 is the initial amount of the metal ions in the solution, C_e is the amount of the metal ions finally left in the solution after attainment of equilibrium, V is the volume of the solution in mL and M is the mass of the exchange material in grams.

2.1.7. Quantitative separations of zinc from various binary mixtures

Various binary separations Hg (II) - Zn (II), Co (II) - Zn (II) and Ni (II) - Zn (II) were achieved on a SDS-ZrP column. An exchanger in H^+ form (1.0g) was packed in a glass column (0.6 cm internal diameter) with a sintered disc supported at the end. The column was washed thoroughly with dematerialized water and the binary mixtures of metal ions were loaded on it and allowed to pass through the column at a flow rate of 2–3 drops min^{-1}. The process was repeated two or three times to ensure the complete adsorption of one of the metal ions on the bead. The separation of metal ions was achieved by collecting the effluent in 10 mL fractions and titrated against the standard solution of 0.01 M di-sodium salts of EDTA.

3. Results and discussions

In this study, various samples of sodium dodecyl sulphate zirconium phosphate (SDS-ZrP) as new composite intercalated cation exchanger were prepared by the sol–gel method. To confirm the presence of various functional groups and composite nature of the synthesized material, the IR spectrum of the composite material was recorded at room temperature **(Fig.1).** The IR spectrum confirms the presence of SDS, external water molecules in addition to the O-H groups and the metal oxide in the material. The sharp peaks at 567.00 and 1097.62cm^{-1} are due to the presence of a phosphate group [41]. Metal - oxygen and metal- hydroxide bonds were confirmed by a weak band at 723 cm^{-1}.Peaks at 1598 and 3469.30 cm^{-1} were due to the presence of water at crystallization [42]. The lower values of υO-H$_{sym}$ in this composite compared to that of free OH groups (>3650 cm^{-1}) indicates that all the OH groups are involved in hydrogen bonding with SDS. The band at 3500cm^{-1} is also indicative of the strongly hydrogen-bonded OH or extremely strongly coordinated H_2O. Peaks at 2927.40, 2883.60,1462.52 and 1352.62 cm^{-1} reveal the presence of methyl and/or methylene groups in material.

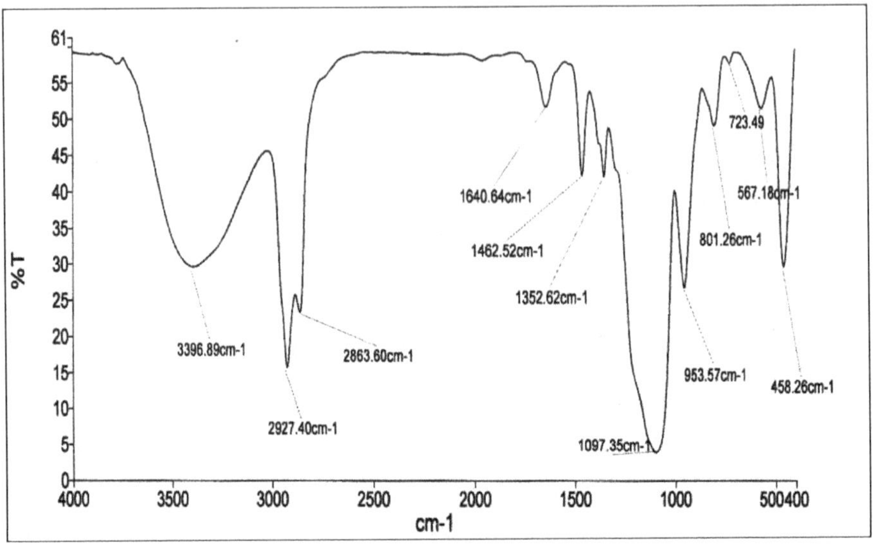

Figure 1.FTIR spectrum of sodium dodecyl sulphate zirconium phosphate.

In order to study thermal characteristics of the sample, thermo gravimetric analyses (TGA) as well as differential thermal gravimetry (DTG) were performed.TG-DTG curves of the SDS-ZrP shown in **Fig.2** depict two distinct weight loss stages. Within the temperature range of 50°C to 125°C weight loss of 16.37% has been observed with a sharp endothermic peak in its DTG plot at 77°C.This weight loss is due to the evaporation of external water molecules [43] and partial degradation of SDS moiety of composite material. At the temperature of 460°C a second weight loss of approx. 45% occurs with a sharp endothermic peak at 458.1°C. This is attributed to the loss of strongly coordinated water molecule and complete decomposition of surfactant molecules. After temperature of 550°C a small weight loss occurs due to condensation of phosphate groups in pyrophosphate groups until it become constant after 620°C due to formation of zirconia (ZrO_2).

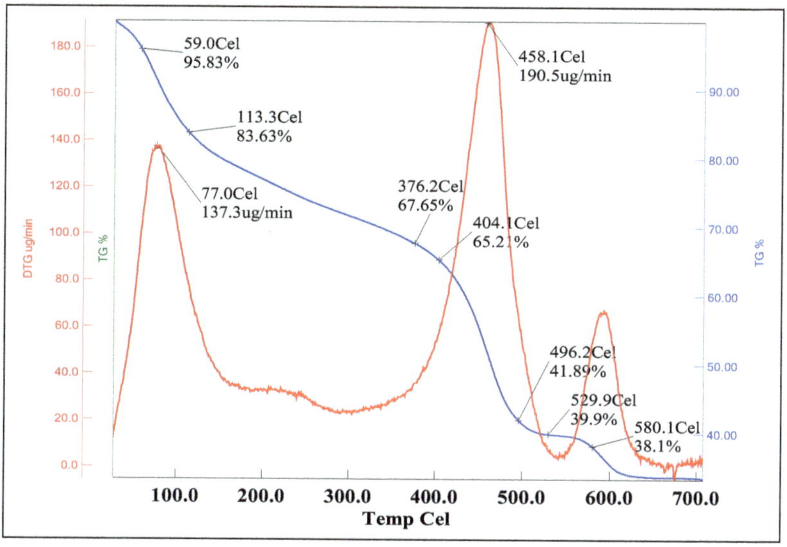

Figure 2. TGA/DTG plot of sodium dodecyl sulphate zirconium phosphate .

XRD analysis was carried out for structural identification of the composite material in the 2θ range between 05 and 60 degrees **(Fig. 3a).** The X- ray diffraction spectrum of the composite polymer material confirms the amorphous nature of the material. Scanning electron microscopy technique (SEM) was used to study the sample morphololology. It confirmed that the material possess a nearly stacked layered type structure (**Fig. (3b)**).

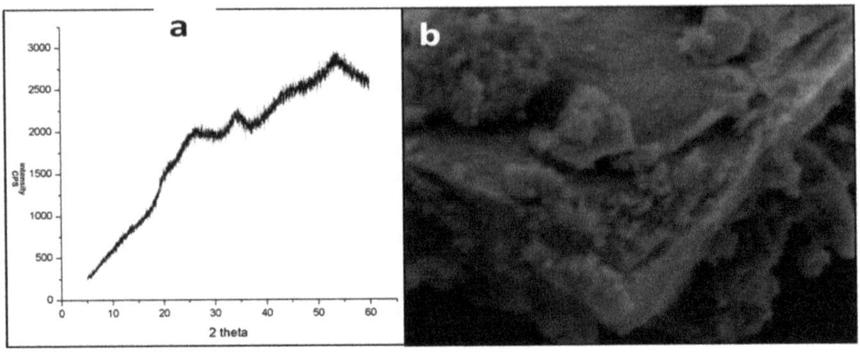

Figure 3. (a) XRD spectrum of sodium dodecyl sulphate zirconium phosphate. (b) SEM images of sodium dodecyl sulphate zirconium phosphate.

In order to explore the potentiality of composite polymer material as an ion-exchanger, the ion exchange capacities of some monovalent and divalent cations were investigated and the results are summarized in **Table 1**.The material shows good exchange capacity of 3.11 mequiv g^{-1} for Sr^{2+} ions in aqueous medium. This may be attributed to the enhanced interlayer distances in the material as a result of adsorption of surfactant molecules into the layer of zirconium phosphate matrix. The material shows a good reproducibility of Na^{+} ion exchange capacity. In accordance to the decreasing order of the hydrated ionic radii of selected alkali and alkaline earth metal ions the order of their ion exchange capacity is: $Li^{+} < Na^{+} < K^{+} < Mg^{2+} < Sr^{2+}$. From **Table 1** it is clear that the size and charge of the exchanging ions influence the ion exchange capacity of this material. Ions with smaller hydrated ionic radii, results in faster exchange and hence greater ion exchange capacity. The ion exchange capacity increases down the group because of decrease of hydrated ionic radii. Our group and others [14, 44-46] have reported similar results, for exchange of alkali and alkaline earth metals on zeolites. It has been shown that the entropy contributions to the free energies of exchange are favorable to the uptake of cations with smaller hydrated radii [22].

Table 1.Ion exchange capacity of composite ion exchange material sodium dodecyl sulphate zirconium phosphate.

S. No.	Cation	Salt used	Ionic radii (A^0)	Hydrated ionic radii (A^0)	pH	IEC in mequiv g^{-1}
1	Li^+	$LiNO_3$	0.68	10	1	1.43
2	Na^+	$NaNO_3$	0.97	2.76	1	2.17
3	K^+	KNO_3	1.33	2.32	1	2.39
4	Mg^{2+}	$Mg(NO_3)_2$	0.78	7.00	1	2.75
5	Sr^{2+}	$Sr(NO_3)_2$	1.27	6.10	1	3.11

The efficiency of the exchange process using this material in column chromatography is studied by evaluating the elution behavior of 1g column bed. It was found that the exchange is quite fast at the beginning of the process, as most of the exchangeable H^+ ions are eluted out in the first 50 ml of the effluent **(Fig.4.)**

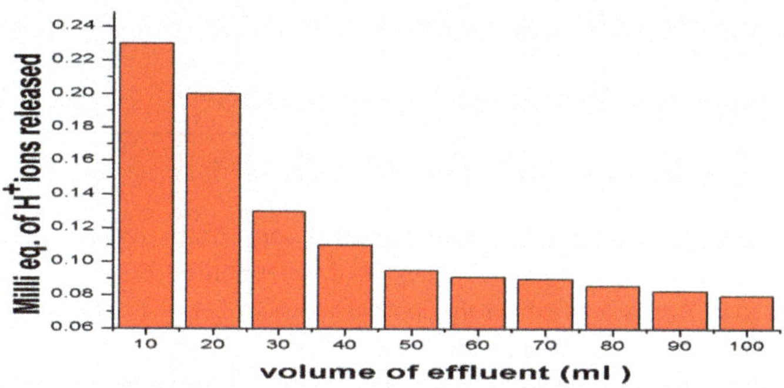

Figure 4.Histogram showing elution behavior of sodium dodecyl sulphate zirconium phosphate.

In order to explore the potentiality of this hybrid cation exchange material in the separation of metal ions, distribution studies for various heavy metal ions was performed in aqueous, acidic media (HNO_3) and in presence of Triton X-100 (**Table. 2).**

Table 2.K_d values of some metal ions on composite ion exchange material Sodium dodecyl sulphate Zirconium Phosphate in different solvent systems including surfactant media

METAL ION	DMW	0.001M SDS	0.01MHNO_3	Aqueous triton X (1%)
Zn^{2+}	3233	3900	1900	4900
Co^{2+}	800	900	680	1000
Cd^{2+}	1329	1900	890	2400
Hg^{2+}	650	800	233	1000
Pb^{2+}	450	850	150	900
Ni^{2+}	300	800	150	1200
Sn^{2+}	571	800	500	1250
La^{3+}	1000	1329	800	1900
Ce^{3+}	450	650	255	900

This behavior can be useful for the uptake of metal ions from the industrial and laboratory wastes and for the environmental analysis of water bodies. Effect of non ionic surfactant (Triton X-100) was studied on the ion exchange of heavy metals and it was found that in the presence of surfactants ion exchange rate is very much enhanced as compared to an aqueous and an acidic medium **Fig. 5a.** Triton X-100 surfactant increases the ion exchange rate by forming complex with heavy metal ions due to its polyoxyethylene shell. Conversely, by lowering the pH of the ion exchange medium, Kd values for various heavy metal ions are decreased in comparison to that in DMW **Fig. 5b**.This observation could be partly attributed to competitive ion exchange between metal ions and H^+ ions in the exchange medium and partly due to repulsion offered by the H^+ ions present at the ionogenic sites to incoming metal ions.

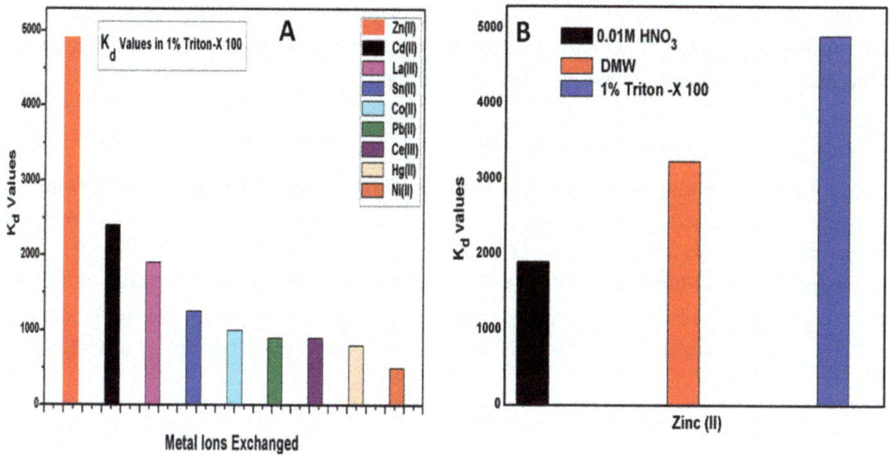

Figure 5.(a) K_d of various metal ions in presence of Triton-X 100 and (b) Comparison of K_d values for Zinc (II) in various media.

When K_d value of the synthesized material SDS-ZrP was compared with other surfactant intercalated hybrid ion exchangers [47], it was justified that our synthesized material provides better ion exchange (**Table 3**) in terms of its distribution coefficient especially towards Zn (II).

Table 3. Comparison of distribution coefficient (K_d) of various surfactant intercalated cation exchanger towards Zn (II)

Metal ion	**Kd values**					
	SDBS-CeP		AOT-SnP		**SDS-ZrP (Present study)**	
	DMW	0.01M HNO_3	DMW	0.01M HNO_3	DMW	0.01M HNO_3
Zn(II)	57	87	633	340	3233	1900

Ion exchange materials are used to separate metal ions from each other, which are having a large difference in their K_d values in a particular media, in which separation has to be carried out. On the basis of K_d values of the metal ions of interest, separation factor provides a better indicative parameter and we invoke separation factor for the binary separation of metal ions. Separation factor is defined as

$$SF = \frac{K_d^A}{K_d^B} \quad (2)$$

where A and B are the two metal ions of interest. The SF is a scale used to know whether a media may be used to separate out two metal ions from one another. The higher the values of SF, usually > 10,the more efficient is the separation. In case of SDS-ZrP, it was arrived at SF Zn^{2+}/Ni^{2+} value of 12.66 in 0.01M HNO_3. Significant separation factors calculated were used in achieving some important binary separations **(Table 4)** such as Hg (II) - Zn (II), Co (II) - Zn (II) and Ni (II) - Zn (II) and the binary plots are shown in **Fig. 6.**

Table 4.Quantitative separation of metal ions from a binary mixture using composite ion exchange material Sodium dodecyl sulphate Zirconium Phosphate columns at room temperature.

S. No.	Metal ions separation	Amount loaded (mg)	Amount found (mg)	Recovery (%)	Volume of eluent used (mL)	Eluent used
1	Zn^{2+}	6.53	6.37	97.55	80	1M HNO_3
	Hg^{2+}	20.06	19.81	98.75	80	0.0IM HNO3
2	Zn^{2+}	6.53	6.35	97.24	90	1M HNO_3
	Ni^{2+}	5.84	5.73	98.11	100	0.01M HNO3
3	Zn^{2+}	6.53	6.41	98.16	80	1M HNO_3
	Co^{2+}	5.89	5.72	97.11	90	1%Triton– X

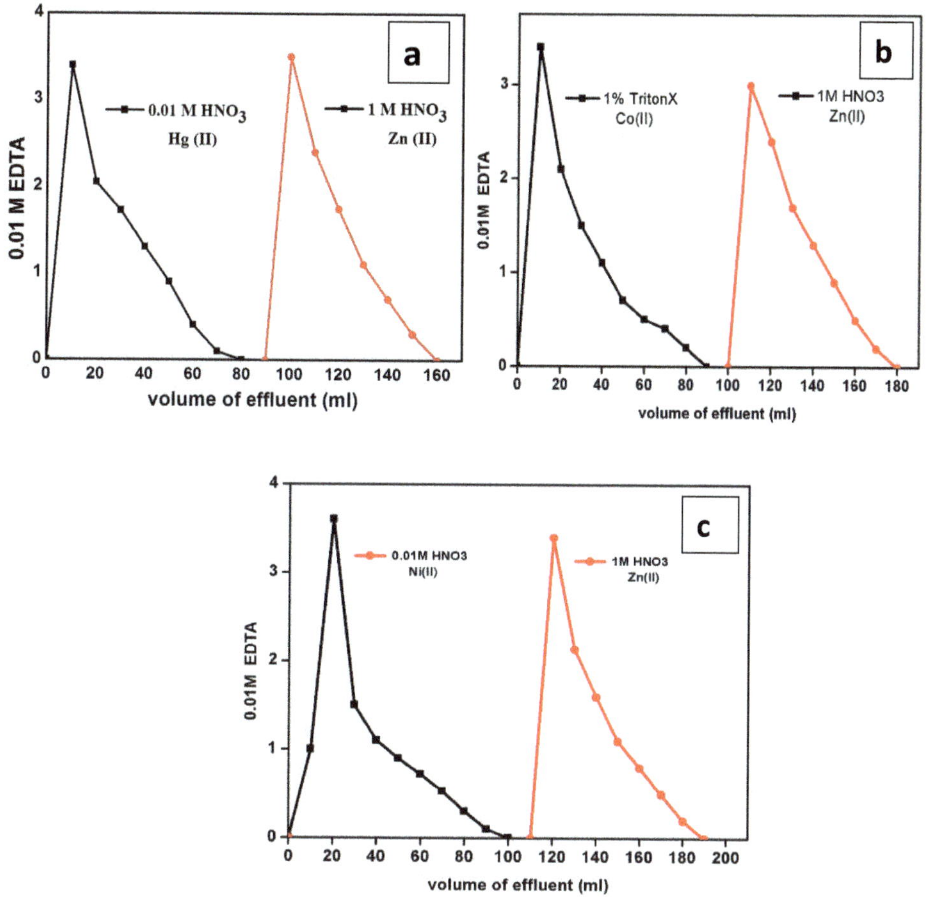

Figure 6.Binary separation plots of (a) Hg (II)- Zn (II) (b) Co (II) - Zn (II) (c) Ni (II) - Zn(II) on sodium dodecyl sulphate zirconium phosphate loaded column.

4. Conclusion

The new surfactant intercalated hybrid material, SDS-ZrP showed good reproducibility, and it has been found highly selective for Zn (II). It is most suitable to separate Zn^{2+} from various metal ions in the binary mixture column chromatography. The addition of surfactant enhances the ion exchange performance of the material.

Acknowledgment

We are grateful to the University Grants Commission (UGC), Government of India, for providing financial assistance vide research grant no. F.No.42-305/2013. A.B. would like to thank Council of Scientific and Industrial Research (CSIR) for financial assistance in the form of a Junior Research Fellowship (JRF).

References

[1] D.V. Marinin, G.N. Brown, Studies of sorbent/ ion exchange materials for the removal of radioactive strontium from liquid radioactive waste and high hardness ground waters, Waste Manag. 20 (2000) 545–553. *http://dx.doi.org/10.1016/S0956-053X(00)00017-9*

[2] B. Preetha, C. Jonardanan, Ion exchange method for the detection of trace amounts of Mn2+ using nano cerium zirconium phosphate cation exchanger, Ion Exch. Lett. 3 (2010) 12-18.

[3] R. Chitrakar, S. Tezuka , A. Sonoda, K. Sakane, K. Ooi, T. Hirotsu, Selective adsorption of phosphate from sea water and waste water by amorphous zirconium hydroxide, J. Coll. Interf. Sci. 297 (2006) 426-433. *http://dx.doi.org/10.1016/j.jcis.2005.11.011*

[4] V.N. Lebedev, N.A. Melnik, A.V. Rudenko, Sorption of cesium on titanium and zirconium phosphates, Radiochemistry. 45 (2003) 149-151. *http://dx.doi.org/10.1023/A:1023829125035*

[5] K.G. Varshney , A. H. Pandith ,U. Gupta, Synthesis and characterization of zirconium alumino phosphate. A new cation esxchanger,Langmuir. 14 (1998) 7353-58. *http://dx.doi.org/10.1021/la970464j*

[6] K.G. Varshney, A.H. Pandith, Foreward and reverse ion –exchange kinetics for some alkali and akaline earth metal ions on amorphous zirconium (IV)alumino phosphate, Langmuir. 15(1999) 7422-25. *http://dx.doi.org/10.1021/la990009i*

[7] S.D. Ajagekar, Z.R. Turel, Selective adsorption and radiochemical separation of Sr (II) using zirconium phosphate ion exchanger, J. Ind. Council. Chem. 26 (2010) 187-189.

[8] M.A. Hafez, I.M. Kenway, M.A. Akl, R.R. Lashein, Preconcentration and separation of total mercury in environmental samples using chemically modified chloromethylated polystyrene-PAN (ion-exchanger) and its determination by cold vapour atomic absorption spectrometry,Talanta. 53(2001) 749-760. *http://dx.doi.org/10.1016/S0039-9140(00)00524-5*

[9] M. Nushad, R. Mitra, J, Raguvanshi, Use of neutral red modified strong acid cation exchange resin for separation of heavy metal ions, Ion Exch. Lett. 2 (2009) 31-34.

[10] Inamuddin, S.A. Khan, W.A. Siddiqui, A.A. Khan, Synthesis, characterization and ion-exchange properties of a new and novel organic–inorganic hybrid cation exchanger : Nylon-6,6, Zr(IV) phosphate, Talanta. 71 (2007) 841–847. *http://dx.doi.org/10.1016/j.talanta.2006.05.042*

[11] K.G. Varshney, A.H. Pandith, Synthesis and ion exchange behavior of acrylonitrile based zirconium-phosphate; a new hybrid ion-exchanger, J. Indian Chem. Soc. 78 (2001) 250-253.

[12] A.Somya, M.Z.A. Rafiquee, K.G. Varshney, Synthesis, characterization and analytical applications of sodium dodecyl sulphate cerium (IV) phosphate: A new Pb (II) selective, surfactant-based intercalated fibrous ion exchanger, Colloid Surf. A: Physiochem. Eng. Asp. 336 (2009) 142–146. *http://dx.doi.org/10.1016/j.colsurfa.2008.11.036*

[13] A.A Khan, Inamuddin, Cation-exchange kinetics and electrical conductivity studies of an organic-inorganic composite cation-exchanger: polypyroleTh(IV) phosphate, J. Appl. Polym. Sci. 105(2007) 2806 - 2815. *http://dx.doi.org/10.1002/app.25183*

[14] S. Rehman, N. Islam, S.Ahad, S.Z. Fatima, A. H.Pandith, Preparation and characterization of 5-sulphosalicylic acid doped tetraethoxysilane composite ion-exchange material by sol–gel method. J. Hazard. Mat. 260 (2013) 313 – 322. *http://dx.doi.org/10.1016/j.jhazmat.2013.05.036*

[15] H. Patel, A. Parikh, U. Chudasama, A comparative study of proton transport properties of metal (IV) tungstates and their organic derivatives, Bull. Mater. Sci. 28(2005) 137-144. *http://dx.doi.org/10.1007/BF02704233*

[16] A. Khan, A. M. Asiri, M.A. Rub, N. Azum, A.S.P. Khan, Khan, P.K. Mondal, Review on composite cation exchanger as interdisciplinary materials in analytical chemistry, Int. J. Electrochem. Sci. 7(2012) 3854-3902.

[17] A. Shahat, M.R. Awual, M. Naushad, Functional ligand anchored nanomaterial based facial adsorbent for cobalt(II) detection and removal from water samples. Chem Eng J 271 (2015) 155-163.

[18] S.A. Nabi, A.H. Shalla, EDTA-stannic (IV) iodate: preparation, characterization and its analytical applications for metal content determination in real and synthetic samples, J Porous Mater. 16 (2009) 587-597. *http://dx.doi.org/10.1007/s10934-008-9236-5*

[19] B.C. Pan, Q.R. Zhang, W. Du, Q.J. Zhang, Selective removal of heavy metals from water by zirconium phosphate; behavior and mechanism, Water Res. 41(2007) 3103-3111. *http://dx.doi.org/10.1016/j.watres.2007.03.004*

[20] T. Kijima and Y.Matsui, A new type of host compound consisting of α-zirconium phosphate and an amminatedcyclodextrin, Nature, 322 (1986) 533 – 534. *http://dx.doi.org/10.1038/322533a0*

[21] R. Thakkar, H. Patel, U. Chudasma, A comparative study of proton transport properties of zirconium phosphate and its metal exchanged phases, Bull. Mater. Sci. 30 (2007) 205-209. *http://dx.doi.org/10.1007/s12034-007-0036-3*

[22] A.A. Khan, T. Akhtar, Preparation, physio-chemical characterization and electrical conductivity measurement studies of an organic-inorganic nano-composite cation exchanger: poly-o-toluidine Zr(IV) phosphate, Electrochim. Acta. 53(2008) 5540-5548. *http://dx.doi.org/10.1016/j.electacta.2008.03.002*

[23] J. Wang, Y. Hua, B. Li, Z. Guia, Z. Chen, Preparation of polyacrylamide and gamma-zirconium phosphate nanocomposites by intercalative polymerization, UltrasonicsSonochemistry. 11 (2004) 301–306. *http://dx.doi.org/10.1016/j.ultsonch.2003.06.001*

[24] L. Wang, X. Wu, W. Xu, X. Huang, J. Liu, and A. Xu, Stable Organic–inorganic hybrid of polyaniline/α-zirconium phosphate for efficient removal of organic pollutants in water environment, Appl. Mater. Interfaces.4 (2012)2686–2692. *http://dx.doi.org/10.1021/am300335e*

[25] V.K. Gupta, Deepak Pathania, N.C. Kothiyal, G. Sharma, Polyaniline zirconium(IV) silicophosphate nanocomposite as absorbent for removal of methylene blue dye from water system, J. Mol. Liq. 190 (2014) 139-145. *http://dx.doi.org/10.1016/j.molliq.2013.10.027*

[26] K.G. Varshney and A. H.Pandith, Synthesis and characterization of styrene supported zirconium phosphate; A new ion- exchange material, J. Chem and Env. Res. 5 (1996)1-4.

[27] M.H.H. Alhendawi, Synthesis and structural characterization of a new chiral porous hybrid organic–inorganic material based zirconium phosphates and L-(+)-phosphoserine, J.Solid State Chem.201 (2013) 24-28. *http://dx.doi.org/10.1016/j.jssc.2013.02.025*

[28] W. A. Siddiqui, S.A. Khan, Inamuddin, Synthesis, characterization and ion-exchange properties of a new and novel 'organic–inorganic hybrid'cation-exchanger Poly (methylene methacrylate) Zr(IV)phosphate, Colloid Surf. A: Physiochem. Eng. Asp. 295 (2007) 193–199. *http://dx.doi.org/10.1016/j.colsurfa.2006.08.053*

[29] Y.Zhou, R. Huang, F.Ding, D. A. Brittain, J. Liu, M. Zhang, M. Xiao, Y. Meng, and L. Sun, Sulfonic acid-functionalized α-zirconium phosphate single-layer nanosheets as a strong solid acid for heterogeneous catalysis applications, J.Appl. Mater.Interfaces.6 (2014) 7417–7425. *http://dx.doi.org/10.1021/am5008408*

[30] V.K. Gupta, D. Pathania, P. Singh, B. S. Rathore, P. Chauhan, Cellulose acetate–zirconium (IV) phosphate nano-composite with enhanced photo-catalytic activity, Carbohydrate Polym. 95 (2013) 434–440. *http://dx.doi.org/10.1016/j.carbpol.2013.02.045*

[31] Q. Zhang, B. Pan, W.M. Zhang and K.Jia, Selective sorption of lead, cadmium and zinc ions by a polymeric cation exchanger containing nano-Zr (HPO3S)2, J. Environ. SciTechnol . 42(2008) 4140–4145. *http://dx.doi.org/10.1021/es800354b*

[32] T. K. Sen and M. V. Sarzali, Removal of cadmium metal ion Cd2+ from its aqueous solution by aluminium oxide: A kinetic and equilibrium study, J. Chem. Eng. 142 (2008) 256-262. *http://dx.doi.org/10.1016/j.cej.2007.12.001*

[33] F. Arias and T. K. Sen, Removal of zinc metal ion (Zn2+) from its aqueous solution by kaolin clay mineral: A kinetic and equilibrium study, Colloid Surf. A, 348 (2009) 100-108. *http://dx.doi.org/10.1016/j.colsurfa.2009.06.036*

[34] C. H. Weng and C. P. Huang, Adsorption characteristics of Zn (II) from dilute aqueous solutions by fly ash, Colloids Surf. A. 247 (2004) 137-143. *http://dx.doi.org/10.1016/j.colsurfa.2004.08.050*

[35] D. Mohan and K. P. Singh, Single and multi-component adsorption of cadmium and zinc using activated carbon derived from bagasse-an agricultural waste, Water Res.36 (2002) 2304-2318.
http://dx.doi.org/10.1016/S0043-1354(01)00447-X

[36] A. K. Bhattacharya, S. N. Mandal and S. K. Das, Adsorption of Zn (II) from aqueous solution by using different adsorbents, J. Chem. Eng. 123 (2006) 43-51.
http://dx.doi.org/10.1016/j.cej.2006.06.012

[37] S. A. Nabi, R. Bushra, Z. A. Al-Othman & Mu.Naushad, Synthesis, characterization, and analytical aplications of a new composite cation exchange material acetonitrile stannic(IV) selenite: adsorption behavior of toxic metal ions in nonionic surfactant medium, J. Sep. Sci. and Tech. 46 (2011)847–857.
http://dx.doi.org/10.1080/01496395.2010.534759

[38] K.G. Varshney, M.Z.A. Rafiquee, A Somya, M. Drabik, Synthesis and characterization of Hg(II) selective n-butylacetate cerium(IV) phosphate as a new intercalated fibrous ion exchanger: Effect of surfactants on the adsorption behavior, Ind. J. Chem. A. 45 (2006) 1856-1860.

[39] J.A. Loughlin and L. S. Romsted. A new method for estimating counter-ion selectivity of a cationic association colloid: trapping of interfacial chloride and bromide counter-Ions by reaction with micellar bound aryldiazonium Salts, Colloids and Surf. 48 (1990) 123-137.
http://dx.doi.org/10.1016/0166-6622(90)80223-Q

[40] M. Naushad, Z.A. ALOthman, Inamuddin, H. Javadian, Removal of Pb(II) from aqueous solution using ethylene diamine tetra acetic acid-Zr(IV) iodate composite cation exchanger: Kinetics, isotherms and thermodynamic studies.J Ind Eng Chem 25 (2015) 35-41.

[41] E.V. Bakhmotava, O. Zhiang, D.G. Medevdev, A. Clearfield, Cobalt phosphonates: an unusual cobalt polymeric phosphonate containing a clathrated phosphonate anion and a layered bisphosphonate, J. Inorg. Chem. 42 (2003) 7046-7051.
http://dx.doi.org/10.1021/ic0301425

[42] H.C. Yang, K. Aoki, H.G. Hong, D.D. Sackett, Growth and characterization of metal(II) alkane bisphosphonate multilayer thin films on gold surfaces, J. Am. Soc. 115 (1993)11855-11862.
http://dx.doi.org/10.1021/ja00078a025

[43] G. Alberti, U. Constantino, R. Millini, G. Perego, R.Vivan, Preparation, characterization and structure of α-zirconium hydrogen phosphate hemihydrates, J. Solid State Chem.113(1994) 289-295.

http://dx.doi.org/10.1006/jssc.1994.1373

[44] F.C. Nachod, W. Wood, The reaction velocity of ion exchange, J. Am. Chem. Soc. 66(1944) 1380-1384.
http://dx.doi.org/10.1021/ja01236a050

[45] S.A. Nabi, A.H. Shalla, A.M. Khan, S.A. Ganie, Synthesis, characterization and analytical applications of titanium(IV) molybdosilicate: a cation ion-exchanger, J. Coll. Surf. 203(2007) 241-250.
http://dx.doi.org/10.1016/j.colsurfa.2007.02.034

[46] M. Naushad, Z.A. ALOthman, M.M. Alam, M.R. Awual, Sol-gel synthesis of sodium dodecyl sulphate supported nanocomposite cation exchanger: Removal and recovery of Cu^{2+} from synthetic, pharmaceutical and alloy samples.Iran J Chem12 (2015) 1677-1686

[47] M. Naushad, Inamuddin, T.A. Rangreez, Potentiometric determination of Cd(II) ions using PVC based polyaniline Sn(IV) silicate composite cation exchanger ion selective membrane electrode. Des & Water Treat 55 (2015) 463-470.

CHAPTER 6

Artificial Intelligence (AI) Based Tools for Predicting the Removal Efficiency of Heavy Metals by Adsorption

Nusrat Parveen[1], Sadaf Zaidi[2*] and Mohammad Danish[3]

Department of Chemical Engineering, Aligarh Muslim University, India

[1]nusratalig@gmail.com, [2*]sadaf63in@yahoo.com and [3]mohddanish.chem@gmail.com

Abstract

The two most popular artificial intelligence (AI) techniques namely, artificial neural networks (ANN) and support vector machines (SVM) have been applied for predicting the removal efficiency of heavy metals like Copper (II), Arsenic (III), Lead (II), etc. in an adsorption process using low cost biosorbents. A comparison has been made between ANN, SVM and multiple linear regression (MLR) models based on the statistical parameters such as: correlation coefficient (R), average absolute relative error (AARE) etc. SVM is found to be the better predictive model than the commonly used MLR model and ANN model. Hence, its results are more accurate and generalized.

Keywords

Artificial Intelligence (AI), Artificial Neural Networks (ANN), Support Vector Machines (SVM), Adsorption, Biosorption.

Contents

1. Introduction

Pollution due to heavy metals has become a major issue in many countries and this problem is becoming more serious every as the concentration of heavy metals is found to continually exceed the permissible standards. This is caused by heavy metals which are not biodegradable and toxic and thus are a great threat to the environment and human life. The major cause of increased heavy metals concentration in the environment is the effluents from many industries. Some of the prominent industries are: fuels related, fertilizer, iron and steel, leather, mining, petroleum refining etc. [1]. Most commonly found heavy metals in the effluents are: toxic metals (such as Hg, Cr, Pb, Zn, Cu, Ni, Cd, As, Co, Sn, etc.), precious metals (such as Pd, Pt, Ag, Au, Ru etc.) and radio nuclides such as U, Th, Ra, Am, etc. [2, 3]. These heavy metals are responsible for various serious health problems like brain damage, liver and kidney damage, break down of nervous system, lung cancer, bronchitis etc. Due to the severe impact of these metals on both environment and human health, removal and recovery of heavy metals from effluent streams is essential.

Several methods are being employed for the removal of these heavy metals coming from industrial effluents such as chemical precipitation, membrane filtration, ion exchange, activated carbon adsorption, evaporation etc. Recently, the biosorption process which is adsorption on a biosorbent has been found to offer an attractive substitute to these existing methods [4, 5]. Besides being economical it has several advantages as compared to other conventional techniques such as high metal removal efficiency, minimization of chemical and toxic sludge or biological sludge which can be easily disposed of, minimum requirement of additional nutrients, easy regeneration of biosorbent and metal recovery [6, 7, 3].

Biosorption of heavy metal removal is greatly affected by various parameters such as nature of biomass and its concentration, contact time between solution containing metal and the biomass, temperature, surface area to volume ratio, initial metal ion concentration, metal affinity to biosorbent and pH of the solution [8]. Heavy metal removal efficiency predictions are made by using several techniques like MLR, ANN, genetic algorithm (GA) etc. [9, 10, 11]. Recently, support vector machines (SVM) have been developed that are being applied for classification and regression termed as a support vector classification (SVC) and support vector regression (SVR). SVM possesses many advantages over traditional ANN. While ANN suffers from multiple local solution, SVM gives unique, optimal and global solution for quadratic problems. Only two parameters need to be selected, upper bound and kernel parameter. The use of a structural risk minimization (SRM) principle in SVM provides a good generalization performance. There are many past studies which show the superiority of SVM over ANN [12, 13, 14, 15].

SVR now finds many applications for the prediction of many real world problems like time series prediction, predicting thermal–hydraulic performances in compact heat exchangers, predicting the auto-ignition temperatures of the organic compounds, atmospheric temperature prediction, predicting heat transfer coefficient and circulation rate in a thermosiphon reboiler etc. [16, 17, 18, 19, 20, 21].

This chapter mainly focuses on the application of artificial intelligence tools, primarily SVR, to predict the removal efficiency of heavy metals during the adsorption process using low cost biosorbents. The first section of this chapter introduces the adsorption process as used for heavy metal removal from waste water. The second section discusses the basic principle and theory of the two most popular artificial intelligence techniques i.e. ANN and SVM. This is followed by a third section which presents three different case studies for the prediction of the removal efficiency of heavy metals: Cu (II), As (III) and Pb (II) using SVR for all the three cases and ANN for predicting the removal efficiency of Pb (II) only. Comparison of the results is made with the conventional MLR method. Further, for the case of Pb (II), the prediction performance of these three models is assessed in the light of existing theory and the experimental findings. The same along with the conclusions are presented in the last section.

1.1 Adsorption

Adsorption is essentially a surface phenomenon, the substance which gets adsorbed on to the surface is called adsorbate which may either be a liquid or gas on the other hand the surface usually a solid on to which this adsorbate gets adsorbed is referred as adsorbent. As adsorption being the surface phenomenon entirely different from absorption which is a bulk phenomenon, adsorbate permeates deeply into the adsorbent i.e. the whole volume of the material is involved in the absorption process. The term sorption includes both the process namely absorption and adsorption. The adsorption process is basically used to remove individual component from liquid or gas mixture. The component to be removed is physically or chemically bonded to a surface. Thus, adsorption may be categories into physical adsorption and chemical adsorption.

Physical Adsorption or Physisorption

When the force of attraction between adsorbate and adsorbent is of the order of weak Vanderwaal force, the process is called physical adsorption or physisorption. In physisorption the enthalpy change is very low, typically 20-40 KJ/mol. It takes place at low temperature and as the temperature increases, the physisorption process decreases.

Chemical Adsorption or Chemisorption

When the force of attraction between adsorbate and adsorbent is similar to those present in chemical bonding, the process is called chemical adsorption or chemisorptions. As compared to physisorption, the enthalpy change associated with chemisorptions is very high i.e. around 200-400 KJ/mol. Chemisorption first increases with temperature and then decreases.

The major difference between physisorption and chemisorption include [22]:

- Physisorption is basically a reversible process and does not require any activation energy while chemisorption is an irreversible process and requires high activation energy.
- Chemisorption is highly selective and occurs only between certain active surface and adsorbent where as physisorption occurs on all surfaces if the conditions are favorable.
- Physisorption results in forming multimolecular layer while chemisorptions process proceeds in a monomolecular layer.
- Heat of adsorption for physisorption process is low on the other hand for chemisorption heat of adsorption is high.

The most commonly used industrial adsorbents are activated carbon, silica gel, molecular sieves, alumina, clay mineral and synthetic polymers etc. as they provide large surface areas per unit weight. A good adsorbent should have the following properties [23]:

- High porosity, high selectivity and large surface area per unit weight.
- Thermal, mechanical and chemical stability.
- Low pressure drop over the adsorber bed.
- Low fouling, no side reaction and easy regeneration.
- Low cost.

Applications of adsorption

Some of the important applications of adsorption are listed below [24, 25]:

- Separation of gaseous mixture, chemicals and air.
- Purification of liquids and water, softening of hard water, removal of impurities from liquid and gas etc.
- Removal of coloring matter from solution.
- Volatile organic compound (VOCs) removal from the polluted air.
- Production of high vacuum.
- Using activated charcoal for the preparation of gas masks to avoid poisonous gases like CO, CH_4 etc.
- It plays a vital role in controlling the environmental pollution especially the control of 'greenhouse gases' like CH_4, CO_2, N_2O etc., thus, helps in reducing global warming.

Factors affecting the adsorption process

There are various factors which affect the adsorption process [26]:

- Nature of the adsorbent: The type of adsorbent used in the adsorption process plays an important role in the adsorption rate such as the chemical nature of adsorbent surface, its specific area and the extent of porosity. Since, adsorption is a surface phenomenon, the larger the surface area of the adsorbent, the larger is its capacity for adsorption.
- Temperature and Pressure: Adsorption is an exothermic process, therefore with the decrease in temperature, the rate of adsorption increases. Further adsorption of a gas is accompanied by decrease in pressure, the rate of adsorption increases as the pressure increases. Whereas decrease in pressure leads to desorption. Thus, the decrease in temperature and increase of pressure both leads to an increased rate of adsorption.
- Nature of the gas: The type of gases used also affects the adsorption process such as chlorine, sulphur dioxide and ammonia are more easily adsorbed because of weak Vander wall forces compared to oxygen, hydrogen and nitrogen. The rate of adsorption is in direct proportion to the critical temperature of gases and these gases have high critical temperature.

The amount of heavy metal removed during adsorption process is evaluated either in terms of adsorption capacity/metal uptake or heavy metal removal efficiency. In order to obtain the adsorption capacity, the amount of metal ions adsorbed per unit mass of adsorbent is determined using the following equation [27, 28]:

$$q_e = \frac{V(C_o - C_f)}{m}$$

Where, q_e, V, m, C_o and C_f represent the metal ions adsorbed per adsorbent at equilibrium (mg/g), volume of solution (L), mass of adsorbent (g), the initial metal ions concentration in the solution (mg/L), final metal ions concentration in the solution (mg/L) respectively.

Heavy metal ion removal efficiency can be calculated with the amount of metal ions which gets adsorbed on the adsorbent and the available amount of metal ions in the solution as given by the equation below:

$$\text{Heavy metal removal efficiency (\%)} = \frac{\text{Heavy metal ions removed}}{\text{Heavy metal ions remaining}} * 100$$

2. An Overview of Artificial Intelligence (AI) Techniques

Artificial intelligence is the simulation of human intelligence processes by machines i.e. making the computer behave like humans. These processes include learning, reasoning, problem solving, perception, planning etc. Particular applications of AI include expert systems, speech recognition, and machine vision.

This section describes two very popular computational intelligence techniques namely, artificial neural networks (ANN) and support vector machines (SVMs).

2.1 Artificial Neural Networks (ANN)

Artificial neural networks are parallel processing computer algorithms inspired by the working of the human brain. The basic components of neural networks is composed of neurons and weights (bias). The neural networks architecture is made up of an input layer, one or more hidden layers and the output layer. The input layer receives the information and conveys this information/data to the hidden layer and does not carry any activation or transfer function. Hidden layers are sometimes called as "black boxes" as they are within the networks and no direct connection exists to the external environment [29]. The output layer processes the outcome of the hidden layer using some activation function [30] and then finally it sends the calculated output to the user. The most widely used activation functions are: linear function, piecewise linear function, tangent hyperbolic function, threshold function and sigmoidal (S shaped) function. The neural networks structure can be of three types [31, 32, 33, 34, 35]: if the input layer is directly connected to the output then it is often called a single-layer network, and if some hidden layer exists between the input and the output layer then it is called a multi-layer network as shown in Figs. 1 and 2. A feed forward network with one or more hidden layers as well as with feedback loop is termed as a recurrent network i.e. output of the network is fed back to its input layer as shown in Fig 3. The number of neurons at the input and output layers is decided by the nature of the problem and the no. of neurons as the hidden layer is generally decided by the trial and error method. Too few neurons lead to an inadequately fit-to the data and problem of under fitting arises while too many neurons lead to an over fit-to the data and have poor generalizability [36].

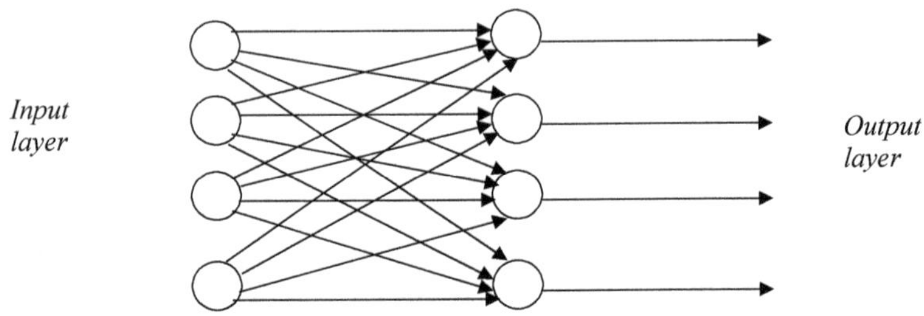

Figure 1. Single layer feed-forward NN (FFNN)

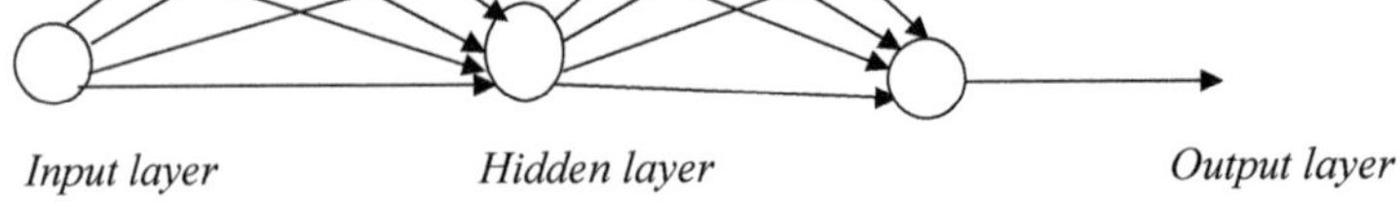

Figure 2. Multi layer feed-forward NN (FFNN)

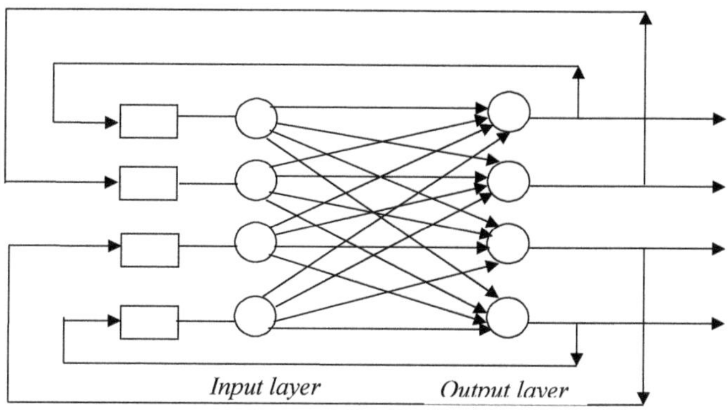

Figure 3. A recurrent network

Neural learning

Learning in neural networks basically refers to the method of modifying the weight of the connections between the nodes. Learning may be classified as supervised learning, unsupervised learning and reinforced learning. In supervised learning, the desired set of output (target) is available for each set of input. The weights of the connection are modified till the output of the response and target matches, otherwise the network produces an error. Whereas in unsupervised learning, no desired output (target) is available and the network adjusts its weight according to the input, and reinforced learning is in between the supervised and unsupervised learning. The desired output of the response is not compared with the target and it provides the information of the computed result, whether it is correct or not. Network weights are adjusted to have a good performance (based on certain criteria) and are rewarded while the other parts of ANN are penalized [37, 38].

The various steps involved for the development of ANN model are listed below [39, 40, 41, 42]:

1. Collection of data as input and output variables for a specified problem.
2. Data preprocessing i.e. scaling or normalization of data.
3. Dividing the whole data set into training, testing and validation sets.

4. Selecting ANN architecture i.e. the type of ANN, no. of hidden layers, no. of neurons at the input layer, hidden layer and the output layer, type of activation function at the hidden layer and the output layer.
5. Neural networks training:
   (a) A training set is used to adjust the weight and bias.
   (b) A Validation set is used to check the generalization performance and when to stop the training process.
   (c) A test set is used to check the performance of the developed ANN model based on some statistical parameters.

2.2 Support Vector Machines (SVMs)

Support vector machines (SVMs) are a set of related supervised learning methods [43, 44] originally developed for classification purposes i.e. support vector classification (SVC). However, it's application has now been extended to regression problems [45, 46, 47]. SVM is based on machine learning theory and enhances the accuracy of prediction while avoiding an over-fit to the data. Basically, SVM is superior than the conventional techniques like ANN on the following counts [48, 49, 12]: (a) only two parameters need to be selected i.e. upper bound and the kernel parameter, (b) unique, optimal and global solution for quadratic programming problem, and (c) implementation of structural risk minimization (SRM) principle provides good generalization performance [50]. Thus, SVM overcomes the problem of over fitting explicitly. Robustness of SVM is achieved by considering absolute, instead of quadratic values of the errors. As a consequence, the influence of outliers is less pronounced [51]. Because of these advantages, SVM has found many applications [52, 53].

Theory and Basic Principles

Fitting a regression function, $y = f(x)$ in a ε-SVR model is the basic goal so as to accurately predicts the targets $\{y_i\}$ corresponding to a set of input samples, $\{x_i\}$, With the given training data set as D= $\{(x_1, y_1), (x_2, y_2),\ldots, (x_N, y_N)\}$, where $x_i \in R^N$ is a vector of input variables and $y_i \in R^N$ is the corresponding scalar output (target) value.

For the case of a linear function, SVR approximates the relationship between the output and input parameters given by:

$$f(\mathbf{x}, \mathbf{w}) = (\mathbf{w}.\mathbf{x} + b) \tag{1}$$

Where, **w** expresses the weight vector and **b** refers to a constant called bias.

However, for non-linear problems of the real world where the input data cannot be correlated to the required output linearly, the following linear model can be constructed for high-dimensional feature space using a nonlinear mapping function $\phi(\mathbf{x})$ [54]:

$$f(\mathbf{x},\mathbf{w}) = (\mathbf{w} \bullet \phi(\mathbf{x}) + b) \tag{2}$$

Where, $\phi(\mathbf{x})$ is the function termed feature and $(\mathbf{w} \bullet \phi(\mathbf{x}))$ the dot product in the feature space F.

The regression problem is equivalent to minimizing the following regularized function:

$$\text{Minimize: } R(f) = C\frac{1}{N}\sum_{i=1}^{N} L_\epsilon\left(y, f(x,w)\right) + \frac{1}{2}\|w^2\| \tag{3}$$

$$L_\varepsilon(y, f(\mathbf{x},\mathbf{w})) = \begin{cases} 0 & \text{if } |y - f(\mathbf{x},\mathbf{w})| \le \varepsilon \\ |y - f(\mathbf{x},\mathbf{w})| - \varepsilon & \text{otherwise} \end{cases} \tag{4}$$

In Eq. (3), the first term represents the empirical error and the term C gives a measure of the trade-off between the training or empirical error and the model flatness given by the second term of the said equation. Eq. (4) defines the Ɛ- insensitive loss function proposed by Vapnik [56, 57]. The loss function assumes a zero value in the ε -insensitive zone and as a result it does not penalize the prediction errors with magnitudes smaller than ε. However, practically it is not possible to establish a function, f(x,w) with errors less than ε. In that case, some allowance for errors has to be considered, which can be represented in terms of slack variables ξ and ξ^*. These slack variables have non-zero values outside the ε -insensitive zone and measure the deviation from the boundaries of the ε -insensitive zone (penalty).
Thus after introducing the slack variables ξ and ξ^*, SVR is formulated as minimization of the following convex optimization problem:

$$\text{Minimize} \quad \frac{1}{2}\left\|\mathbf{w}^2\right\| + C\sum_{i=1}^{N} \xi_i + \xi_i^{(*)} \tag{5}$$

Subject to

$$(y_i - \mathbf{w} \bullet \phi(\mathbf{x}) - b) \le \varepsilon + \xi_i, \tag{6}$$

$$(\mathbf{w} \bullet \phi(\mathbf{x}) + b - y_i) \le \varepsilon + \xi_i^{(*)}, \quad \xi_i, \xi_i^* \ge 0 \tag{7}$$

The above optimization problem can be transformed into a dual problem by incorporating Lagrangian multipliers α and α^* as follows:

Maximize

$$L_d(\alpha_{i,j}^{(*)}) = -\frac{1}{2}\sum_{i=1}^{N}\sum_{j=1}^{N}(\alpha_i-\alpha_i^*)(\alpha_j-\alpha_j^*)\times(\phi(\mathbf{x}_i)\bullet\phi(\mathbf{x}_j)) - \varepsilon\sum_{i=1}^{N}(\alpha_i-\alpha_i^*) - \sum_{i=1}^{N}(\alpha_i+\alpha_i^*)y_i \quad (8)$$

Subject to constraints

$$C \ge \alpha_i,\ \alpha_i^* \ge 0 \text{ and } \sum_{i=1}^{N} y_i(\alpha_i-\alpha_i^*) = 0 \quad (9)$$

By solving this convex quadratic programming problem (QP), the coefficients α and α^* in Eq. (8) are obtained. Only the non-zero coefficients, and the corresponding input vectors, $\mathbf{x}_i$, are called support vectors (SVs). These SVs are the most informative data points that compress the information content of the training set, thereby representing the entire SVR function.

The expanded form of the SVR is thus given as:

$$f(\mathbf{x},\alpha_i,\alpha_i^*) = \sum_{i=1}^{N_{sv}}(\alpha_i-\alpha_i^*)(\phi(\mathbf{x}_i)\bullet\phi(\mathbf{x}_j)) + b \quad (10)$$

However, for the above-mentioned problem [Eq. (10)], the inner product in the feature space needs to be calculated. But with the increase in the input dimensions, the computation in the high-dimensional feature space becomes a computationally complicated problem (curse of dimensionality). Such a problem of contradiction between high dimensions and computational complexity can be overcome by using the kernel trick or defining appropriate kernel functions in place of the dot product of the input vectors in high-dimensional feature space. The kernel function is used to compute directly the inner product from the input space, rather than in the high dimensional feature space. Kernel functions provide a way to avoid the curse of dimensionality.

The most commonly used kernel functions include linear function, polynomial function, Gaussian function, and sigmoid function, spline, etc. [58]. The Gaussian function is a widely used kernel function as it can map the input space sample set into a high dimensional feature space effectively and is good for representing the complex nonlinear relationship between the input and output samples. Furthermore, only one variable (the width parameter) is there to be defined [21, 19]. Considering the above advantages, the Gaussian radial basis function (RBF) is chosen as the kernel function in this work and the expression is given as:

$$\mathrm{K}(x,x_i) = \exp\left(-\frac{\| x-x_i \|^2}{\sigma^2}\right) \quad (11)$$

Where, σ denotes the width of the RBF and $\gamma = \frac{1}{2\sigma^2}$

Thus by applying the kernel function, the basic SVR function modeling the data obtained by using the above-mentioned maximization function is then given by:

$$f(\mathbf{x}, \alpha_i, \alpha_i^*) = \sum_{i=1}^{N_{sv}} (\alpha_i - \alpha_i^*) K(\mathbf{x}, \mathbf{x}_i) + b \tag{12}$$

The bias parameter, b, can be obtained by applying Karush–Kuhn–Tucker (KKT) conditions.

SVR modeling procedure

The computational procedure for SVR modeling can be broadly divided into five stages [59, 60, 61]:

1. The first step is to collect the data set as independent input variables and the target output variable.
2. Thereafter, the data set is normalized and then divided into a training data set and test data set in the ratio of 8:2.
3. Once the training data set is selected and normalized, the next step is to optimize the model and kernel parameters using grid search methodology with K-fold cross validation.
4. Establishment of final SVR model is carried out using the obtained optimal parameters.
5. The performance of the obtained SVR model is evaluated and validated in terms of the statistical evaluation parameters such as average absolute relative error (AARE), root mean square error (RMSE), coefficient of determination (R^2) etc., comparing the prediction performance with models existing in literature and in the light of existing theory.

3. Application of AI Techniques

The following sections discuss three case studies for the prediction of the removal efficiency for Cu (II), As (III) and lead (II) by applying AI techniques like ANN and SVM and the comparison of the results with MLR.

CASE STUDY – I

PREDICTION OF Cu (II) BIOSORPTION EFFICIENCY

In this study, an effort has been made to assess the performance of a developed SVR-based model for predicting the biosorption efficiency of Cu(II) ions on the litter of a natural trembling poplar (*Populus tremula*) forest (LNTPF) as a low-cost and abundant biosorbent vis-à-vis an MLR model using the same data set extracted from published literature [62].

Analysis of the MLR model

As per the existing theory, it has been found that the Cu(II) biosorption efficiency (%) depends on pH (a_1), temperature (a_2), particle size (a_3), agitating speed (a_4), initial Cu(II) concentration (a_5), adsorbent concentration (a_6), and the contact time (a_7). The following MLR model equation has been obtained:

$$y = -8.3357 + 11.0513\,(a_1) + 0.3574\,(a_2) - 30.3253\,(a_3) + 0.0608\,(a_4) - 0.2541\,(a_5) + 5.9614\,(a_6) + 0.06619\,(a_7)$$

The independent variables are found to be significant into the regression.

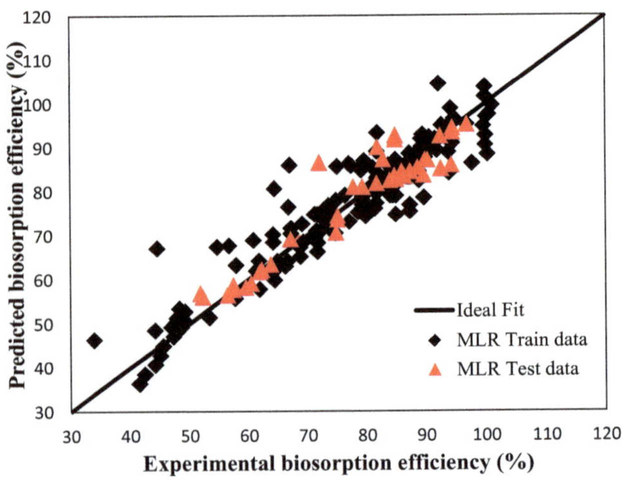

Figure 4. Performance of MLR model using training data and test data for prediction of Cu(II) biosorption efficiency (%).

Performance assessment of the MLR model has been carried out based on statistical parameters namely, AARE, R2, RMSE, SD and MRE. A comparison between the predicted values of Cu(II) biosorption efficiency (%) by the MLR model and those calculated from the experimental data using the training dataset and the test dataset is shown in the Fig. 4. An AARE of 5.514 %, R2 of 0.8708, RMSE of 0.0733, SD of 14.0785 and MRE of 0.0527 on the training data are obtained while the corresponding values for the test data are: 5.327 % AARE, 0.8965 R2, 0.0665 RMSE, 12.94 SD and 0.0525 MRE. It can be seen that the MLR model has consistent prediction ability. In fact, the model gives an improved generalized performance on the test data set.

Analysis of the SVR-based model

The entire dataset of 197 samples is divided into two non-repetitive groups as the training dataset and the testing dataset using the simple random sampling (SRS) technique, in the ratio of 8 (158 data points): 2 (39 data points), respectively. Thereafter, using the same training data set, a SVR-implementation known as "ε-SVR" in the open software LIBSVM library [63] on MATLAB platform has been used to develop the SVR-based model. The generalization ability of the SVR model has been assessed using the virgin test data set.

RBF kernel has been used in this study because of its good general performance and the requirement of few parameters to be adjusted [58]. For achieving the optimal values of the SVR model hyper-parameters (C and ε) and the kernel parameter (γ), grid search methodology with 10-fold cross validation has been carried out on the training data set by considering a wide range of these parameters: C $[2^5, 2^{15}]$, γ $[2^{-15}, 2^2]$ and ε $[2^{-15}, 2^4]$. The optimal values of the model parameters for the SVR model are given in Table 1.

Table 1. Optimal parameters for SVR-based model for Cu (II) biosorption efficiency in a biosorption Process.

Property to be predicted	C	$\gamma = 1/2\sigma^2$	ε	Kernel type	Type of Loss function	Number of support vectors	Number of training points
Biosorption efficiency	32768	0.01195	0.0625	RBF	ε - insensitive	104	158

After optimization of the SVR parameters, the model output for the training and test data has been plotted in Fig. 5. It is evident that both the predicted values for the training data as well as the test data lie close to the ideal fit line. Table 2 exhibits the model evaluation parameters of the SVR-based model for both the training data as well as the test data. The close proximity to each other of the model evaluation parameters for the training data and test data testify to the excellent prediction performance and also to the high generalization ability of the SVR-based model.

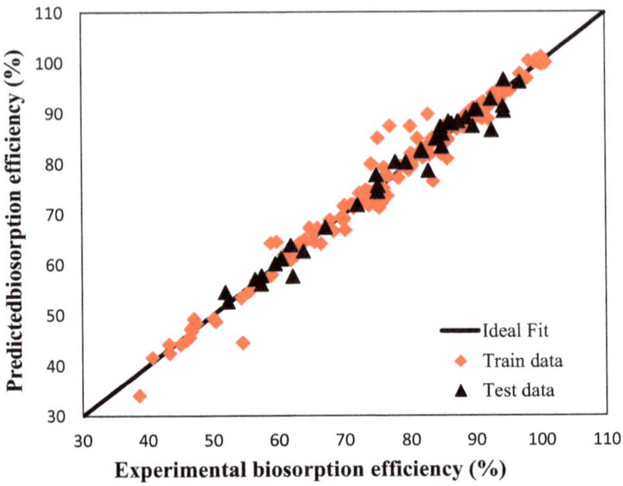

Figure 5. SVR simulation of Cu(II) biosorption efficiency(%) in a biosorption process with optimal parameters using training data and test data.

Table 2. Model evaluation parameters for SVR-based model using the training data and the test data.

SVR Model evaluation parameter	Train data	Test data
AARE (%)	2.07	1.97
R	0.9881	0.9879
RMSE	0.0011	0.0007
SD	14.837	13.2689
MRE	0.0206	0.0199
Q^2LOO (Train data),Q^2ext(Test data)	0.9763	0.9759

Comparison of SVR-based model with the MLR model.

A comparison was made of the prediction performances of the developed MLR model with the SVR-based model. Fig. 6 clearly illustrates the superior prediction performance of the SVR-based model. Table 3 exhibits the relevant model evaluation parameters for the two modeling methods on the test data set. It can be easily surmised that the SVR-based model is a vastly improved model than the MLR model. Further, in terms of absolute deviation (AD) values, the SVR-based model again scores over the MLR model as shown in Table 4 and 5.

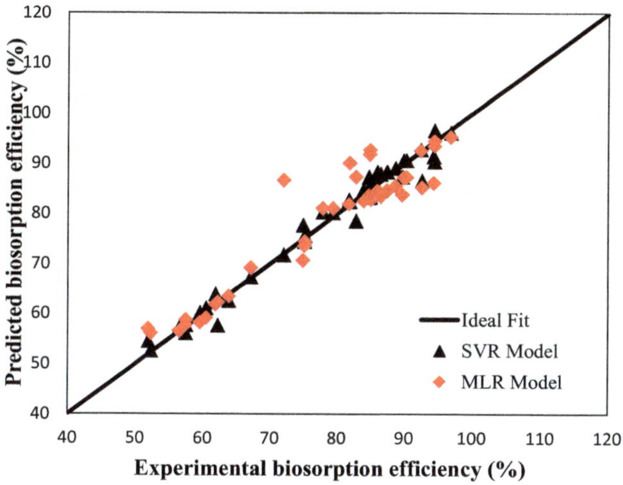

Figure 6. Comparison of SVR model with MLR model for prediction of Cu(II) biosorption efficiency (%) in a biosorption process using test data.

Table 3. Model evaluation parameters for SVR-based model and MLR model using the test data.

SVR Model evaluation parameter	SVR-based model	MLR model
AARE (%)	1.97	5.33
R^2	0.9760	0.8965
RMSE	0.0007	0.0665
SD	13.2689	12.9434
MRE	0.01994	0.0525

Table 4 illustrates the distribution of predicted data points of Cu (II) biosorption efficiency by the SVR-based model in terms of absolute deviation (AD) for the training data. The SVR-based model was found to predict nearly 92.4 % data points within an absolute deviation of less than 5 % and a total of 97.5 % data points within an absolute deviation of less than 10%. Only 2.5 % data points have an AD of more than 10%. Whereas, the MLR-based model was observed to predict only 65.1% data points within an absolute deviation of less than 5 % and a total of 86.2% data points within an absolute deviation of less than 10%. Only 13.8% data points have an AD of more than 10%.

Table 5 depicts the distribution of predicted data points of Cu (II) biosorption efficiency by MLR and SVR-based model in terms of absolute deviation for the test data set. It is observed that the SVR based model predicts nearly 92.3% data points within an absolute deviation of less than 5% and all the data points fall within an absolute deviation of no more than 10%. While, the MLR-based model predicts nearly 56.4 % data points within an absolute deviation of less than 5 % and a total of 84.6 % data points within an absolute deviation of less than 10%. Only 15.4 % data points have an AD of more than 10%. This shows that the SVR-based model is highly accurate and has good generalization ability.

Table 4. Percentage distribution of predicted data points of Cu (II) biosorption efficiency by MLR model and SVR- based model in terms of absolute deviation (AD) for training data.

Absolute deviation (AD) (%)	% of SVR model predicted values	Cumulative score	% of MLR model predicted values	Cumulative score
AD< 2	76.6	76.6	27.2	27.2
2<AD< 5	15.8	92.4	37.9	65.1
5<AD< 10	5.1	97.5	21.1	86.2
AD>10	2.5	100	13.8	100
Total	100		100	

Table 5. Percentage distribution of predicted data points of Cu (II) biosorption efficiency by MLR model and SVR- based model in terms of absolute deviation (AD) for test data.

Absolute deviation (AD) (%)	% of SVR model predicted values	Cumulative score	% of MLR model predicted values	Cumulative score
AD< 2	61.5	61.5	23.1	23.1
2<AD< 5	30.8	92.3	33.3	56.4
5<AD< 10	7.7	100	28.2	84.6
AD>10	00.00		15.4	100
Total	100		100	

CASE STUDY II

PREDICTION OF ARSENIC (III) BIOSORPTION EFFICIENCY

In this study biosorption efficiency of As (III) ions has been predicted with the developed SVR-based model and the MLR model. Biosorption data for the present work have been taken from available published literature [64] where the input parameters are taken as: initial As (III) concentration, biosorbent dosage, contact time and temperature. The whole dataset of 81 samples are divided into a training dataset and a testing dataset using the simple random sampling (SRS) technique, in the size ratio of 80% (65 data points) and 20% (16 data points) respectively.

Analysis of the MLR model

From the literature [64] we found that biosorption of As (III) (y) depends on initial As (III) concentration (m_1), biosorbent dosage (m_2), contact time (m_3) and temperature (m_4) and the following MLR model equation has been obtained:

$$y = 54.5550 - 0.3314 (m_1) + 1.3083 (m_2) + 0.2120 (m_3) + 0.2442 (m_4)$$

These independent variables are found to play a significant role into the regression.

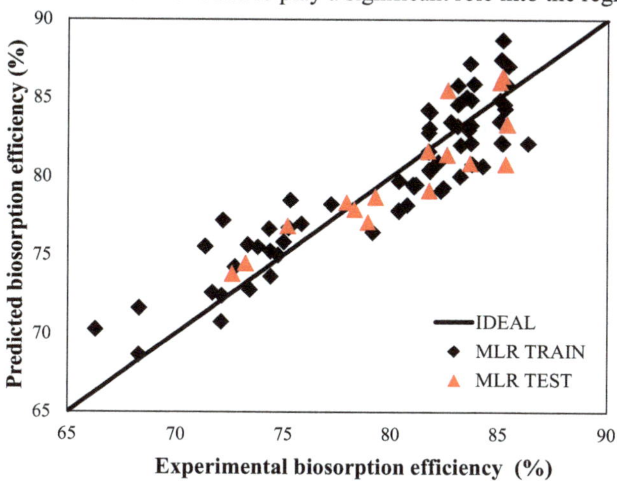

Figure 7: Performance of MLR model using a training data and a test data for predicting the biosorption efficiency of As(III).

Performance assessment of the MLR model for the prediction of biosorption efficiency As (III) based on statistical parameters like AARE, R, RMSE, SD and MRE. A comparison between the predicted values of As(II) biosorption efficiency by the MLR model and those calculated from the experimental data using the training dataset and the test dataset is shown in the Fig. 7 and Table 6.

Table 6: Model evaluation parameters for MLR model using the training data and the test data

MLR Model evaluation parameter	Train data	Test data
AARE (%)	2.3206	1.74
R	0.9066	0.8775
RMSE	0.0275	0.0233
SD	4.7056	4.1988
MRE	0.0231	0.0177
Q^2LOO (Train data),Q^2ext(Test data)	0.8217	0.7533

Analysis of the SVR-based model

Similar procedure as discussed in "Case Study I" were used to develop the SVR model by grid search methodology with a 10-fold cross validation. The obtained optimal values of the model parameters are given in Table 7.

Table 7. Optimal parameters for SVR-based model for biosorption efficiency of As (III).

Property to be predicted	C	$\gamma = 1/2\sigma^2$	ε	Kernel type	Type of loss function	Number of support vectors	Number of training points
Biosorption efficiency(%)	32768	0.055	0.0848	RBF	ε - insensitive	46	65

The training and test course has been constructed using the training data set and test data set as shown in the Fig. 8 and Fig. 9 respectively. One can observe that the predicted data points for the training as well as for testing set lies very close to the experimental data. Further, it can be clearly seen that the SVR prediction results are much improved over the test data set.

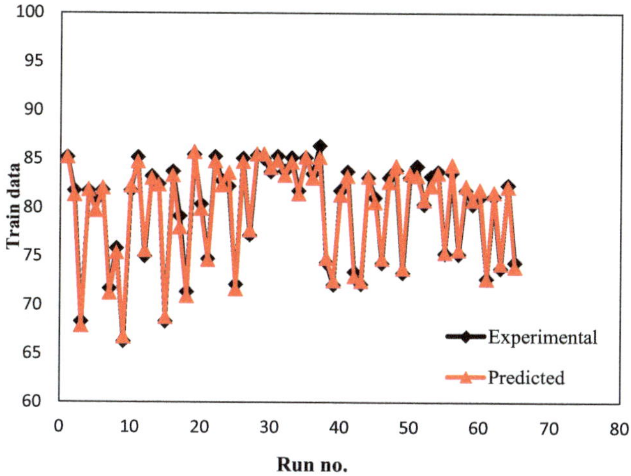

Figure 8. Training course curve for the Prediction of As(III) biosorption efficiency(%)

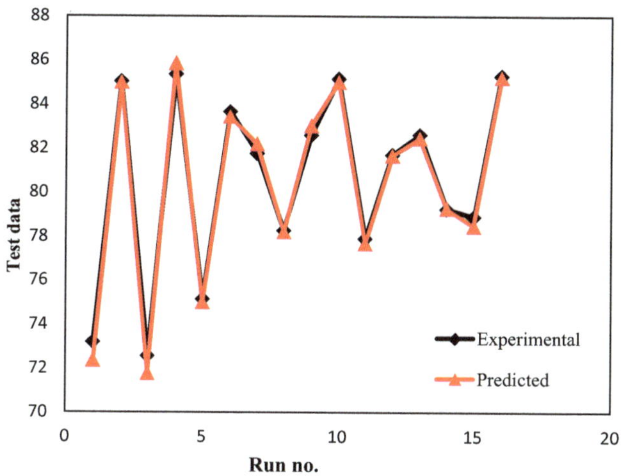

Figure 9. Test course curve for the Prediction of As(III) biosorption efficiency(%)

Comparison of SVR-based model with the MLR model

A comparison of the prediction performances of the developed MLR model with the SVR-based model is shown in Table 8 and Fig. 10.

Table 8. Model evaluation parameters for SVR-based model and MLR model using the test data.

SVR Model evaluation parameter	SVR-based model	MLR model
AARE (%)	0.5591	1.74
R	0.9905	0.8775
RMSE	5.6418E-05	0.0233
SD	3.6245	4.1988
MRE	0.00562	0.0177

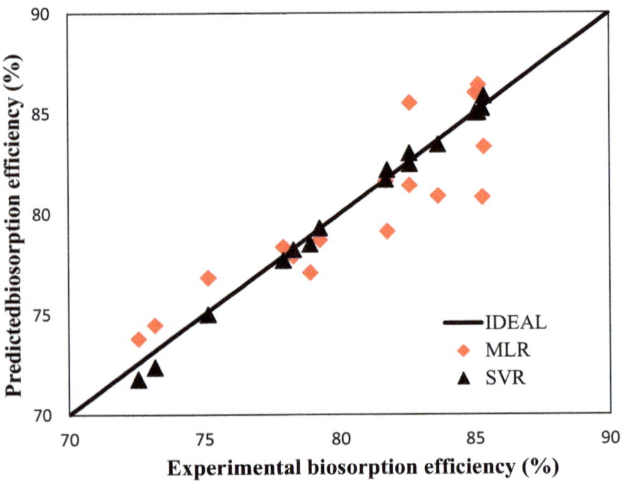

Figure 10. Comparison of SVR model with the MLR model using the test data set.

Table 9 illustrates the distribution of predicted data points of As (III) biosorption efficiency by the SVR-based model in terms of absolute deviation for the training data set. It has been found that the SVR-based model predicts nearly 90.7 % data points within an absolute deviation of less than 1 %. Only 9.23 % data points within an absolute deviation of less than 5% and no deviation have been found above an AD of more than 5%. Whereas, it is observed that the MLR-based model predicts nearly 95.38% data points within an absolute deviation of less than 5 % and a total of 4.6% data points have an AD of more than 5%.

Table 10 depicts the distribution of predicted data points of As (III) biosorption efficiency by MLR and SVR-based model in terms of absolute deviation for the test data set. It is observed that the SVR based model predicts nearly 56.25% data points within an absolute deviation of less than 1% and all the data points fall within an absolute deviation of no more than 5%. While, the MLR-based model predicts nearly 25 % data points within an absolute deviation of less than 1 % and a total of 93.75 % data points within an absolute deviation of less than 5%. Only 6.25 % data points have an AD of more than 5%.

Table 9. Percentage distribution of predicted data points of the sorption efficiency of As (III) by MLR model and SVR- based model in terms of absolute deviation (AD) for the training data set.

Absolute deviation (AD) (%)	% of MLR model predicted values	Cumulative score	% of SVR model predicted values	Cumulative score
AD< 1	20	20	90.7692	90.7692
1<AD< 5	75.3846	95.3846	9.2308	100
AD>5	4.6154	100	0.00	
Total	100		100	

Table 10. Percentage distribution of predicted data points of the sorption efficiency of As (III) by MLR model and SVR- based model in terms of absolute deviation (AD) for test data.

Absolute deviation (AD) (%)	% of MLR model predicted values	Cumulative score	% of SVR model predicted values	Cumulative score
AD< 1	25	25	56.25	56.25
1<AD< 5	68.75	93.75	43.75	100
AD>5	6.25	100	0.00	
Total	100		100	

CASE STUDY - III

PREDICTION OF THE SORPTION CAPACITY OF LEAD (II)

To predict the sorption capacity of Lead (II) using an agricultural by-product 'tree fern', as a low cost biosorbent, similar procedure has been adopted as discussed in earlier case studies. Biosorption data for SVR, MLR and ANN model have been taken from the available published literature [65, 66]. The entire dataset consisting of 145 samples are divided into training dataset and testing dataset using the simple random sampling (SRS) technique, in the size ratio of 80% (116 data points) and 20% (29 data points) respectively. Finally developed SVR model has been evaluated and validated on the basis of statistical parameters as well as by comparing it with the commonly used multiple linear regression (MLR) model and ANN model.

Analysis of the MLR model

The MLR model was developed for predicting the sorption capacity of lead (II) ions (y) with the independent input parameters being: initial lead ion concentration (x_1), temperature (x_2), pH (x_3) and contact time (x_4).

The following MLR model equation has been obtained through MS-Excel (2007):

$$y = -141.8444 + 0.0398\,(x_1) + 0.4826\,(x_2) + 2.8756\,(x_3) + 0.1336\,(x_4)$$

The independent variables are found to be significant into the regression.

In this study, performance of the developed MLR model for the training and test data set is shown in the Fig. 11 as well as also, given in the Table 11. It can be seen from Table 11 that the performance of the MLR model for test data set (unseen data) is poor as compared to the training data set.

Table 11. Model evaluation parameters for MLR model using the training data and the test data.

MLR Model evaluation parameter	Train data	Test data
AARE (%)	21.1178	25.0056
R	0.78661	0.7613
RMSE	0.0469	0.1188
SD	8.3075	8.3246
MRE	0.1745	0.1799
Q^2LOO (Train data),Q^2ext(Test data)	0.6153	0.5764

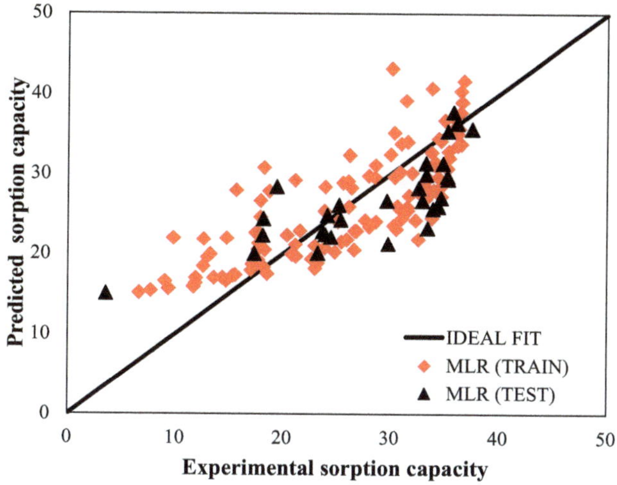

Figure 11. Performance of MLR model using training data and test data for predicting the sorption capacity of lead(II).

Analysis of ANN model

The schematic representation of neural networks architecture shown in Fig. 12. comprises of: (1) input layer (independent variables), (2) hidden layers, (3) and the output layer (dependent variables). Three layered multiple feed forward neural networks (4:12:1) have been used for modeling of the sorption capacity of lead (II) with 4 – neuron for initial lead ion concentration (mg/L), temperature (°C), pH of the solution and contact time (min) at the input layer, 12 – neuron at hidden layer and 1- neuron for the sorption capacity at the output layer. The entire data set of 145 samples is partitioned into three subsets consisting training dataset, 65% (data), validation dataset, 15% (data) and test dataset, 20% (data). The numbers of input and output neurons are set according to the nature of the problem whereas an appropriate number of neuron in the hidden layer has been chosen by trial and error method as insufficient neurons may cause under fitting while too many neurons might lead to over fitting [67, 68, 69]. The training of the ANN model is carried out with the complete input data set to the network till the MSE is minimized [70, 71, 72, 73]. A model should be trained properly in order to have its high correlation coefficient (R) close to 1 and low MSE. The ANN toolbox present in MATLAB has been used in this study to develop the model.

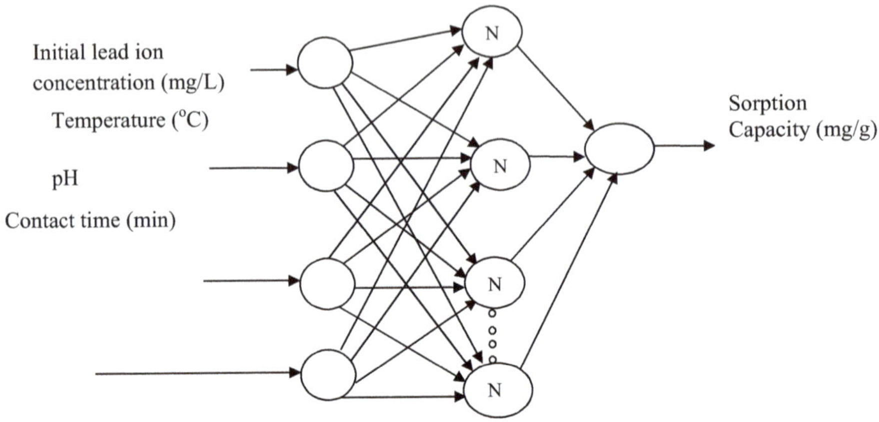

Figure 12. Schematic representation of ANN architecture.

Fig. 13 demonstrates that both the predicted values for the training data as well as the test data lie close to the ideal fit line. Table 12 exhibits the model evaluation parameters of the ANN model for both the training data as well as the test data.

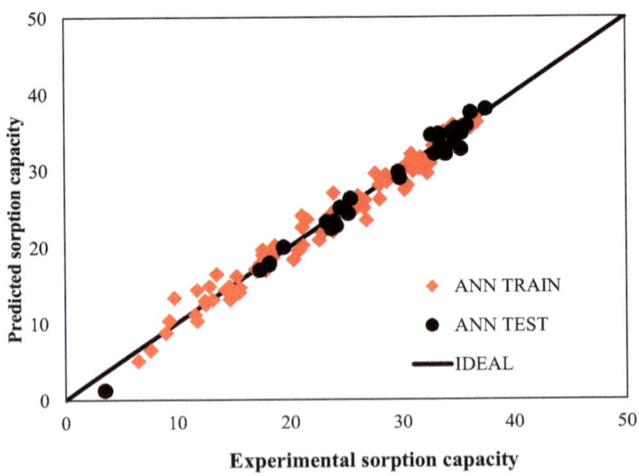

Figure 13. Performance of ANN model using training data and test data for predicting the sorption capacity of lead(II).

Table 12. Model evaluation parameters for ANN model using training and the test data.

Model evaluation parameter	Train data	Test data
AARE (%)	4.4545	4.941
R	0.9889	0.9845
RMSE	0.0048	0.0587
SD	8.2293	7.9204
MRE	0.0444	0.0912
Q^2LOO(Train data),Q^2ext(Test data)	0.9826	0.9779

Analysis of the SVR-based model

An SVR model with the same data set of 145 samples is developed using the procedure described in the previous case studies, the obtained optimal values of the SVR model parameter is given below in Table 13.

Table 13. Optimal parameters for SVR-based model for sorption capacity of Lead (II).

Property to be predicted	C	$\gamma = 1/2\sigma^2$	ε	Kernel type	Type of loss function	Number of support vectors	Number of training points
Sorption capacity [mg/g]	32768	0.058	0.09	RBF	ε-insensitive	70	116

After optimizing the SVR parameters, training course curve and test course has been constructed as shown in Fig. 14 and Fig. 15 respectively.

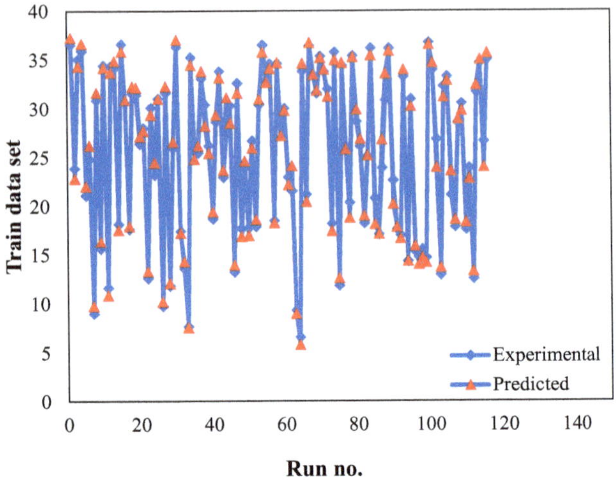

Figure 14. Training course curve for predicting the sorption capacity of lead (II) using SVR.

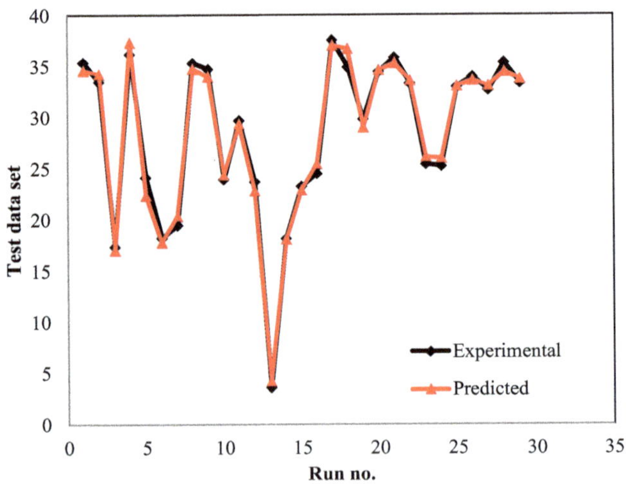

Figure 15. Test course curve for predicting the sorption capacity of lead (II) using SVR.

A comparison between the predicted values of the sorption capacity of lead (II) by the SVR-based model and those calculated from the experimental data using the training dataset and the test dataset has been shown in the Fig. 16 and the model evaluation parameters are given in the Table 14.

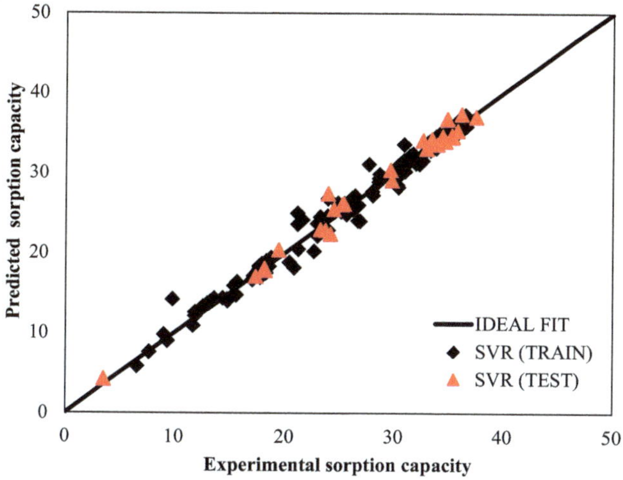

Figure 16. SVR simulation of sorption capacity (mg/g) of lead (II) ions with optimal parameters using training data and test data set.

Table 14. Model evaluation parameters for SVR-based model using training and the test data.

SVR Model evaluation parameter	Train data	Test data
AARE (%)	3.9874	3.3816
R	0.9901	0.9914
RMSE	0.0041	0.0025
SD	8.2529	7.8962
MRE	0.0399	0.0338
Q^2LOO(Train data),Q^2ext(Test data)	0.9800	0.9845

Comparison of SVR-based model with ANN and MLR model

Prediction performance of the developed SVR-based model has been compared with the ANN and the MLR model as shown in Fig. 17. Table 15 clearly depicts the superior prediction performance of the SVR-based model.

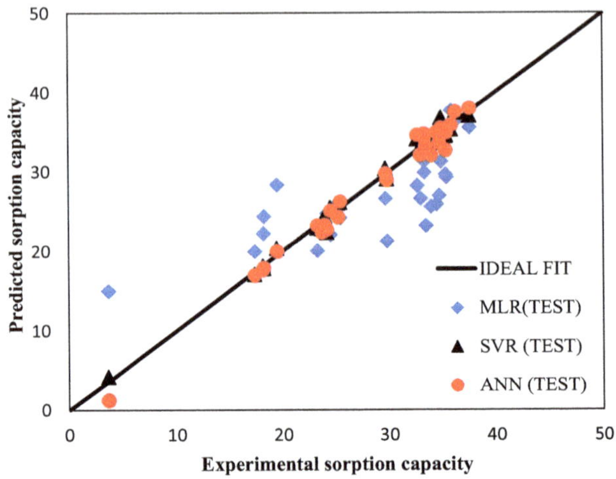

Figure 17. Comparison of SVR model with MLR and ANN model for predicting the sorption capacity (mg/g) of lead (II) ions using the test data set.

Table 15. Model evaluation parameters for SVR-based model, ANN and MLR model using test data set.

Model evaluation parameter	SVR-based model	ANN model	MLR model
AARE (%)	3.38	4.941	25.01
R	0.9915	0.9845	0.7613
RMSE	0.0025	0.0587	0.1188
SD	7.8962	7.9204	8.3246
MRE	0.0338	0.0912	0.1799

Table 16 illustrates the distribution of predicted data points of lead (II) sorption capacity by SVR-based model in terms of absolute deviation for the training data set. It has been found that the SVR-based model predicts nearly 77.58 % data points within an absolute deviation of less than 5 % and a total of 91.37 % data points within an absolute deviation of less than 10%. Only 0.86 % data points have an AD of more than 20%. Whereas, it is observed that the ANN model have 69.83 percent data points within an absolute deviation of less than 5 % and a total of 86.21 per cent data points within an absolute deviation of less than 10%. Only 96.55% data points have an AD of more than 20%.

MLR-based model predicts nearly 14.65% data points within an absolute deviation of less than 5 % and a total of 36.2% data points within an absolute deviation of less than 10%. Only 31.03% data points have an AD of more than 20%.

Table 17 depicts the distribution of predicted data points of lead (II) sorption capacity by MLR and SVR-based model in terms of absolute deviation for the test data set. It is observed that the SVR based model predicts nearly 86.2% data points within an absolute deviation of less than 5% and a total of 93.10 % data points within an absolute deviation of less than 10% and all the data points fall within an absolute deviation of not more than 20%. While, it is found that ANN model have 79.31 percent data points within an absolute deviation of less than 5 % and a total of 96.55 % data points within an absolute deviation of less than 10%. Only 3.44% data points have an AD of more than 20%.

MLR-based model predicts nearly 20.68 % data points within an absolute deviation of less than 5 % and a total of 37.93 % data points within an absolute deviation of less than 10%. Only 31.03 % data points have an AD of more than 20%. This shows that the SVR-based model is highly accurate and has good generalization ability for the unseen test data set.

Table 16 Percentage distribution of predicted data points of the sorption capacity of lead (II) by MLR model and SVR- based model in terms of absolute deviation (AD) for training data.

Absolute Deviation AD (%)	% of MLR model predicted values	Cumulative score	% of ANN model predicted values	Cumulative score	% of SVR model predicted values	Cumulative score
AD< 5	14.6552	14.6552	69.8276	69.8276	77.5862	77.5862
5<AD< 10	21.5517	36.2069	16.3793	86.2069	13.7931	91.3793
10<AD< 20	32.7586	68.9655	10.3448	96.5517	7.7586	99.1379
AD>20	31.0345	100	3.4483	100	0.8621	100
Total	100		100		100	

Table 17 Percentage distribution of predicted data points of the sorption capacity of lead(II) by MLR model, ANN and SVR- based model in terms of absolute deviation (AD) for test data set.

Absolute Deviation AD (%)	% of MLR model predicted values	Cumulative score	% of ANN model predicted values	Cumulative score	% of SVR model predicted values	Cumulative score
AD< 5	20.6897	20.6897	79.3103	79.3103	86.2069	86.2069
5<AD< 10	17.2414	37.9310	17.2414	96.5517	6.8965	93.1035
10<AD< 20	31.0345	68.9655	0.00	96.5517	6.8965	100
AD>20	31.0345	100	3.4483	100	0.00	
Total	100		100		100	

Performance Assessment of the Commonly Used MLR Model, ANN And SVR-Based Model in the Light of Existing Theory and Experimental Findings

A. *Effect of initial conc. of lead (II)*

Experiments were performed at constant condition (pH 4.9, temperature = 20° C, agitating speed = 300 rpm) to show the effect of various initial concentration of lead (II) on the sorption capacity of lead (II). It was observed that the sorption capacity of lead (II) initially increases due to ion exchange with the surface of cations on the tree fern and afterwards, it slows down representing a gradual uptake of cation exchange at the inner surface. The equilibrium sorption capacity increases from 18.2 to 41.5 mg/g, as initial lead (II) concentration vary from 74.1 to 350 mg/l.

Prediction of sorption capacity of lead (II) by SVR-based model, ANN and MLR model is shown in the Fig. 18 and it is found that the SVR-based model shows fair agreement with the experimental results and thus, it can be said that SVR-based has the best prediction ability and generalizability than ANN and MLR model.

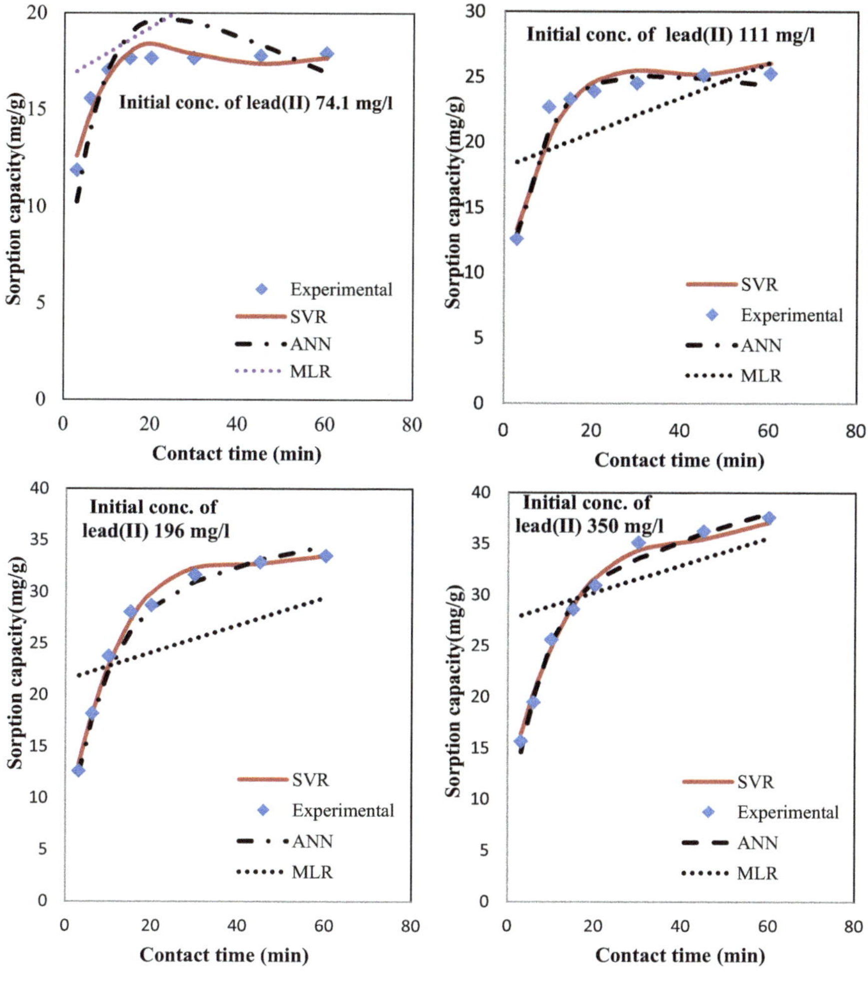

Figure 18 (a)-(e). Effect of initial lead (II) ion concentrations on the sorption capacity (mg/g) of lead against contact time (pH 4.9, temperature = 20° C, agitating speed = 300 rpm).

B. *Effect of temperature on sorption capacity of lead (II)*

The effect of various temperatures on sorption capacity of lead (II) shown in Fig. 19 has been obtained by experiments performed with a constant condition (pH 4.9 initial lead concentration 200 mg/L, agitating speed = 300 rpm). Sorption capacity increases with an increase in solution temperature from 278 K to 318 K. However, the equilibrium sorption capacity was hardly affected by increased temperature. The values of rate constant were found to increase from $9.18 * 10^{-4}$ to 0.103 g/mg.min with an increase in the solution temperature from 278 K to 318 K.

The variation in temperature influenced the time required to reach saturation with excellent agreement between the experimental and the predicted curves. It is observed that the SVR-based follows the same trends of the experimental results and is the better predictive model than the other two.

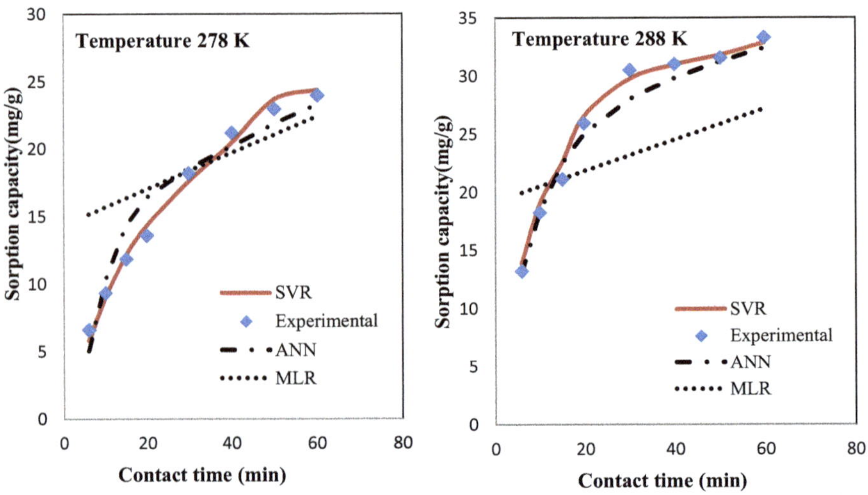

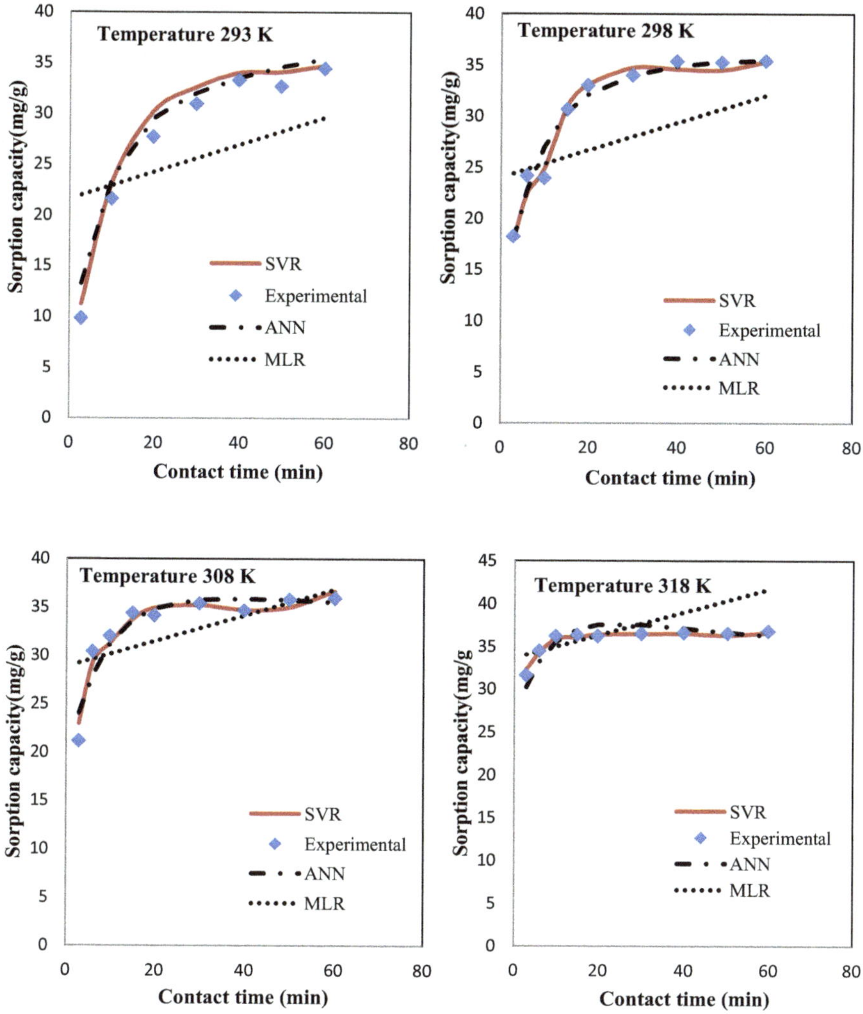

Figure 19 (a)-(e). Effect temperatures on the sorption capacity (mg/g) of lead against contact time (pH 4.9 initial lead concentration 200 mg/L, agitating speed = 300 rpm).

C. *Effect of pH on sorption capacity of lead (II)*

Ho [66] had performed the experiments at constant condition (initial lead concentration 200 mg/L, temperature = 20° C, agitating speed = 300 rpm) to show the effect of pH on sorption capacity of lead (II). The obtained results shown in Fig. 20 demonstrate that the sorption was initially fast and then slowed until equilibrium was reached, it is believed that ion exchange may have occurred in the beginning state of sorption. It was also found that at low pH, surface of the tree fern becomes positively charged and as pH increased more ligands (neutral molecule) of tree fern gets attracted to the positively charged lead ion.

Predicting the effect of pH on sorption capacity of lead (II) by the SVR-based model, ANN and MLR and the data showed good compliance with the SVR model.

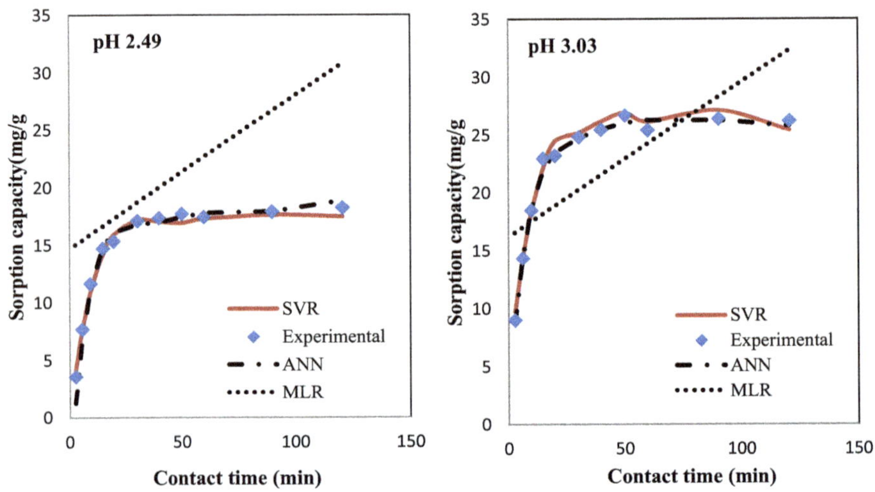

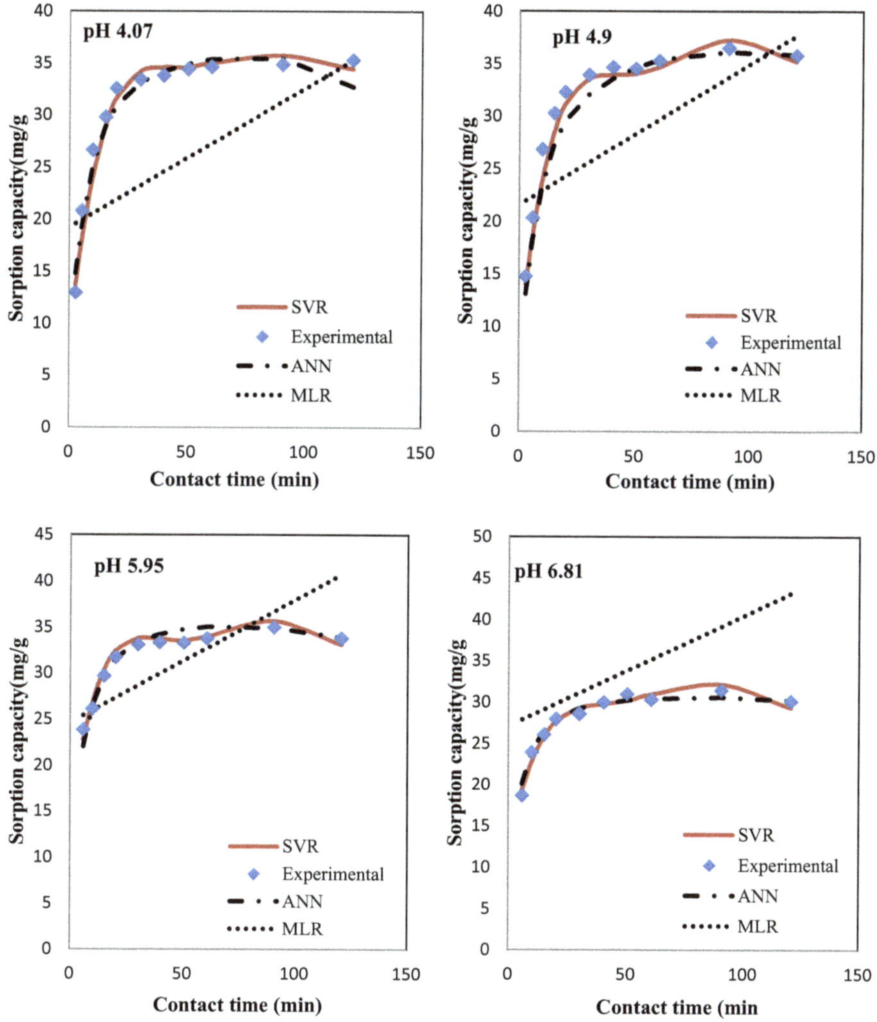

Figure 20 (a)-(e). Effect pH on the sorption capacity (mg/g) of lead against contact time (initial lead concentration 200 mg/L, temperature = 20° C, agitating speed = 300 rpm).

4. Conclusions

Artificial intelligence techniques such as ANN and SVM have been applied to predict the heavy metals removal efficiency in adsorption process and the results thus obtained have been compared with those obtained by MLR. Three case studies have been studied for the removal of heavy metals like Cu (II), As (III) and lead (II). It has been observed that the prediction performance of SVR is the best as compared followed by ANN and MLR, in that order. It is also found that the SVR model most effectively predicts the behavior of the complex interaction process of lead (II) removal from waste water under various experimental conditions, followed by the ANN model. Good prediction results from these AI techniques can be helpful in a more efficient design and fabrication of an efficient waste water treatment plant for treating heavy metal rich effluent.

Acknowledgements

We wish to gratefully acknowledge and thank all the authors whose works have been referred to in this study, especially those whose data have been used to develop the models.

Nomenclature

b	Bias term
C	Cost function
C_0	Initial Cu(II) concentration, mgL^{-1}
F	High-dimensional feature space
$f(\mathbf{x})$	Regression function
$K(\mathbf{x}_i,\mathbf{x}_j)$	Kernel function
L	Lagrangian function (dual form)
Q^2ext	Leave-one-out cross validation on the test set
Q^2LOO	Leave-one-out cross validation on the training set
W	Weight vector
x_i	Input vector
y_i	Output vector

Greek symbols

σ	Width parameter of RBF kernel
ε	Loss function
γ	Regularization parameter
α and α^*	Lagrange multipliers
$\varphi(x_i)$	Mapping function to high dimensional feature space for input vector $\mathbf{x}$

Abbreviations

AARE	Average absolute relative error
AD	Absolute deviation

ANN	Artificial neural networks
MSE	Mean square error
RMSE	Root mean square error
SD	Standard deviation
MRE	Mean relative error
MLR	Multiple linear regression
SVM	Support vector machine
SVR	Support vector regression

References

[1] N. Das, R. Vimala, P. Karthika Biosorption of heavy metals- A review, Indian J. Biotechnol. 7 (2008) 159-169.

[2] J. Wang, C. Chen, Biosorption of heavy metals by Saccharomyces cerevisiae: A review, Biotechnol. Adv. 24 (2006) 427–451. *http://dx.doi.org/10.1016/j.biotechadv.2006.03.001*

[3] S. H. Abbas, I. M. Ismail, T. M. Mostafa, A. H., Sulaymon, Biosorption of Heavy Metals: A Review, J. Chem. Sci. Technol. 3 (2014) 74–102.

[4] A. Kapoor, T. Viraraghavan, Fungal biosorption- an alternative treatment option for heavy metal bearing wastewater: a review, Bioresour. Technol. 53 (1995) 185–206.

[5] D. J. Sweetly, K. Sangeetha, B. Suganthi, Biosorption of Heavy Metal Lead from Aqueous Solution by Non-living Biomass of Sargassum myriocystum, International Journal of Application on Innovation in Engineering & Management (IJAIEM) 3 (4) (2014) 39-45.

[6] V. K. Gupta, A. Nayak, S. Agarwal, Bioadsorbents for remediation of heavy metals: Current status and their future prospects, Environ. Eng. Res. 20 (2015) 1-18. *http://dx.doi.org/10.4491/eer.2015.018*

[7] M. A. Ashraf, K. Mahmood, A. Wajid, Study of low cost biosorbent for biosorption of heavy metals, Int. Conf. Food Eng. Biotechnol. 9 (2011) 60–68.

[8] M. Zabochnicka-Świątek, M. Krzywonos, Potentials of Biosorption and Bioaccumulation Processes for Heavy Metal Removal, Pol. J. Environ. Stud. 23 (2) (2014) 551-561.

[9] V. Eyupoglu, B. Eren, E. Dogan, Prediction of ionic Cr (VI) extraction efficiency in flat sheet supported liquid membrane using artificial neural networks (ANNs), Int. J. Environ. Res. 4(3), (2010) 463-470.

[10] Y. Vasseghian, M. F. Ahmadi, Dolati, A. Heydari, Modeling and Experimental Prediction of Wastewater Treatment Efficiency in Oil Refineries Using Activated Sludge Process, Journal of Chemical and Petroleum Engineering, 48 (1) (2014) 69-79.

[11] T. Brey, A. Jarre-Teichmann, O. Borlich, Artificial neural networks versus multiple linear regressions: predicting P/B ratios from empirical data, Mar Ecol Prog Ser. 140 (1996) 251-25. *http://dx.doi.org/10.3354/meps140251*

[12] K.O. Akande, T.O. Owolabi, S. Twaha, S.O. Olatunji, Performance Comparison of SVM and ANN in Predicting Compressive Strength of Concrete, IOSR Journal of Computer Engineering IOSR-JCE). 16 (5) (2014) 88-94. *http://dx.doi.org/10.9790/0661-16518894*

[13] A. A. Hasseim, R. P. Sudirman, I. Khalid, N. Tabatabaey-Mashadi, Comparison of ANN and SVM to Identify Children Handwriting Difficulties, Engineering 5 (2013) 1-5. *http://dx.doi.org/10.4236/eng.2013.55B001*

[14] A. Eskandari, R. Nouri, S.H. Meraji, A. Kiaghadi, Developing a Proper Model for Online Estimation of the 5-day Biochemical oxygen demand based on Artificial neural networks and support vector machine, Journal of Environmental Studies 38 (61) (2012) 22-24.

[15] E. A. Zanaty, Support Vector Machines (SVMs) versus Multilayer Perception (MLP) in data classification, Egyptian Informatics Journal 13, (2012) 177-183. *http://dx.doi.org/10.1016/j.eij.2012.08.002*

[16] U. Thissen, R. V. Brakel, A. P. D. Weijer, W.J. Melssen, L.M.C. Buydens, Using support vector machines for time series prediction, Chemometr. Intell. Lab. 69, (2003) 35– 49. *http://dx.doi.org/10.1016/S0169-7439(03)00111-4*

[17] Y. Pan, J. Jiang, R. Wang, H. Cao, Y. Cui, Predicting the auto-ignition temperatures of organic compounds from molecular structure using support vector machine, J. Hazard. Mater. 164 (2009) 1242–1249. *http://dx.doi.org/10.1016/j.jhazmat.2008.09.031*

[18] Y. Radhika, M. Shashi, Atmospheric Temperature Prediction using Support Vector Machines, International Journal of Computer Theory and Engineering 1(1) (2009) 1793-8201.
http://dx.doi.org/10.7763/ijcte.2009.v1.9

[19] H. Peng, X. Ling, Predicting thermal–hydraulic performances in compact heat exchangers by support vector regression, Int. J. Heat Mass Transf. 84 (2015) 203–213.
http://dx.doi.org/10.1016/j.ijheatmasstransfer.2015.01.017

[20] S. Zaidi, Development of support vector regression (SVR)-based model for prediction of circulation rate in a vertical tube thermosiphon reboiler, Chem. Eng. Sci. 69 (2012) 514–521.
http://dx.doi.org/10.1016/j.ces.2011.11.005

[21] S. Zaidi, Novel application of support vector machines to model the two phase boiling heat transfer coefficient in a vertical tube thermosiphon reboiler, Chem. Eng. Res. Des. 98 (2015) 44–58.
http://dx.doi.org/10.1016/j.cherd.2015.04.002

[22] P. A. Webb, Introduction to Chemical Adsorption Analytical Techniques and their Applications to Catalysis, Micromeritics Instrument Corp., Norcross, Georgia, (2003)1-12.

[23] A. Buekens, N. N. Zyaykina, Adsorbents and Adsorption Processes for Pollution Control, Pollution control technologies 2 (2009) 99-123.

[24] A. Dabrowski, Adsorption - from theory to practice, Advances in Colloid and Interface Science 93 (2001) 135-224.
http://dx.doi.org/10.1016/S0001-8686(00)00082-8

[25] F. I. Khan, A. K. Ghoshal, Removal of Volatile Organic Compounds from polluted air, Journal of Loss Prevention in the Process Industries 13 (2000) 527–545.
http://dx.doi.org/10.1016/S0950-4230(00)00007-3

[26] J. Kipling, Adsorption from solutions of Non-Electrolytes, Academic Press. Inc., London, New York, (1965).

[27] Ju, Okoli, I. Ezuma, Adsorption Studies of Heavy Metals by Low-Cost Adsorbents, J. Appl. Sci. Environ. Manage. 18 (3) (2014) 443-448.

[28] B. Yasemin, T. Zeki, Removal of heavy metals from aqueous solution by sawdust adsorption, Journal of Environmental Sciences 19 (2007) 160–166.
http://dx.doi.org/10.1016/S1001-0742(07)60026-6

[29] C. A. Comrie, Comparing Neural Networks and Regression Models for Ozone Forecasting, J. Air & Waste Manage. Assoc. 47 (1997) 653-663. *http://dx.doi.org/10.1080/10473289.1997.10463925*

[30] M. A. Doori, B. Beyrouti, Credit Scoring Model Based on Back Propagation Neural Networks Using Various Activation and Error Function. International Journal of Computer Science and Network Security (IJCSNS), 14(3) (2014) 16-24.

[31] C. -h., Kung, W. -S. Yang, C. -M. Kung, A Study on Image Quality Assessment using Neural Networks and Structure Similarty, Journal of computers, 6 (10) (2011) 2221-2228.

[32] V. Sharma, S. Rai, A. Dev, A Comprehensive Study of Artificial Neural Networks, International Journal of Advanced Research in Computer Science and Software Engineering, 2 (10) (2012) 278-284.

[33] M. Maraqa, F. Al-Zboun, M. Dhyabat, R. A. Zitar, Recognition of Arabic Sign Language (ArSL) Using Recurrent Neural Networks, Journal of Intelligent Learning Systems and Applications, 4 (2012) 41-52. *http://dx.doi.org/10.4236/jilsa.2012.41004*

[34] B. Rajkumar, T. Gopikiran, S. Satyanarayana, Neural Networks Design in Cloud Computing, International Journal of Computer Trends and Technology- 4 (2) (2013) 63-67.

[35] D. K. Sonar, S. S. Gupta, Prediction and Estimation of Gyroscopic Couple by Analytical and Neural Networks, International Journal of Engineering Science Invention, 4 (6) (2015) 62-71.

[36] F. S. Panchal, M. Panchal, Review on Methods of Selecting Number of Hidden Nodes in Artificial Neural Networks, International Journal of Computer Science and Mobile Computing (IJCSMC), 3 (11) (2014) 455-464.

[37] M. Alhaz uddin, M. Jameel, H. A. Razak, Application of artificial neural networks in fixed offshore structure, Indian J. Mar. Sci. 44 (3) (2015) 1-10.

[38] A. D. Dongare, R.R. Kharde, A. D. Kachare, Introduction to Artificial Neural Networks, International Journal of Engineering and Innovative Technology (IJEIT). 2 (1), (2012)189-194.

[39] M. Madić, M. Radovanović, Methodology of developing optimal BP-ANN model for the prediction of cutting force in turning using early stopping method, Facta universitatis series: Mechanical Engineering. 9 (1) (2011) 21 - 32.

[40] S. Mallar, R. M., T. Thyagarajan, Modelling of continuous stirred tank reactor using artificial intelligence techniques, Int. J. Simul. Model. 8 (3) (2009) 145-155. *http://dx.doi.org/10.2507/IJSIMM08(3)2.128*

[41] V. K. Devabhaktuni, M. C. E. Yagoub, Y. Fang, J. Xu, Q.-J. Zhang, Neural Networks for Microwave Modeling: Model Development Issues and Nonlinear Modeling Techniques, John Wiley & Sons, Inc. (2001) 4-21.

[42] F. Musharavati, A.S.M. Hamouda, Application of artificial neural networks for modelling correlations in age hardenable aluminium alloys, Journal of achievements in materials and manufacturing engineering, 41 (2010) 140-146.

[43] R. Herbrich, Learning Kernel Classifier: Theory and Algorithm, Massachusetts Institute of Technology, MIT Press Cambridge, MA, USA (2002).

[44] S. G. Anantwar, R. Shelke, Simplified Approach of ANN: Strengths and Weakness, International Journal of Engineering and Innovative Technology (IJEIT), 1 (2012) 73–77.

[45] S. Gunn, Support Vector Machines for Classification and Regression, ISIS Technical Report, university of Southampton, (1997) 1–42.

[46] A. J. Smola, B. Scholkopf, A tutorial on support vector regression, Statistics and Computing 14 (2004) 199–222. *http://dx.doi.org/10.1023/B:STCO.0000035301.49549.88*

[47] W. Wang, Z. Xu, A heuristic training for support vector regression, Neurocomputing 61 (2004) 259 – 275. *http://dx.doi.org/10.1016/j.neucom.2003.11.012*

[48] A. M. Deris, A. M. Zain, R. Sallehuddin, Overview of support vector machine in modeling machining performances, Procedia Engineering 24 (2011) 308–312. *http://dx.doi.org/10.1016/j.proeng.2011.11.2647*

[49] W.H. Chen, J.Y. Shih, Comparison of support-vector machines and back propagation neural networks in forecasting the six major Asian stock markets, Int. J. Electronic Finance. 1 (1) (2006) 49-67. *http://dx.doi.org/10.1504/IJEF.2006.008837*

[50] A. Marti, Support vector machines, IEEE Intell. Syst. 13 (1998) 18–28. *http://dx.doi.org/10.1109/5254.708428*

[51] G. Nalbantov, J.F. Patrick, Groenen, J. C. Bioch, Support Vector Regression Basics, Medium Econometrische Toepassingen, 13 (2005) 16–19.

[52] K. S. Shin, T. S. Lee, H. Y. Kim, An application of support vector machines in bankruptcy prediction model, Expect Systems with Application. 28 (2005) 127-135.
http://dx.doi.org/10.1016/j.eswa.2004.08.009

[53] V. Sugumaran, G.R. Sabareesh, K. I. Ramachandran, Fault diagnostics of roller bearing using kernel based neighborhood score multi-class support vector machine, Expert Systems with Applications 34 (2008) 3090–3098.
http://dx.doi.org/10.1016/j.eswa.2007.06.029

[54] A.B. Gandhi, J.B. Joshi, Estimation of heat transfer coefficient in bubble column reactors using support vector regression, Chem. Eng. J. 160 (2010) 302-310.
http://dx.doi.org/10.1016/j.cej.2010.03.026

[55] V.N. Vapnik, S. Golowich, A.J. Smola, Support vector method for function approximation, regression estimation and signal processing. Adv. Neural Inform. Process. Syst. 9 (1996) 281-287.

[56] C. Cortes, V. Vapnik, Support-vector networks, Machine Learning. 20 (1995) 273–297.
http://dx.doi.org/10.1007/BF00994018

[57] S. Nandi, Y. Badhe, J. Lonari, U. Sridevi, B.S. Raoc, S. S. Tambe, B. D. Kulkarni, Hybrid process modeling and optimization strategies integrating neural networks/support vector regression and genetic algorithms: study of benzene isopropylation on Hbeta catalyst, Chem. Eng. J. 97 (2004) 115–129.
http://dx.doi.org/10.1016/S1385-8947(03)00150-5

[58] R. Samant, S. Rao, A study on Comparative Performance of SVM Classifier Models with Kernel Functions in Prediction of Hypertension, International Journal of Computer Science and Information Technologies. 4 (2013) 818-821.

[59] S. X. Zhi, Z. Jian, W. B. Biao, H. D. Wei, Support vector machine approach to mean particle size of rock fragmentation due to blench blasting prediction, Trans. Nonferrous Met. Soc. China. 22 (2012) 432-441.
http://dx.doi.org/10.1016/S1003-6326(11)61195-3

[60] C. Y. Lee, S. G. Chern, Application of a Support Vector Machine for Liquefaction Assessment, J. Mar. Sci. Technol. 21 (2013) 318-324.

[61] V. Cherkassky, Y. Ma, Practical selection of SVM parameters and noise estimation for SVM regression, Neural Networks. 17 (2004) 113–126.
http://dx.doi.org/10.1016/S0893-6080(03)00169-2

[62] M. Dundar, C. Nuhoglu, Y. Nuhoglu, Biosorption of Cu(II) ions onto the litter of natural trembling poplar forest, J. Hazard. Mater. 151 (2008) 86–95. *http://dx.doi.org/10.1016/j.jhazmat.2007.05.055*

[63] C. C. Chang, C. J. Lin, LIBSVM: A Library for Support Vector Machines, Department of Computer Science, National Taiwan University, Taipei, Taiwan, (2001).

[64] A.K. Giri, R.K. Patel, S.S. Mahapatra, Artificial neural networks (ANN) approach for modelling of arsenic (III) biosorption from aqueous solution by living cells of Bacillus cereus biomass, Chemical Engineering Journal 178 (2011) 15– 25. *http://dx.doi.org/10.1016/j.cej.2011.09.111*

[65] Y. S. Ho, W. T. Chiu, C. S. Hsu, C. T. Huang, Sorption of lead ions from aqueous solution using tree fern as a sorbent, Hydrometallurgy. 73 (2004) 55–61. *http://dx.doi.org/10.1016/j.hydromet.2003.07.008*

[66] Y. S. Ho, Effect of pH on lead removal from water using tree fern as the sorbent, Bioresour. Technol. 96 (2005)1292–1296. *http://dx.doi.org/10.1016/j.biortech.2004.10.011*

[67] M. Massinaei, Estimation of metallurgical parameters of flotation process from froth visual features, Int. J. Min. & Geo-Eng. 49(1) (2015) 75-81.

[68] S. Karsoliya, Approximating Number of Hidden layer neurons in Multiple Hidden Layer BPNN Architecture, International Journal of Engineering Trends and Technology. 31 (6) (2012) 714-717.

[69] H. Merdun, Ö. Çinar, Artificial neural networks and regression techniques in modelling surface water quality, Environ. Prot. Eng. 36(2) (2010) 95-109.

[70] S. Suganthi, K. Murugesan, S. Raghavan, ANN Model of RF MEMS Lateral SPDT Switches for Millimeter Wave Applications, Journal of Microwaves, Optoelectronics and Electromagnetic Applications 11(1) (2012) 130-143. *http://dx.doi.org/10.1590/S2179-10742012000100011*

[71] A. Verikas, M. Bacauskiene, Using artificial neural networks for process and system modeling, Chemometr. Intell. Lab. 67 (2003) 187 – 191. *http://dx.doi.org/10.1016/S0169-7439(03)00093-5*

[72] M. M. Joseph, O. M. Callistus, A. I. Gabriel, Application of Artificial Neural Networks For Path Loss Prediction In Urban Macrocellular Environment, American Journal of Engineering Research (AJER) 3(2) (2014) 270-275.

[73] S. Aber, A.R. Amani-Ghadim, V. Mirzajani, Removal of Cr(VI) from polluted solutions by electrocoagulation: Modeling of experimental results using artificial neural networks, J. Hazard. Mater. 171 (2009) 484–490. *http://dx.doi.org/10.1016/j.jhazmat.2009.06.025*

CHAPTER 7

Polyacrylamide Grafted Alginic Acid Copolymer Based Programmable Adhesive

Rahul Rahul[1]*, Priti Rani[1], Dipti Thakur[1], Gautam Sen[1], Dhaneshwar Mahto[2]

[1]Department of Chemistry, Birla Institute of Technology, Mesra, Ranchi – 835215, Jharkhand, India

[2]Department of Mechanical Engineering, Birla Institute of Technology, Mesra, Ranchi – 835215, Jharkhand, India

*rahulsharma17@gmail.com

Abstract

In this study, aqueous microwave assisted methodology was utilized to synthesize alginic acid grafted polyacrylamide. The synthesis was optimized by varying acrylamide and ceric ammonium nitrate concentrations in order to get an optimum failure shear load. The structural, thermal and physicochemical changes of alginic acid and its graft copolymers were ascertained with the aid of scanning electron microscopy (SEM), Fourier transform infrared spectroscopy (FTIR), thermogravimetric analysis (TGA), intrinsic viscosity measurement, number average molecular weight determination and elemental analysis (C, H, N & O). The applicability of the synthesized graft copolymer as programmable adhesive has been studied by fabricating a single lap joint of two wooden blocks using an aqueous paste of the graft copolymer as adhesive. After curing, the breaking point of the joint was determined using a universal testing machine (UTM).

Keywords

Alginic Acid, Adhesive, Microwave Assisted Synthesis, Graft Copolymer

Contents

1. Introduction

Alginic acid is a naturally occurring, hydrophilic polysaccharide produced by various species of brown seaweed (Phaeophyceae) [1,2] as well as bacteria (Pseudomonasaeruginosa and Azetobactervinilandii) [3]. It is a linear, anionic block copolymer and its structural framework consists of β-D-mannuronic acid and α-L-guluronic acid units [4,5]. Alginic acid has a broad range of applications in agricultural, pharmaceutical and biomedical fields [6,7]. However, application of alginic acid derivatives as adhesives has not been fully investigated, to the best of our knowledge.

Adhesion is the phenomenon which involves the binding together of two materials. This fascinating field has come a long way since man first realized that the blood caused his hair to cling together quite glutinously [8]. The past decade has resulted in a number of industrial applications for adhesives. In this regard, aromatic polyimides, asphalt, cellulose nitrate and soya bean glue have been of significant importance [9-11]. A wide range of tunable properties like thermal stability, excellent mechanical properties, radiation resistance etc., has made these materials an excellent candidate to be utilized as adhesives.

The polymeric adhesives offer considerable advantage over the conventional methods because of their ability to join dissimilar materials with superior product formation, such as honeycomb sandwich panels [12,13]. Polymer based materials have been used in a number of applications for centuries and play important roles in everyday life [14-17] ranging from household applications, automobiles, railway and military, microelectronics etc. [18-20]. Many renewable materials have been developed that are eco-friendly and have been used to prepare polymer composite materials that possess superior mechanical and physicochemical properties, suited to a wide range of applications [21-26].

The present investigation involves aqueous microwave assisted synthesis [27,28] of polyacrylamide (PAM) chains onto the alginic acid (Alg) framework, resulting in the formation of polyacrylamide grafted alginic acid (Alg-g-PAM). The synthetic protocol involves the use of ceric ammonium nitrate (CAN) as the free radical initiator and optimization in terms of failure shear load by varying the acrylamide and CAN concentrations. The applicability of the synthesized material as adhesive has been studied by fabricating a single lap joint by sticking two wooden blocks using an aqueous paste of the graft copolymer. After curing, the breaking point of the joint was determined using a Universal Testing Machine (UTM).

2. Experimental

2.1 Materials and methods

2.1.1 Materials

Alginic acid was procured from CDH, New Delhi, India. Acrylamide and ceric ammonium nitrate was obtained from E. Merck, Mumbai, India. Acetone was purchased from S.D. Fine Chemicals, Mumbai, India. All the chemicals were used as received, without further purification.

2.2 Synthesis of graft copolymer

2.2.1 Microwave assisted synthesis of polyacrylamide grafted alginic acid (Alg-g-PAM)

Alginic acid (1 g) was dissolved in 40 mL of distilled water and to this was added a desired amount of acrylamide and ceric ammonium nitrate. The reaction mixture was transferred to the microwave reactor (Catalyst^{TM} system CATA 4 R) and irradiated with microwave radiation (700 W) until gelling set in, keeping the irradiation temperature at 70°C. After completion of microwave irradiation, the gel like mass generated in the reaction vessel was cooled and poured in excess of acetone. The precipitated grafted polymer was collected and dried in a hot air oven, pulverized and sieved. The synthetic details of various grades of the grafted product have been shown in Table 1and the mechanism has been illustrated in Scheme 1.

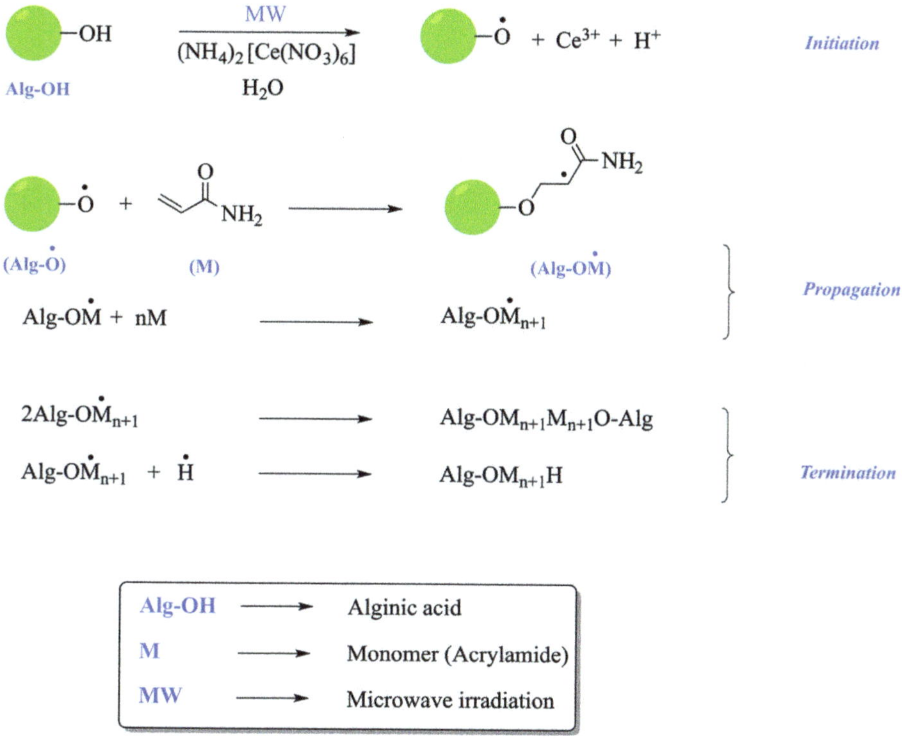

Scheme 1: Schematic representation of mechanism for microwave assisted synthesis of Alg-g-PAM.

Table 1: Synthesis details of Alg-g-PAM grades and their failure shear load.

Polymer Grades	Wt. of alginic acid (g)	Wt.of acrylamide (g)	Wt.of CAN (g)	Percentage grafting	Number average molecular weight (M_n)	Failure Shear load (kg)
Alg-g-PAM 1	1	5	0.1	280	680	100
Alg-g-PAM 2	1	5	0.2	486	875	290
Alg-g-PAM 3	1	5	0.3	598	1350	390
Alg-g-PAM 4	1	5	0.4	548	920	380
Alg-g-PAM 5	1	**10**	**0.3**	**1009**	**1900**	**520**
Alg-g-PAM 6	1	7.5	0.3	752	1875	440
Alg-g-PAM 7	1	15	0.3	1512		80
Alginic acid (Alg)	-	-	-	-		-

2.2.2 Purification of the graft copolymer by solvent extraction method

Any occluded polyacrylamide, formed by competing homopolymer formation reaction was removed from the synthesized graft copolymer by solvent extraction with acetone [29].

2.3 Characterization

2.3.1 Intrinsic viscosity measurement

Viscosity measurements of the polymer solutions were carried out with an Ubbelohde viscometer (Constant: 0.003899) at 25°C. The viscosities were measured in aqueous solutions at neutral pH. The procedure followed was the so called capillary flow method as described in earlier studies [30,31]. The intrinsic viscosities and the relation between intrinsic viscosity and percentage grafting of different grades of the graft copolymer have been illustrated in Table 1 and Fig. 1, respectively.

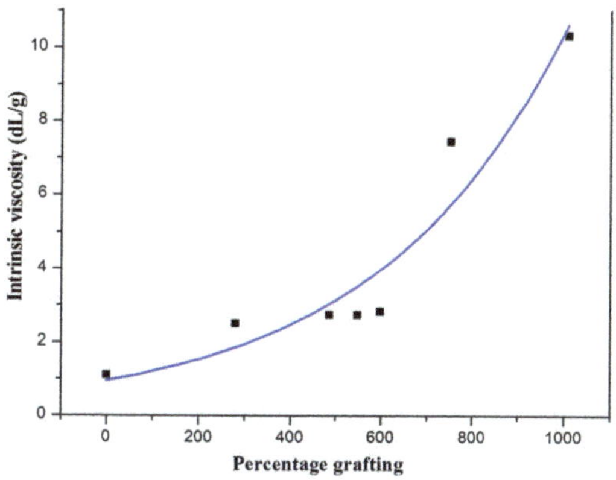

Figure 1: Intrinsic viscosity versus percentage grafting plot for Alg-g-PAM

2.3.2 Determination of number average molecular weight

The number average molecular weights of alginic acid (Alg) and various grades of Alg-g-PAM were determined in aqueous medium by Osmometry (Model: 3320, Osmometer, A^+ Adv. Instruments, INC) in Table 1.

2.3.3 Elemental analysis

Elemental analysis of alginic acid and the best grade of the synthesized graft copolymer (Alg-g-PAM 5) were undertaken with an Elemental Analyzer (Model: Vario EL III, Elementar, Germany). The estimation of the elements: carbon, hydrogen and nitrogen were undertaken. The results have been summarized in Table 2.

Table 2: Elemental analysis of Alg, PAM and best grade of Alg-g-PAM.

Polymer grades	%C	%N	%H	%O
Alginic acid (Alg)	35.45	0.00	5.798	58.38
PAM	50.71	19.17	8.642	21.21
Alg-g-PAM 5	42.58	**15.45**	8.657	33.06

2.3.3 FTIR spectroscopy

The FTIR spectra of alginic acid (Fig. 2a) and of Alg-g-PAM 5 (Fig. 2b) were recorded in solid state by KBr pellet method using a FTIR spectrophotometer (Model: IR-Prestige 21, Shimadzu Corporation, Japan) between 400 and 4000 cm^{-1}.

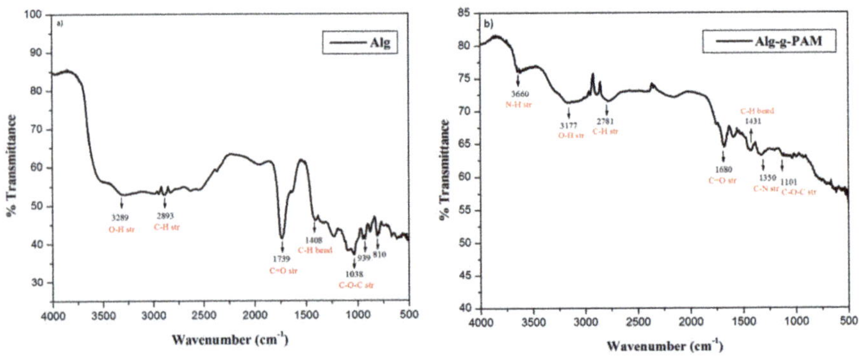

Figure 2: FTIR spectra of a) alginic acid (Alg) and b) Alg-g-PAM 5

2.3.4 Scanning electron microscopy

Surface morphology of patterns alginic acid (Fig. 3a) and Alg-g-PAM 5 (Fig. 3b) were obtained by scanning electron microscopy (SEM) in powdered form (Model: JSM-6390LV, Jeol, Japan).

Figure 3: SEM of a) alginic acid (Alg) and b) Alg-g-PAM 5

2.3.5 Thermo gravimetric analysis

The thermo gravimetric analysis (TGA) studies of Alg and that of the Alg-g-PAM 5 were carried out with TGA instrument (Model: DTG-60, Shimzadu, Japan) under inert atmospheric condition in the temperature range of 25 to 800°C. The temperature rate was maintained at 5°C /min. The TGA curves have been reported in (Fig. 4a-b).

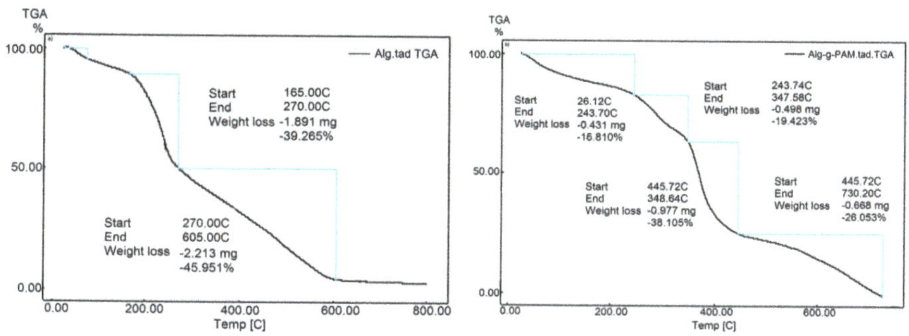

Figure 4: TGA of a) alginic acid (Alg) and b) Alg-g-PAM 5

2.4 Adhesive study

The adhesive properties of various synthesized grades of alginic acid and that of a commercial adhesive were examined by universal testing machine. Adhesive lap joint test experiments were conducted on wooden blocks joined in single lap, using aqueous paste of various materials to be tested. The specimens were left for 2 days, under 0.5 kg pressure at room temperature for curing and were tested in UTM (Alfred J. Amsler & Co., Schaffhouse, Switzerland, range 0-25,000 kg). Failure shear load of various grades of Alg-g-PAM was correlated with their respective percentage grafting, acrylamide and ceric ammonium nitrate concentration and was compared with that of a commercial adhesive. This has been illustrated in Fig. 5a-d.

3. Results and discussion

3.1 Synthesis of Alg-g-PAM by microwave assisted method

Alg-g-PAM has been synthesized by the microwave assisted method in aqueous medium. Various grades of the graft copolymer were synthesized by varying the concentration CAN and acrylamide. In each case, the microwave irradiation of the reaction mixture was continued until it sets into a viscous gel like mass (or up to 3 min if no gelling). The microwave irradiation cut off temperature was maintained at 70°C throughout the process. The synthetic details have been tabulated in Table 1. The optimized grade has been determined through failure shear load of wooden blocks joined by using the graft copolymer grade as adhesive (in adhesive lap joint test experiment) as discussed earlier. The synthetic approach involved optimization with respect to CAN, keeping the acrylamide concentration constant (i.e., Alg-g-PAM 1, Alg-g-PAM 2, Alg-g-PAM 3 and Alg-g-PAM 4) followed by optimization with respect to acrylamide, keeping the CAN concentration as optimized before (i.e., Alg-g-PAM 5, Alg-g-PAM 6 and Alg-g-PAM 7). From Table 1, it is apparent that the optimum percentage grafting and the failure shear load of wooden block were obtained at acrylamide concentration of 10 g and CAN concentration of 0.3 g in the reaction mixture (~50 mL), when the microwave power is maintained at 700 W. The proposed mechanism of microwave assisted grafting has been depicted in Scheme 1, outlined above.

3.2 Characterization

3.2.1 Estimation and interpretation of intrinsic viscosity

The intrinsic viscosity was evaluated for alginic acid and for all the grafted grades, as shown in Table 1. Intrinsic viscosity is practically the hydrodynamic volume of the macromolecule in the solvent solution. As evident from Fig. 1, intrinsic viscosities of all the grades of Alg-g-PAM is greater than that of alginic acid. This is due to the increase in hydrodynamic volume resulting from grafting of the PAM chains on the main polymer backbone (alginic acid). The grafted PAM chains are supposed to increase hydrodynamic volume either by uncoiling of the protein chain through steric hindrance to intermolecular bonding or by contributing their own hydrodynamic volume.

Further, this is in good agreement with Mark–Houwink–Sakurada relationship (intrinsic viscosity $\eta = KM^{\alpha}$where K and α are constants, both related to stiffness of the polymer chains), that the increase in intrinsic viscosity as a result of increase in molecular weight is due to the grafted PAM chains.

3.2.2 Determination of the number average molecular weight (M_n)

Molecular weights of alginic acid and various grades of Alg-g-PAM are listed in Table 1. The number average molecular weight (M_n) of various grades of Alg-g-PAM were greater than that of alginic acid. This illustrates that M_n increases with the increase of percentage grafting.

3.2.3 Elemental analysis

The results of elemental analysis (C, H, N & O) for Alg, PAM and that of best grade Alg-g-PAM 5 are presented in Table 2. The presence of nitrogen in Alg-g-PAM 5 is a clear indication of the grafting of PAM chains onto the polysaccharide backbone, as intended (the nitrogen in Alg-g-PAM were contributed by the grafted PAM chains, by virtue of their amide groups).

3.2.4 FTIR spectroscopy

As evident from (Fig. 2a), alginic acid has a broad O-H stretching peak centered at 3289 cm^{-1} and a peak at 2893 cm^{-1} due to C-H stretching. An intense peak at 1739 cm^{-1} is attributed to the stretching vibration of carbonyl group of carboxylic acid. The peaks at 1408 cm^{-1}, 1038and 939 cm^{-1} are due to C-H bending and C-O-C stretching respectively. In case of Alg-g-PAM 5 (Fig. 2b), there is a clear change in the intensity of O-H stretching peak and a frequency shift from 3289 cm^{-1} to 3177 cm^{-1}. The grafting of PAM is further confirmed by characteristic absorption band at 3660 cm^{-1} (N-Hstretch) and a peak at 1350 cm^{-1}(C-N stretch).

3.2.5 Scanning electron microscopy (SEM) analysis

The SEM micrographs of alginic acid and Alg-g-PAM 5 are shown in (Fig. 3a, Fig. 3b). Grafting of PAM chains have resulted in clear and distinct change in surface morphology with granular pattern of alginic acid being transformed to a flaky structure.

3.2.6 Thermo gravimetric analysis (TGA)

The TGA curves of alginic acid (Fig. 4a) essentially involved three distinct zones of weight loss. The first zone of weight loss (25°C–75°C) is due to the presence of traces of moisture. The second zone (165°C–270°C) is attributed to the degradation of carboxylate groups and consequent loss of carbon dioxide. The third zone of weight loss (270°C–605°C) was due to degradation of the alginic acid backbone (probably due to degradation of C-O-C bonds). In case of Alg-g-PAM 5 (Fig. 4b), in addition to the above mentioned three zones, there is a fourth zone of weight loss (445°C–730°C), as a result of degradation of PAM chains grafted on alginate backbone.

3.3 Adhesive study

3.3.1 Adhesive study and grade optimization by failure shear load

The adhesion activity of the graft copolymer is due to entanglement of PAM chains of neighboring grafted alginic macromolecules with the irregularities of the adhering surface and can be ascribed mainly to the concentration of ceric ammonium nitrate (CAN) and acrylamide (AM) in the reaction mixture.

Effect of CAN concentration

Failure shear load increased on increasing the CAN concentration (0.1 g to 0.3 g) in the reaction mixture (60 mL). This may be due to the fact that as the concentration of initiator increases, it initiates more grafting sites on the backbone of alginic acid, resulting in large number of polyacrylamide chains. These chains are in a better position to intertwine with the neighboring chains/adhering surface. But as the concentration increases further (0.3 g to 0.4 g), there is a decrease in the failure shear load. This may be because beyond certain CAN concentration in the reaction mixture, the high number of grafting sites results in considerable reduction in the length of the PAM chains which fail to entangle.

Effect of monomer concentration

With the increase in concentration of acrylamide (5 to 10 g), failure shear load increases continuously and achieves a maximum value at 10 g. This is then followed by a decrease in the failure shear load value. This behavior can be explained by the fact that with increase in acrylamide concentration in the reaction mixture, there is lengthening of polyacrylamide chains. These longer PAM chains cause a greater extent of entanglement with neighboring PAM chains, leading to higher adhesive strength. However, beyond certain monomer concentration, the PAM chain becomes too long and hence fragile. The synthesis of the graft copolymer has been optimized (in terms of breaking strength during lap joint experiment), first in terms of CAN concentration and then acrylamide concentration.

The optimized grade (Alg-g-PAM 5) is obtained when CAN and acrylamide concentrations in the reaction mixture were 0.3 and 10 g, respectively Table 1. The effects of percentage grafting, acrylamide concentration and CAN concentration on breaking strength of single lap joint experiment are depicted in (Figs. 5a-c). After performing the adhesion test, it was found that Alg-g-PAM 5 showed higher and better failure shear load (i.e., 520 kg) compared to the commercial adhesive (Fig. 5d).

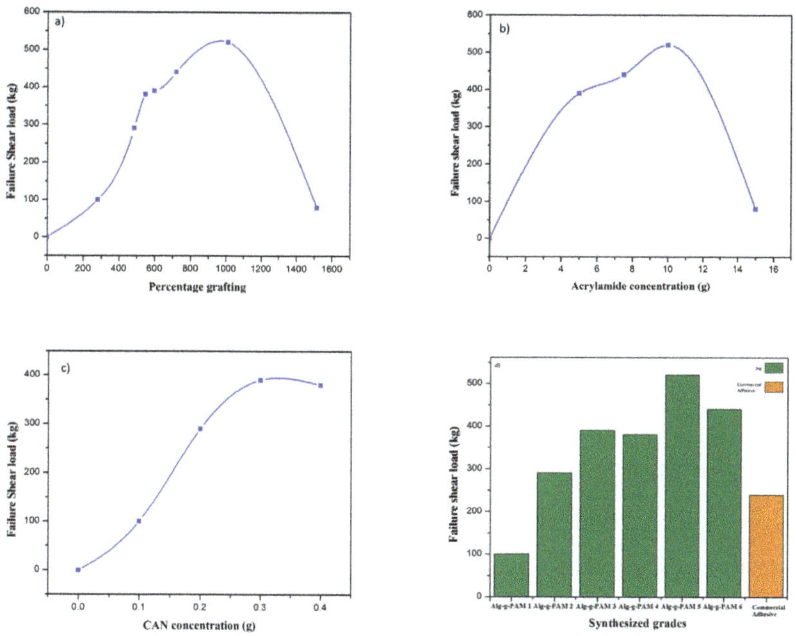

Figure 5: a) Effect of % grafting on failure shear load b) Effect of acrylamide concentration on failure shear load c)Effect of CAN concentration on failure shear load d) Comparison of failure shear load of various grades of Alg-g-PAM and commercial adhesive

Cellulose is the main wood component, so the dangling chains of graft copolymer get entangled with the hydroxyl functionality of the cellulose, creating covalent linkages between the adhesive and the wooden blocks [32]. The interactions that hold the adhesive and the adherents together may also receive contributions from mechanical interlocking, diffusion or electrostatic mechanisms. Since the graft copolymer is a programmable adhesive, it can be tuned according to our requirement and suitability. This adhesive is intended for applications where the joined surfaces are not likely to come in contact with water during the product lifetime and will be programmed to disassemble automatically once disposed.

4. Conclusion

The present work reports the synthesis of graft copolymers of alginic acid and their application as a water soluble adhesive. Acrylamide units have been grafted onto the alginic acid backbone by the microwave assisted technique and the synthesized product (Alg-g-PAM) has been confirmed by standard physicochemical characterization techniques. The graft copolymer has been optimized at CAN concentration of 0.3 g and acrylamide concentration of 10 g in the reaction mixture, leading to a failure shear load of 520 kg (shear stress breaking). This alginate based synthetic graft copolymer (Alg-g-PAM 5) shows superior adhesion property than that of the commercial adhesive.

Acknowledgement

The support and assistance from Central Instrumentation Facility, Birla institute of Technology (BIT), Mesra is earnestly acknowledged.

References

[1] P. Malkaj, E. Pierri, E. Dalas, The crystallization of hydroxyapatite in the presence of sodium alginate, Journal of Materials Science: Materials in medicine, 16 (2005) 733–737.
http://dx.doi.org/10.1007/s10856-005-2610-9

[2] H. Ertesvag, S. Valla, Biosynthesis and applications of alginates, Polymer Degradation Stability, 59 (1998) 85–91.
http://dx.doi.org/10.1016/S0141-3910(97)00179-1

[3] F.A. Johnson, D.Q.M. Craig, A.D. Mercer, Characterization of the block structure and molecular weight of sodium alginates, Journal of Pharmacy and Pharmacology, 49 (1997) 639–643.
http://dx.doi.org/10.1111/j.2042-7158.1997.tb06085.x

[4] F.G. Fischer, H. Dörfel, Polyuronic acids in brown algae, Hoppe Seylers Z Physiological Chemistry, 302 (1955) 186–203. *http://dx.doi.org/10.1515/bchm2.1955.302.1-2.186*

[5] W.L. Nelson, L.H. Cretcher, The alginic acid from Macrocystis Pyrifera, Journal of American Chemical Society, 51 (1929) 1914–1922. *http://dx.doi.org/10.1021/ja01381a045*

[6] W.H. McNeely, D.J. Pettit, Algin, in: R.L. Whistler (Ed.), Industrial gums, Academic Press, New York, 1973, pp. 49–81. *http://dx.doi.org/10.1016/B978-0-12-746252-3.50009-5*

[7] K.I. Darget, O. Smidsrod, G. Skjàk-Bræk, Alginate from algae, in: A. Steinbüchel, & S.K. Rhee (Eds.), Polysaccharide and Polyamides in the food Industry, Wiley-VCH Verlag GmbH & Co. KGaA, Germany, 2005, pp. 1–30. *http://dx.doi.org/10.1002/3527600035.bpol6008*

[8] R.L. Patrick, Treatise on adhesion and adhesives (Vol. 2), Marcel Dekker, New York, 1969.

[9] D. Wilson, H.D. Stenzenberger, P.M. Hergenrother, Polyimides, Chapman and Hall, New York, 1990. *http://dx.doi.org/10.1007/978-94-010-9661-4*

[10] G. Goud, R.N. Rao, The effect of alkali treatment on dielectric properties of Roystonearegia/epoxy composites, International Journal of Polymer Analysis and Characterization, 16 (2011) 239–250. *http://dx.doi.org/10.1080/1023666X.2011.570039*

[11] B. Siroka, J. Siroky, T. Bechtold, Application of ATR-FT-IR single-fiber analysis for the identification of a foreign polymer in textile matrix, International Journal of Polymer Analysis and Characterization, 16 (2011) 259–268. *http://dx.doi.org/10.1080/1023666X.2011.570066*

[12] A.J. Kinloch, Adhesion and adhesives, Science and technology, Chapman and Hall, London, 1987. *http://dx.doi.org/10.1007/978-94-015-7764-9*

[13] A.J. Kinloch, Adhesive in engineering, Proceedings of institution of mechanical engineering, Journal of Aerospace Engineering, 211 (1997) 307–335.

[14] Z.Y. Zhang, Z.Z. Dong, C.F. Xiao, Estimation of ethyl cellulose layer thickness of coated polyethylene oxide particles using differential scanning calorimetry, International Journal of Polymer Analysis and Characterization, 18 (2013) 25–29. *http://dx.doi.org/10.1080/1023666X.2012.719115*

[15] R.P.V Atluri, K.M. Rao, A.V.S.K.S. Gupta, Experimental investigation of mechanical properties of golden cane fiber-reinforced polyester composites, International Journal of Polymer Analysis and Characterization, 18 (2013)30–39. *http://dx.doi.org/10.1080/1023666X.2013.745679*

[16] A. Chandramohan, M. Alagar, Preparation and characterization of cyclohexyl moiety toughened POSS-reinforced epoxy nanocomposites, International Journal of Polymer Analysis and Characterization, 18 (2013) 73–81. *http://dx.doi.org/10.1080/1023666X.2013.747253*

[17] R. Rahul, S. Kumar, U. Jha, G. Sen, Cationic inulin: A plant based natural biopolymer for algal biomass harvesting, International Journal of Biological Macromolecules, 72 (2015) 868–874. *http://dx.doi.org/10.1016/j.ijbiomac.2014.09.039*

[18] J. Zhao, C. Xiao, N. Xu, Surface and physical mechanical properties of polypropylene/poly(butyl methacrylate-co-hydroxyethylmethacrylate) blend fiber, International Journal of Polymer Analysis and Characterization, 17 (2012) 557–567. *http://dx.doi.org/10.1080/1023666X.2012.704557*

[19] S. Mallakpour, A. Zadehnazari, New organosoluble thermally stable and nanostructuredpoly(amide-imide)s with dopamine pendant groups: Microwave-assisted synthesis and characterization, International Journal of Polymer Analysis and Characterization, 17 (2012) 408–416. *http://dx.doi.org/10.1080/1023666X.2012.669646*

[20] N. Niimura, Structural study of a Japanese lacquer film with thermogravimetry-linked scan mass spectrometry, International Journal of Polymer Analysis and Characterization, 17 (2012) 540–546. *http://dx.doi.org/10.1080/1023666X.2012.704560*

[21] A.S Singha, V.K. Thakur, I.K. Mehta, A. Shama, A.J. Khanna, R.K. Rana, A.K. Rana, Surface modified Hibiscus sabdariffa fibers: Physicochemical, thermal and morphological properties evaluation, International Journal of Polymer Analysis and Characterization, 14 (2009) 695–711. *http://dx.doi.org/10.1080/10236660903325518*

[22] X. Ren, M. Soucek, Soya based coatings and adhesives, in: R.P. Brentin, (Ed.), Soy-based chemicals and materials, ACS, 2014, pp. 207–254. *http://dx.doi.org/10.1021/bk-2014-1178.ch010*

[24] K. Ramanaiah, A.V. Ratnaprasad, K. Hema, C. Reddy, Thermal and mechanical properties of sansevieria green fiber reinforcement, International Journal of Polymer Analysis and Characterization, 16 (2011) 602–608. *http://dx.doi.org/10.1080/1023666X.2011.622358*

[25] V.K. Thakur, A.S. Singha, M.K. Thakur, Fabrication and physico-chemical properties of high-performance pine needles/green polymer composites, International Journal of Polymer Analysis and Characterization, 62 (2013) 226–230. *http://dx.doi.org/10.1080/00914037.2011.641694*

[26] S. Berdous, S.N. Amroun, M. Saidi, M. Bendaoud, Trapping effect on electrical behavior of polyester film: I. Unipolar injection, International Journal of Polymer Analysis and Characterization, 16 (2011) 416–430. *http://dx.doi.org/10.1080/1023666X.2011.596648*

[27] D. Dallinger, C.O. Kappe, Microwave-assisted synthesis in water as solvent. Chemical Review, 107 (2007) 2563–2591. *http://dx.doi.org/10.1021/cr0509410*

[28] V. Polshettiwar, R.S. Varma, Aqueous microwave chemistry: Aclean and green synthetic tool for rapid drug discovery, Chemistry Society Review, 37 (2008) 1546–1557. *http://dx.doi.org/10.1039/b716534j*

[29] Z. Yang, B. Yuan, X. Huang, J. Zhou, J. Cai, H. Yang, Evaluation of the flocculation performance of carboxymethylchitosan-g-polyacrylamide, A novel amphoteric chemically bonded composite flocculant, Water Research, 46 (2012) 107–114. *http://dx.doi.org/10.1016/j.watres.2011.10.024*

[30] A.K. Sarkar, N.R. Mandre, A.B. Panda, S. Pal, Amylopectin grafted with poly(acrylic acid): Development and application of a high performance flocculant, Carbohydrate Polymers, 95 (2013) 753–759. *http://dx.doi.org/10.1016/j.carbpol.2013.03.025*

[31] R. Rahul, U. Jha, G. Sen, S. Mishra, A novel polymeric flocculant based on polyacrylamide grafted inulin: Aqueous microwave assisted synthesis, Carbohydrate Polymers, 99 (2014) 11–21. *http://dx.doi.org/10.1016/j.carbpol.2013.07.082*

[32] C.R. Frihart, Adhesive bonding and performance testing of bonded wood products, Journal of ASTM International, 2 (2005) 1–7. *http://dx.doi.org/10.1520/JAI12952*

CHAPTER 8

Organic-TiO_2 Nano Composite Membranes

Dhiraj Sud

Department of Chemistry, Sant Longowal Institute of Engineering and Technology, Longowal, Sangrur-148106, Punjab, India

E-mail: suddhiraj@yahoomail.com

Abstract

The synergy between organic and inorganic materials offers the possibility for the development of new generation innovative organic–inorganic composites with diversified applications such as catalysis, photocatalysis, separation technology, energy storage and drug delivery. Recently there has been an upsurge in the development of polymeric matrix nanocomposite membranes for gas separation, protein exchange membrane fuel cell, direct methanol fuel cell, solvent nanofiltration and water treatment. Membrane separation technology has been proven to be an economical method of water purification due to its high separation efficiency, easy operation and low energy consumption. The incorporation of well-known titanium oxide (TiO_2) photocatalyst into polymeric matrices is known to affect membrane structure and properties. The potential of TiO_2 to impart much desired characteristics such as antimicrobial, super-hydrophilic and photocatalytic properties to the membranes could lead to finding aj energy and cost effective solution for water treatment. The present chapter discusses the scientific and technical advancements in the development of organic polymer based TiO_2 nanocomposite membranes for water treatment.

Keywords

Titanium oxide, Photocatalyst, Nanocomposite Membrane, Organic Pollutants

Contents

1. Introduction

The synergy between organic and inorganic materials give the potential for the development of new generation innovative organic–inorganic composites with diversified applications in various area such as catalysis, photocatalysis, separation technology, energy storage and drug delivery [1-5].The Inorganic materials impart enhanced mechanical strength, thermal and chemical stability into realm of highly flexible organics along with the improved electrical properties, ionic conductivity, permeability and selectivity, chemical or biochemical activity. Inorganic materials such as metal/metal oxide particles- Al_2O_3, Fe_3O_4, ZrO_2, TiO_2, carbon -based nanomaterials, zeolites and silica [6-12] hold the potential to create synergism or multifunctional effects when integrated into the polymer system. The incorporation of a wide range of metal/metal oxide nanoparticles into the polymeric membranes has been the focus of numerous investigations for last decade [13]. Polymeric matrix nanocomposite membranes have been used for gas separation, protein exchange membrane fuel cell, direct methanol fuel cell, solvent nanofiltration and water treatment [14]. Membrane separation technology has been proven to be an economical method for water purification due to its high separation efficiency, easy operation and low energy consumption [15]. Organic polymeric membranes are one of the most commonly employed membranes for water purification and treatment. However, the development of polymeric nanocomposite membranes offers new traits to the membrane separation technology. The metal oxides in particular TiO_2 inclusion has resulted in the addition of remarkable attributes such as antimicrobial, super-hydrophilic and photocatalytic properties to the membranes which are much desired characteristics for finding an energy and cost effective solutions for water purification and treatment. The present chapter discusses the scientific and technical advancements in the development of organic polymer based TiO_2 nanocomposite membranes for water treatment.

2. Classification of Membranes

A membrane may be regarded as a *physical wall which acts as the barrier between two different phases* and restricts the transfer of different chemical species. Membranes can be classified by using various criteria based on their *pore size, homogeneity in composition and chemical nature*. Membranes on the *basis of pore size* have been classified as i) microfiltration membranes (MF) having pore size greater than 100A^o, ii) ultrafiltration membranes(UF) with pore size (0.01-0.1μm) and iii) reverse osmosis membranes with pore size in the range 1-10A^o. Newly emerged membranes are nanofiltration membranes which are also termed as loose RO membranes with pore size equal or less than 10A^o. MF membranes separate or reject particles of about 0.05-0.1 μm and efficiently remove protozoa, bacteria, suspended solids and reduce the turbidity of water [15]. UF are usually used to remove viruses, emulsified oils, metal hydroxides, colloids, proteins and other low molecular weight materials from water [16]. Nanofiltration membranes have properties in between UF and RO and offer significant advantages such as lower osmotic pressure difference, higher permeate flux, higher retention, low operational cost [17]. On the basis of *homogeneity in composition*, membranes have been classified as i) Isotropic- having uniform composition throughout the membrane and ii) anisotropic membranes - consists of layers which vary in structure and composition.

Depending on the *basis of materials employed* for synthesis of membranes, they are classified as organic and inorganic membranes. Recently hybrid membranes which incorporate organic and inorganic materials are being developed.

Organic membranes mostly contain membranes formed from the synthetic or natural polymers which are organic in nature. Synthetic polymers are synthesized by homo or copolymerization of monomers to obtain diverse structures such as linear chain, branched chains and crosslinked polymers. Polymeric membranes are extensively used for waste water treatment; however, the major limitations of the polymeric membranes are membrane fouling, permeability and selectivity.

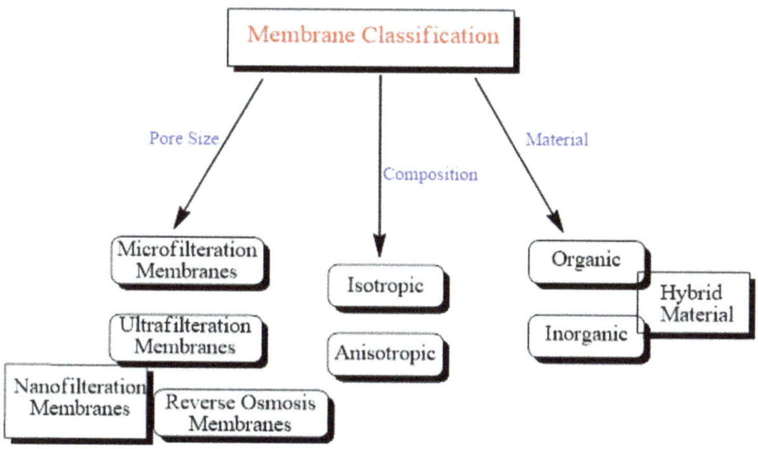

Fig. 1: Classification of Membranes

Inorganic membranes are formed mainly from inorganic materials which commonly include metallic membranes prepared from sintered metal powder such as W, Pd etc. deposited on porous substrates, ceramic membranes, membranes consisting of oxides, nitrides and carbides of metals and zeolite membranes. The inorganic membranes are characterized by high thermal and chemical stability, improved mechanical strength and structural hardness, inertness towards microbial degradation and relative ease of cleaning operations after use. The prominent disadvantages of the inorganic membranes are i) high capital cost ii) predominant surface poisoning iii) sensitivity to temperature variation and membrane cracking. Zeolite membranes have low gas flux and hence thick layers are required to prevent cracks and holes.

Hybrid membranes- Recently there is an upsurge in the development of hybrid/composite membranes to desalinate seawater/brackish water or remove heavy metals, hardness, organic micropollutants such as pesticides, disinfection by- products (DBPs), endocrine disrupting compounds (EDCs) and pharmaceutically active compounds. Thin film composite nano filtration membranes have *excellent permeability and selectivity* over asymmetric nanofiltration membranes. Some examples of inorganic/TiO_2 composite membranes are spiolite/TiO_2 composite, zeolite/TiO_2 membranes and organic/TiO_2 composite membranes PES/TiO_2, PVDF/TiO_2 etc.

3. Organic Polymer /Titanium Dioxide Nanocomposite Membranes

3.1 Polymeric Membranes

They are the most widely used membranes for waste water treatment mainly due to their inherent properties such as pore formation, higher flexibility, easy installation and relatively low cost as compared to inorganic membranes. Table 1 lists the examples of commonly used organic polymers as membranes.

3.2 Unique Properties of TiO_2

TiO_2 is extensively used in diversified applications such as photocatalysis, batteries, photovoltaics, UV blocker, the coating material in textile, paints, cosmetics, paper and biomedical sciences. The unique properties of TiO_2 such as biologically and chemically inertness, resistant to chemical and photo-corrosion, non-toxicity, hydrophilicity and low cost make it a preferred material, particularly in waste water treatment. TiO_2 is more stable than other available photocatalysts in ambient conditions and can even be recycled [18,19]. TiO_2 has a band gap energy of 3.2 eV for anatase phase and 3.0 eV for rutile phase and can use natural UV light due to appropriate “energetic separation between its valence and conduction bands”.

Table 1: Examples of common organic polymers used as membranes

Organic Polymer		Properties	Structure
Polyamide	PA	Thermal mechanical and Chemical Resistance	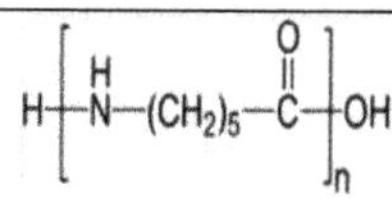
Poly acrylonitrile	PAN	Thermal mechanical and Chemical Resistance	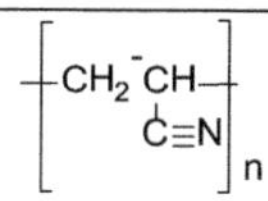
Polyvinylidene fluoride	PVDF	Chemical and thermal stability Hydrophilicity	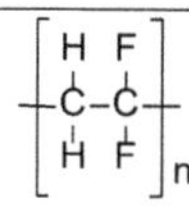
Poly propylene	PP	Thermal and mechanical stability. Less chemical stability	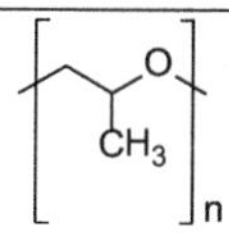
Poly acrylamide	PAA	Highly flexible with lower chemical and mechanical stability	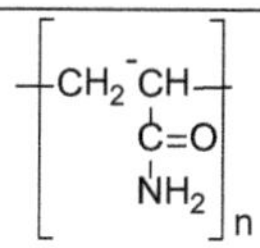
Polyether sulfone	PES	Easy availability and processing Thermal and mechanical stability	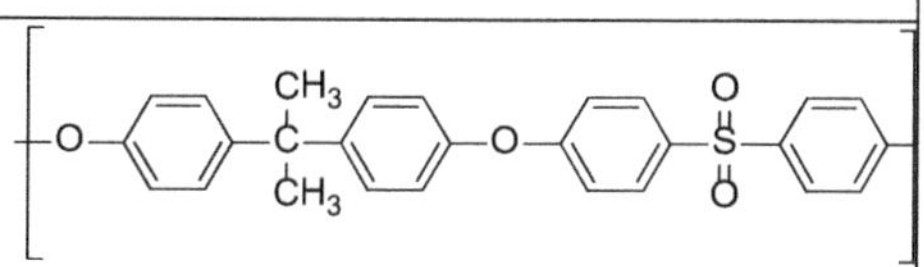

3.3 Structure and Physical Properties of TiO_2

Titanium exists naturally in mineral forms like rutile, ilmenite, leucoxene, anatase, brookite and perovskite. All three prominent forms (brookite, rutile and anatase) contain TiO_6 octahedra, having six oxygen (O_2^-) atoms coordinated to titanium (Ti^{4+}) atom (Figure 2). The anatase form has a tetragonal structure formed by the sharing of corner (vertice) of octahedra. In rutile, the TiO_6 octahedra share two edges to form linear chains with tetragonal structure and are slightly distorted. In brookite, both edges and corners are shared to give an orthorhombic structure [20-23]. The distortion of anatase TiO_6 is slightly larger than the rutile phase and the density of anatase (3.79) is less as compared to the rutile phase (4.23).

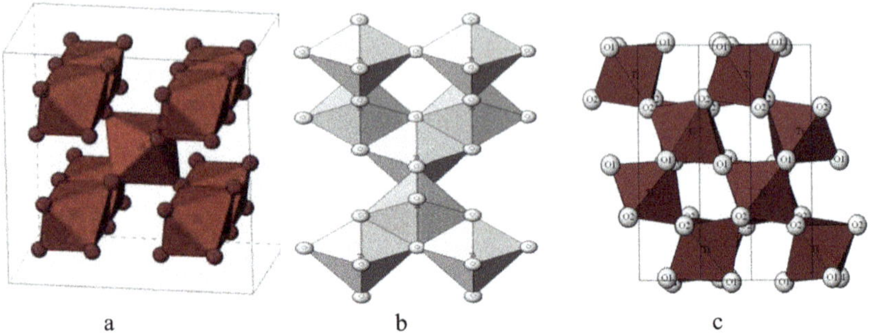

Fig 2: Crystalline structures of titanium dioxide (a)-rutile, (b)-anatase, (c)-brookite [20]

The rutile phase is the thermodynamically favorable at all temperatures and pressures because of its linear chains than the zig zag anatase form which is more stable at lower temperature [24]. Phase transformation (anatase to rutile) occurs at higher temperature 700–1000° C depending on the crystallite size and impurity content. Band gap energies for anatase and rutile have been estimated and found to be 3.2 and 3.0 eV, respectively. The valence band of TiO_2 consists of 2p orbital of hybridized oxygen with 3d orbital of titanium, while the conduction band consists of 3d orbital of titanium. The smaller band gap of rutile, however, implies that a higher fraction of the solar radiation can induce band gap excitation (threefold increase in photocurrents). The structural, physical and chemical properties of anatase and rutile forms are given in Table 2.

The titanium dioxide has the desirable properties of being chemically stable, readily available and widely used as photocatalyst as it absorbs the UV light because of its larger band gap (3.2 eV)[25]. The major disadvantage of the wide band gap is that it hinders the practical application of titanium dioxide in visible light.

Table 2: Polymeric nanocomposite membranes prepared by different methods

Membrane material	Method	Performance	Reference
PVDF/SPES-TiO_2	Immersion precipitation	Lower flux decline	[33]
TFC-TiO_2	Interfacial polymerization	Self cleaning	[34]
PVDF/TiO_2	Phase inversion	Flux reduction and antifouling	[35]
PVDF/TiO_2	Hydrothermal	Antifouling	[36]
PVDF/TiO_2	Atomic layer	Enhance	

	deposition	permeability and selectivity	[37]
PVDF/TiO_2	Sol gel	Membrane distillation performance	[38]
PES/ TiO_2	Sol gel	Improvement in hydrophilicity	[39]
PES/ TiO_2	Chemical and mechanical modification	Improvement pure flux	[40]
PVDF/SPES/TiO_2	Casting solution	Lower flux and antifouling	[41]
PES/ TiO_2	Casting solution	Thermal and mechanical stability Flux enhancement	[42]
PES/ TiO_2	Precipitation	Lower flux decline	[43]
PAI/TiO_2	Interfacial polymerization	High and stable rejection value	[44]
PAI/TiO_2	Phase inversion method	Removal of humic substances	[45]
PAN/TiO_2	Self crosslinked	High thermal and solvent resistance Excellent separation	[46]
PVDF/PAA/TiO_2	Sol-gel grafting	Dye degradation and antifouling properties	[47]
PSf- TIO_2	Coating	Self Cleaning	[48]

3.4 TiO_2 as Photocatalyst for Degradation of Pollutants

(i) Photo excitation of the Semiconductor

The excitation of TiO_2 with the photon of energy higher or equal to the threshold energy resulted in the promotion of an electron from the valence band (VB) to the conduction band (CB) with the simultaneous generation of a hole in the VB (h_{VB}^+). The energy level at the bottom of the CB is actually the reduction potential of photoelectrons and the energy level at the top of the VB determines the oxidizing ability of photo holes, and reflects the capability to promote reduction and oxidation reactions[26-28].

$$TiO_2 + h\vartheta\,(UV) \rightarrow TiO_2\,(e_{CB}^- + h_{VB}^+) \qquad (1)$$

where $h\vartheta$ is photon energy required to excite the semiconductor electron from the valence band (VB) region to the conduction band (CB) region.

(ii) Formation of hydroxyl radical

Both the e_{CB}^- and the h_{VB}^+ can migrate to the catalyst surface and undergo redox reaction with the other species present on the surface. The photogenerated holes can oxidize the (i) organic molecule or (ii) react with OH^- or H_2O to generate $OH^\bullet$ radicals.

$$TiO_2\,(h_{VB}^+) + H_2O \rightarrow TiO_2 + H^+ + OH^\bullet \qquad (2)$$

$$TiO_2\,(h_{VB}^+) + OH^- \rightarrow TiO_2 + OH^\bullet \qquad (3)$$

The photogenerated electrons can also react with oxygen to produce superoxide radical anion ($O_2^{-\bullet}$). The generated $O_2^{-\bullet}$ can also lead to the additional formation of $OH^\bullet$ radicals [29-30] .

$$TiO_2\,(e_{CB}^-) + O_2 \rightarrow TiO_2 + O_2^{\bullet-} \qquad (4)$$

$$O_2^{\bullet-} + H^+ \rightarrow HO_2^\bullet \qquad (5)$$

$$HO_2^\bullet + H^+ + TiO_2\,(e_{CB}^-) \rightarrow H_2O_2 + TiO_2 \qquad (6)$$

$$H_2O_2 \rightarrow 2^\bullet OH \qquad (7)$$

(iii) Recombination reaction

The recombination of e_{CB}^- and the h_{VB}^+ on the surface or in the bulk of the catalyst in a few nanoseconds results in the dissipation of energy as heat and this process must be prevented to have favorable photocatalytic reactions [31-32].

$$e_{CB}^- + h_{VB}^+ \rightarrow e_{CB} + \text{heat} \qquad (8)$$

(iv) Degradation of organic pollutants

Organic pollutants adsorbed on TiO_2 catalyst surface undergo oxidation by an $^\bullet OH$ radical, which is being a very strong oxidizing agent (standard redox potential +2.8V) can oxidize organic pollutant to the mineralized products **(Figure 3)**.

$$\text{Organic pollutant} + h_{VB}^+ \rightarrow \text{oxidation products} \qquad (9)$$

$$\text{Organic pollutant} + e_{CB}^- \rightarrow \text{reduction products} \qquad (10)$$

Organic pollutant + $OH^{\bullet} \rightarrow$ degradation products (11)

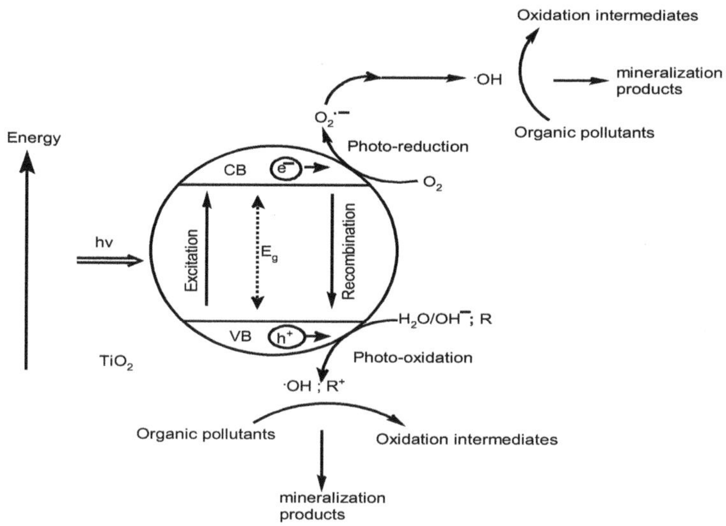

Fig. 3: Schematic diagram of the photocatalytic process over TiO_2 [adapted from 30]

4. Polymer-TiO_2 Nanocomposite Membranes

Polymer –TiO2 nanocomposite membranes have been prepared by using organic polymers namely polyamide (PA), polyvinylidene fluoride(PVDF), polyether sulfone (PES), polyacrylonitrile (PAN) etc. TiO_2 –Polymer composite membranes been prepared mainly by two different approaches to introduce TiO_2 nanomaterials as additive in the polymer matrix i.e. i) phase inversion method- incorporation of the nanomaterials to the casting solution and ii) immersion of porous membrane into aqueous solution of nanoparticles which leads to coating of nanoparticles on membrane surface via chemical bonding or self-assembly(34,35). The different characteristics with desired purposes can be induced through the interaction between the nonmaterial's surface and polymeric chains and/or solvent during the membrane fabrication process.

4.2 Preparation and Mode of Bonding of TiO_2 to the Polymeric Nanocomposite Membranes

PA-TiO_2 membranes have been prepared either by surface deposition or interfacial polymerization. Surface deposition method involves the dipping of the PA membranes in

TiO_2 colloidal solution for a certain time to allow the deposit ion of TiO_2 nanoparticles on the membrane surface. However, the method is vulnerable to easier detachment of TiO2 from the PA matrix, leading to the decrease of TiO_2 concentration in the membranes. The prime goal of the synthesis of nanocomposite membranes is to have stable and sufficient amount of TiO_2 incorporated into the polymeric membrane. The modification of the membrane surface via the functionalization of membrane to provide more hydrophilic groups for surface deposition of TiO_2 particles has been carried out.

A novel polyamide (PA) nanocomposite membrane containing high loading of TiO_2 nanoparticles was synthesized via *in-situ* the interfacial polymerization (44). Interfacial reaction occurred from the aqueous phase of m-phenyl diamine (MPD) and the organic phase of trimesoyl chloride (TMC) in which TiO_2 nanoparticles were homogeneously dispersed (Fig. 4).

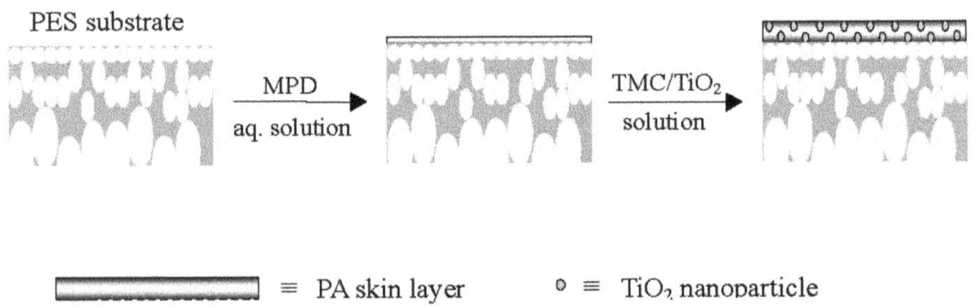

Fig 4 Procedure for preparation of PA-TiO_2 nanocomposite membrane [44]

PES ultrafiltration membrane was prepared via the phase inversion method [42] and TiO_2 nanoparticles were deposited on the neat PES membrane with an area of 38.5 cm^2 .The initial structure of the UF membrane changed due to the self-assembly of TiO_2 nanoparticles and resulted in the formation of the ridge-and-valley structure on the membrane surface which increases the roughness of the membrane surface. The mechanism of self-assembly between TiO_2 nanoparticle onto polymer involves the bonding through ether group and sulfone group present on PES membrane (Fig. 5). Ti^{4+} get attached to the electronegative oxygen atoms present in the ether and sulfone groups of the PES (41)

I

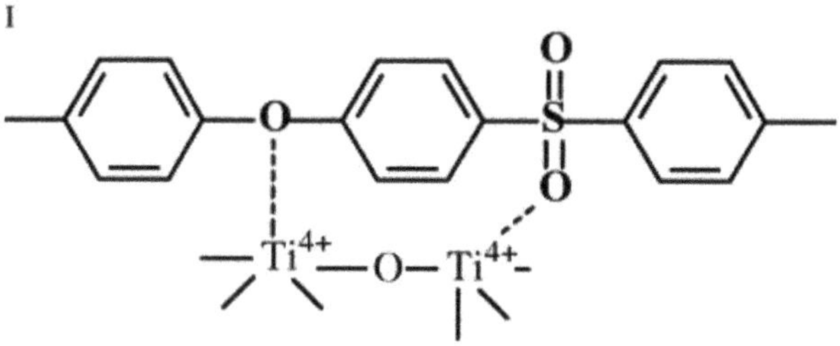

I

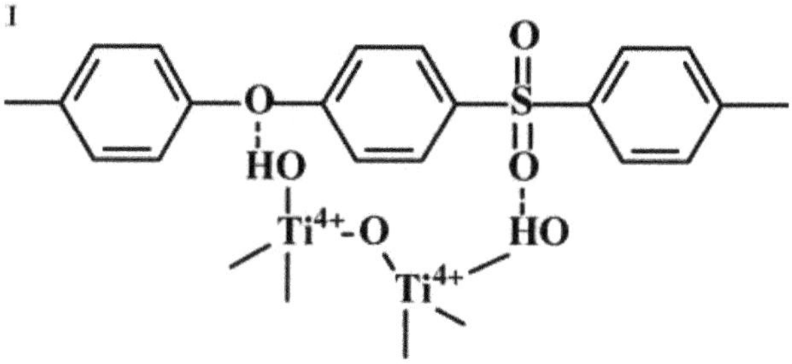

Fig.5: Mechanism of self-assembly of TiO_2 nanoparticles (I) coordination of sulfone group and ether group to Ti^{4+} (II) H - bond between sulfone group and ether bond and surface hydroxyl group of TiO_2 (adapted from 41)

The self-assembly mechanism of TiO_2 on SMA/PVDF blend membranes was proposed [42]. The carboxyl groups present on the blend SMA/PVDF membranes are attached with TiO_2 particles through covalent bonding. The amount of TiO_2 bound on membrane surface increases with increase in the number of –COOH group formed by the hydrolysis of maleic anhydride groups in case of blend membranes.

TiO_2 can be bound with two oxygen atoms of COOH group by a bidentate coordination to Ti^{4+} cation (scheme 1). Alternatively, H-bond is formed between carbonyl group and the surface hydroxyl group of TiO_2 (scheme2) [49-50](Fig 6).

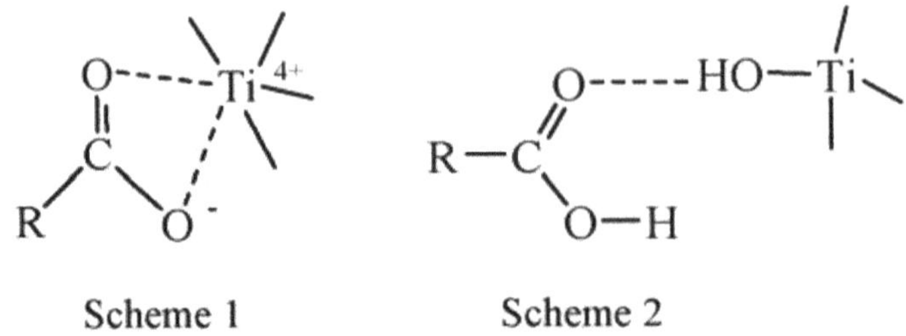

Fig. 6: Mechanism of self assembly of TiO_2 nanoparticles [34]

Mansourpanah et al. prepared the modified PES/PI blend membrane with increased hydrophilicity using reagents such as ethylenediamine, polyethylene glycol triazine and diethanolamine (43). The diethanolamine DEA modified –OH functionalized PES/PI blend membranes were used to prepare hybrid membranes via dipping membranes in TiO2 colloidal suspension [51]

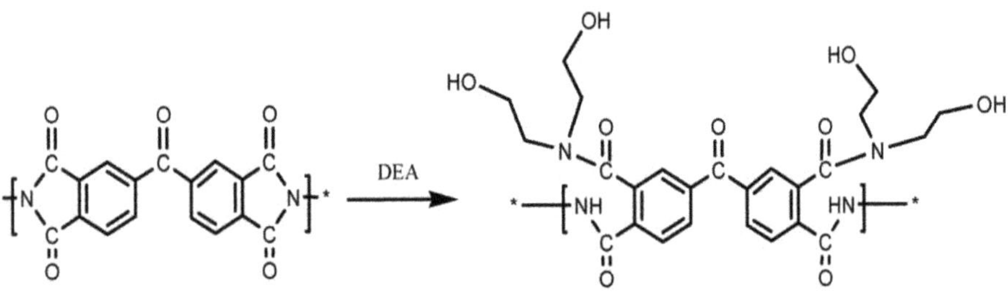

Fig. 7 Membrane modification by DEA

The proposed binding of TiO_2 on to the DEA modified membrane surface of PI has been represented in Fig.8. The –OH groups present on the TiO_2 surface get attached to Ti^{4+} through oxygen atoms of the OH group. In case of blended DEA membranes, the coordination can take place through the oxygen atom of –OH group, CO group of P I and nitrogen groups of the polymer. Further, the roughness of the membrane and surface hydrophilicity increases and DEA modified membrane surface causes additional and superior accumulation of TiO_2 on the membrane surface. Moreover, the hydroxyl groups present on the membrane surface results in a uniform deposition and more accumulation of TiO_2.

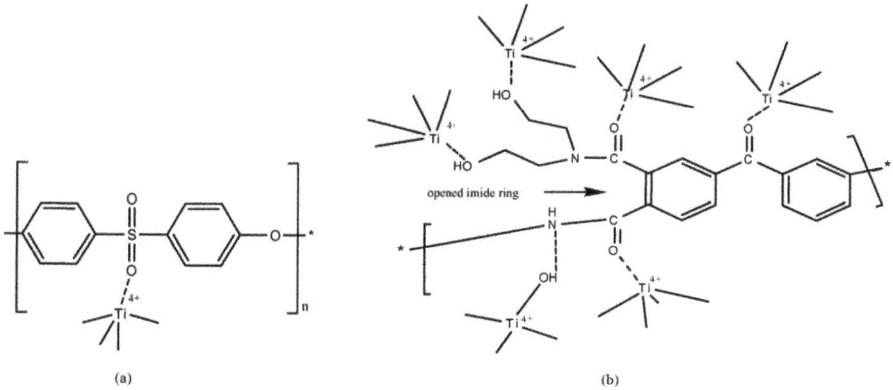

Fig. 8: Mechanism of self - assembly of TiO2 nanoparticles a) on PES surface ii) DEA modified surface of PI [51]

Thin film composite nanofilteration (TFC-NF) membranes are mainly prepared by depositing of TiO2 particles on the thin film membrane surface. Thin film membrane consisting of a very thin polyamide active surface layer and prepared by dry-wet phase inversion technique on the porous support layer may be polysulfone (PSf) or polyether sulfone (PES) [52]. The porous support layer is formed via the interfacial polymerization technique.

Thin film composite (TFC) membranes are made by interfacial polymerization of m-phenylenediamine (MPD) in the aquous phase (2wt%) and trimesoyl chloride (TMC) in the organic phase on the nonwoven fabric reinforced polysulphone supports [34]. The hybrid TFC membrane is devised by the self-assembly between TiO2 nanoparticle and polymer with COOH groups. The schematic diagram of hybrid membrane is given in Fig.9.

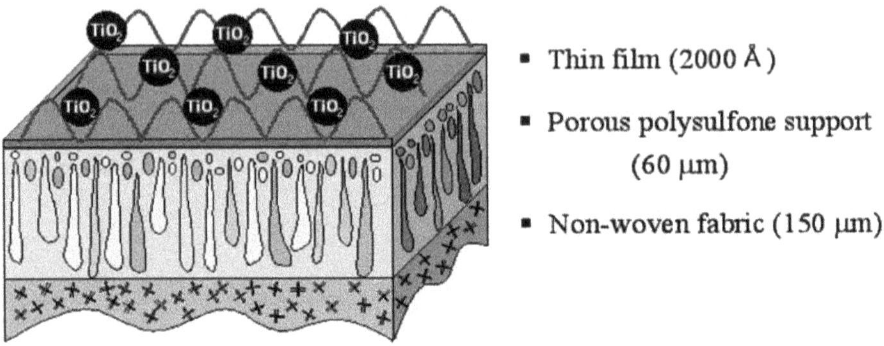

Fig. 9: Schematic diagram of hybrid membrane [34]

5. Membrane Fouling

Fouling is a general problem associated with any membrane separation but in nanofilteration it involves the complex interactions at a nanoscale level. Fouling of membrane results in the irreversible deposition of suspended or dissolved substances on the surface of the membrane or within its pores which lead to flux decline. Substances such as inorganics, suspended solids, colloids, metal oxides, organics and biological species present in feed water can foul the membranes. Depending on the nature of molecules present in the separation medium, the major types of membrane fouling (Fig. 10) includes organic fouling, inorganic fouling, colloidal fouling and biofouling [53-54]. Organic fouling occurs due to the adsorptive interactions between the organic solutes and membrane materials. The octanol-water partition coefficient and adsorption characteristics of the membrane such as nature of residual charge on surface i.e. electrostatic forces and hydrophobicity are the main factors responsible for deciding the flux rate [55,56].

Inorganic fouling results from precipitation of salts on the membrane surface due to the presence of scalants i.e. Scale producing substances such as $CaCO_3$, $BaSO_4$, $SrSO_4$, gypsum, silica etc in the feed water. Nanofiltration membranes can retain inorganic ions which cause, an increase of concentration at the membrane surface [57].

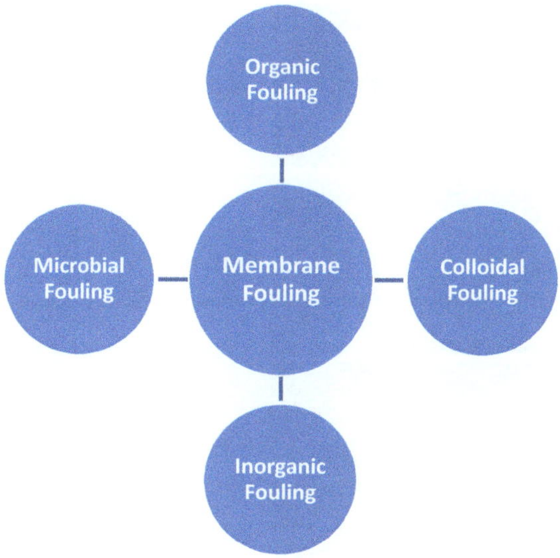

Fig. 10: Type of membrane fouling.

Collodial fouling as the name suggests happens due to the accumulation of colloidal particles of clay or silica on the membrane surface or by their penetration within membrane pores. The preferential movement of colloids into the valleys results in valley clogging. Collodial fouling is related to the membrane **roughness besides surface hydrophobicity and permeability** [58-60**].**

Biofouling is a dynamic process which involves the formation and growth of biofilm at the membrane due to the presence of biologically active microorganisms mainly bacteria and fungi [61]. Biofouling may reduce the water flux and could even prevent the water movement completely. Fouling mechanism of porous membrane is depicted in **Fig 11**. Fouling due to pore blocking or cake layer formation results in degradation of membrane performance.

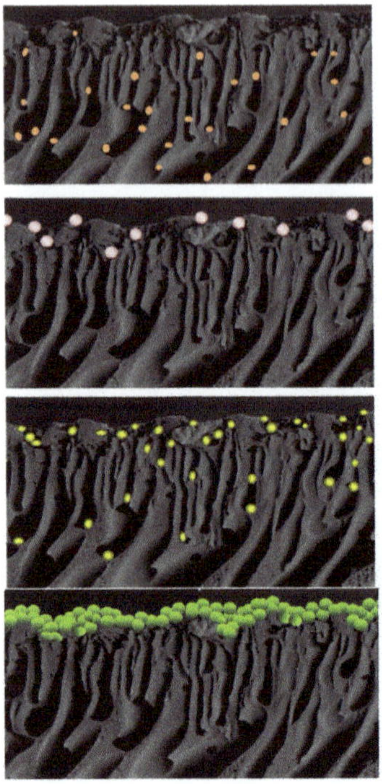

Fig. 11: Fouling behavior of membrane.

5.2 Quantification of Fouling

Fouling can be quantified by measuring the resistance offered to the membrane during the filtration [62]. The resistance is due to the formation of a cake or gel layer on the membrane surface. The flux (J), through the cake or gel layer formed and the membrane may be described by Darcy's law:

$$J = \Delta P/\mu\ R_t \tag{12}$$

where P is the trans membrane pressure(driving force), μ is viscosity of the fluid and R_t is the sum of the resistances .

Rt is the total filtration resistance (m^{-1}) and it is assumed to be the sum of intrinsic membrane resistance (Rm), cake resistance by cake layer formed on membrane surface

(*R*c) and fouling resistance caused by pore plugging and irreversible adsorption of foulants on the membrane pore wall or surface (*R*f).

$$Rt = Rm + Rc + Rf \tag{13}$$

Membrane resistance (Rm) can be estimated from the initial water flux:

$$Rm = \Delta P/\mu \; Jwi \tag{14}$$

Fouling resistance (Rf) can be calculated from the water flux after washing with water:

$$Rf = (\Delta P/\mu \; Jww) - Rm \tag{15}$$

(Rc) can be calculated from the water flux after chemical cleaning

$$Rc = (\Delta P/\mu \; Jwc) - Rm - Rf \tag{16}$$

*J*wi the initial water flux and *J*ww is the final water flux after removing the cake layer. The cake layer was removed by flushing with tap water.

In order to evaluate the fouling resistant of membranes, **flux recovery (FR)** which indicates the recycling property of membrane was calculated as follows:

$$FR\,(\%) = Jww/Jwi \times 100 \tag{17}$$

*J*ww and *J*wi are the pure water flux of fouled and pure membrane respectively.

5.3 Implications of Membrane Fouling

Membrane fouling results in i) decreased membrane permeability thereby affecting membrane performance ii) reduced permeate quality iii) serious damage to the membrane **iv)** diminished production and excessive operational cost in terms of energy cost[63-65].

The classical approaches to control membrane fouling include the following:

- pretreatment of the feed solution
- periodic cleaning of the membranes –back flushing and chemical cleaning
- surface modification of the membrane
- optimization of module arrangement and process conditions

Currently ozonation or UV/H_2O_2 oxidation, adsorption and flocculation techniques are also employed [66, 67]. The most commonly used techniques such as back-flushing and chemical cleaning for removing foulant are not appropriate [68-70] as these involve a filtration process which require regular cleaning or back-flushing to regain membrane permeability. Moreover most of these processes are energy intensive. Surface modification of the polymeric membrane is the most sustainable solution to get fouling resistant membranes. The fouling behavior of membrane strongly depends upon the hydrophobicity of the membranes, pore size and morphology. The increase in membrane hydrophicity appears to be one of the viable approaches to improve the membrane resistance to fouling. Different methods investigated to impart hydrophilicity to the membrane surfaces are

- addition of hydrophilic polymers to the membrane casting solution/ polymeric blend membranes [71-72]
- the immobilization of polymers with hydrophilic segments by photo or plasma polymerization[73-75]
- binding of TiO_2 on to the polymeric membrane

Recently established method to improve the membrane antifouling properties is the incorporation of TiO_2 onto the polymeric membranes and numbers of researchers have tried to modify the polymeric membranes by blending or by treatment with other molecules to introduce polar functional moieties which offer, the sites for binding of TiO_2 as discussed in **section1.4.2.**

6. Effect of TiO_2 Loading on Membrane Structure and Physico-Chemical Properties of TiO_2-Polymer Nanocomposite Membranes

For separation operation, the surface pore size and porosity of the membrane are important parameters for determining the permeability and rejection, especially for MF and UF membranes. Other physico-chemical properties are surface hydrophilicity and roughness which mainly control the membrane fouling characteristics. The membrane performance under various environmental conditions depends on chemical, thermal and mechanical stabilities.

Most of the studies demonstrated the potential of incorporation of TiO_2 into polymeric matrices to affect membrane structure and properties. Hydrophilicity, permeability, flux rate, anti fouling property and remediation of various organic pollutants are some of the properties get improved with the addition of TiO_2 nanoparticles.

6.2 Effect of TiO_2 on Surface Hydrophilicity of Membranes

Surface hydrophilicity is one of the most important factors which control the fouling properties of the filtration membrane and much attention has been diverted towards this aspect [76, 77]. The relative hydrophilicity of the polymer membrane surface is measured by a contact angle which is inversely proportional to the hydrophilicity. Most of the studies reported in the literature reflected the decrease in contact angle in case of TiO_2 based hybrid membranes. In general, increasing the surface hydrophilicity of a membrane decreases the membrane fouling because the hydrogen bonding formation between the surface hydrophilic groups and water molecules creates a water layer on the membrane surface that could prevent foulant adsorption. The self-assembly of TiO_2 nanoparticles on the SMA/PVDF blend membrane gives these membranes an excellent hydrophilicity as the initial contact angle (85°) for SMA/PVDF blend membrane (M1-M3) decreases to 54.1° in case of SMA/PVDF /TiO_2 membrane M1(Ti), M2 (Ti), and M3(Ti) [Fig.12] .

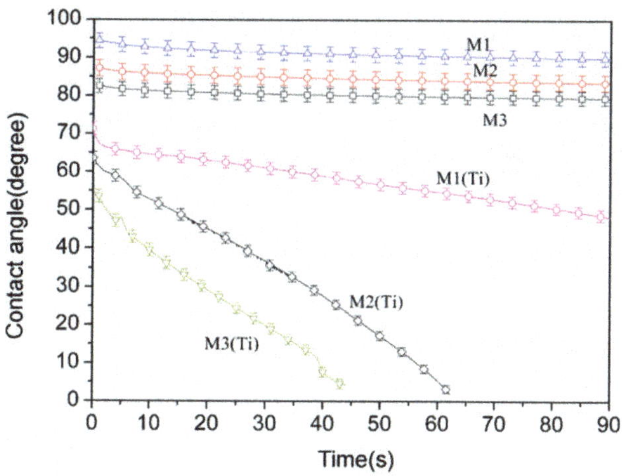

Fig 12: Time dependence of water contact angle of blend membranes and self assembly membranes with different ratio of SMA/PVDF (M1,M2 and M3) corresponds to SMA/PVDF blend membrane and M1(Ti), M2 (Ti), and M3(Ti) corresponds to self-assembly of TiO_2 nanoparticles on SMA/PVDF blend membrane [43]

The hydrophilicity of membrane surface, the pore size and wettability of internal pore channels are three main factors that influence the contact angle change. The change in contact angle has been explained on the basis of formation of three - dimensional tiny voids between nanoparticles due to aggregation of TiO_2 nanoparticles into snowflakes on the membrane surface [78-79]. Thus, when a water droplet is dribbled on the membrane surface, it would spread instantly due to the capillary effects of the three-dimensional tiny voids as well as the hydrophilic effect of anatase TiO_2 nanoparticles in nature [80]. As a result, the TiO_2 self-assembly membranes have smaller initial contact angle and faster contact angle decay rate than blend membranes. Further the self-assembly of TiO_2 nanoparticles on SMA/PVDF blend membrane provides excellent stability as a very small change in contact angle was observed even after one month of shaking of the M3(Ti) membrane in a water bath. Table 3 lists data of few studies carried out to examine the effect of TiO_2 loading on the hydrophilicity of the prepared nanocomposite membranes.

Table 3: Summary of different work carried out for the contact angle measurements by different workers

Hybrid membrane	Contact angle (°)	References
SMA/PVDF blend membrane (M3)	85	[43]
SMA/ PVDF /TiO_2membrane (M3Ti)	54.1	
Neat polyether sulphone	39.6	[52]
PES/TiO_2 composite filteration membrane	19.2	
Neat PES	66.2	[43]
Blend PES membrane	53.5(6wt% TiO_2)	

Similar results for the decrease in the contact angle (from 39.6 to 19.2°) of the membrane treated by colloidal TiO_2 solution to form TiO_2 self assembled polyethersulfone ultrafiltration membrane has been reported. The increase in the roughness of the membrane surface is due to the formation of ridge and valley structure. Generally, hydrophilicity increases with the roughness of the membrane surface [81]. The increase of hydroxyl content on the membrane surface introduces polarity and thus enhances the water permeability of the composite membrane. Although the increase in hydrophilicity can be well correlated with contact angle and indicates the higher affinity of the membrane to water due to the presence of hydrophilic sites. This could facilitate water solubilization and diffusion through the membrane, thus improve water permeability. However, the increase in hydrophilicity does not always raise the membrane permeability [43]. The permeability of TiO_2 –entrapped PES membrane did not increase as compared to neat PES membrane but rather decrease was noticed which may be due to plugging of the membrane pores by nanoparticles due to immersion precipitation. The membranes with large pore size and higher hydrophilicity exhibit higher pure water flux.

6.3 Mechanism of Super Hydrophilicity of Polymer-TiO_2 Nanocomposite Membranes

The mechanism of increase in hydrophilicity by addition of TiO2 particles has been explained on the basis of interaction of the polar –OH groups with water molecules through the van der Waals force and hydrogen bond. It is well established fact that charge separation i.e. formation a pair of holes and electrons on the surface of TiO_2 particles takes place on its irradiation by photons of energy equal to or greater than the band gap energy in ordinary conditions. The photo-generated electrons tend to reduce Ti (IV) cations to the Ti(III) state and the holes oxidize O^{2-} anions. During this process, the oxygen atoms are moved out and thereby generating oxygen vacancies on the surface. The water molecules in the environment can occupy these vacant sites and adsorbed (OH) groups are created on the surface which considerably increases the hydrophilicity of the surface. The mechanism of super hydrophilicity of the membranes is shown in (Fig. 13) [82-87].

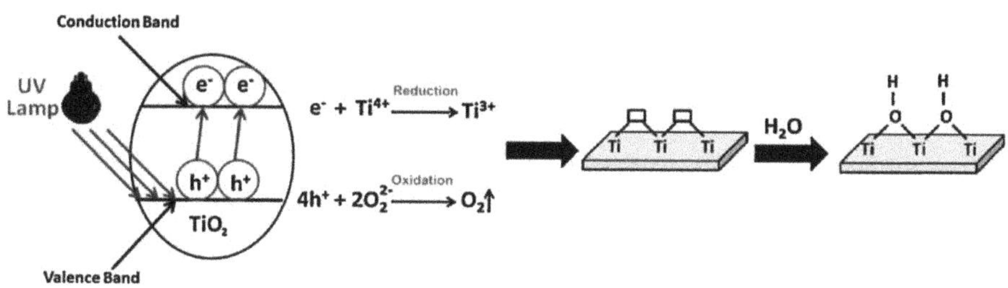

Fig. 13: Mechanism of superhydrophilicity[88]

Madeni et al. [88] also supported the mechanism of super hydrophilicity and further elaborated the theory. The irradiation by UV light and due to high hydrophilicity of the membrane, a layer of water is chemically adsorbed on the surface of the membrane. When this surface is further exposed to the water molecules by keeping in the water, more layers of water get adsorbed through the hydrogen bonds and van der Waals forces. This process leads to the formation of water layers on the surface that causes a higher level of hydrophilicity (Fig. 14) [88].

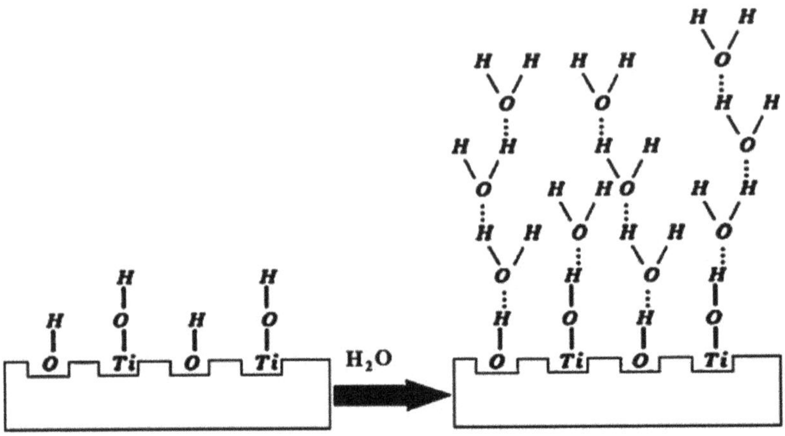

Fig. 14: Mechanism of increasing hydrophilicity by UV-irradiation [92]

The adsorbed layers of water, not only increase the hydrophilicity of the membrane surface but also helpful in the decontamination process by spreading over the surface. Due to the creation of super hydrophilicity, the membrane allows the passes much more water and the water contact angle of both TiO_2 (0.01 wt%) coated and UV radiated membranes with and without immersion in the water before usage, were measured as 5° and 12°, respectively. This decline in the water contact angle (high hydrophilicity) clearly indicates the creation of super hydrophilicity on the membrane surface.

6.4 Effect of TiO_2 Loading on the Other Characteristics Properties of the Membranes

Table 4 presents the data corresponding to change in the polyether sulfone membrane properties on addition of TiO_2 nanoparticles at different concentrations. The properties of PSf substrate are changed with respect to pure water flux, contact angle, overall porosity, pore size and S value changes with variation of TiO_2 loadings

Table 4 Changes in the PSf substrate properties upon addition of TiO_2 nanoparticles at different loadings [89].

Membranes	Pure water flux(L/m^2h) [a]	Contact angle (o_)	Overall porosity (%)	Average pore size (nm)	S value (mm)
Substrate (control)	162.5	71	71	54	0.98
Substrate 0.5	412.5	64	75	47	0.42
Substrate0.75	742.4	61	78	47	0.33
Substrate1	855.2	58	81	46	0.26

[a]Pure water flux of membrane was determined at operating pressure of 2.5 bar.

From the Table 4, it can be inferred that the average pore size decreased while overall porosity increased with increased TiO_2 loadings. In the case of FO and PRO membranes, internal concentration polymerization, ICP (occurring inside the porous support layer) lead to the significant reduction in the available osmotic driving force. ICP is controlled by the hydrophilicity and structural parameter, S.

Structural parameter(S) = thickness and x, tortuosity / porosity

The small S value is desirable to minimize ICP during the FO process. Thus the increase in porosity minimizes ICP in porous support layer which in turn lead to better water permeability. Similarly, the increase in the surface pore size for the TiO_2/PVDF hollow fiber membranes (HFMs) with increasing TiO_2 concentration initially but decrease with further increase inTiO_2 loading was reported [91]

Generally, the addition of hydrophilic substance in the dope solution leads to an accelerated solvent and non-solvent exchange (caused by thermodynamic immiscibility) and thus encourages formation of a more porous structure. However, further increase in the filler loading, however, the viscosity of the dope solution would increase, leading to a diffusion delay and resulted in the formation of the dense and thick with low porosity top layer and low degree of pore interconnectivity. Thus, an optimum dose of TiO_2 is required to create a positive impact.

6.5 Effect of TiO_2 Loadings on Pure Water Flux of Membranes

The pure water flux of substrate membranes (PSF, PSF0.30, PSF0.60 and PSF0.90 corresponding to TiO_2 loading from 0, 30, 60 and 0.90 wt%) was significantly improved

(Fig.14) with increase in TiO_2 loading, mainly due to improved membrane hydrophilicity. The enhancement in substrate hydrophicility can be inferred from decrease in contact angle value from 67° in the pure PSf substrate to 51° in the PSf substrate embedded with 0.90 wt% TiO_2 nanoparticle (Table 4).

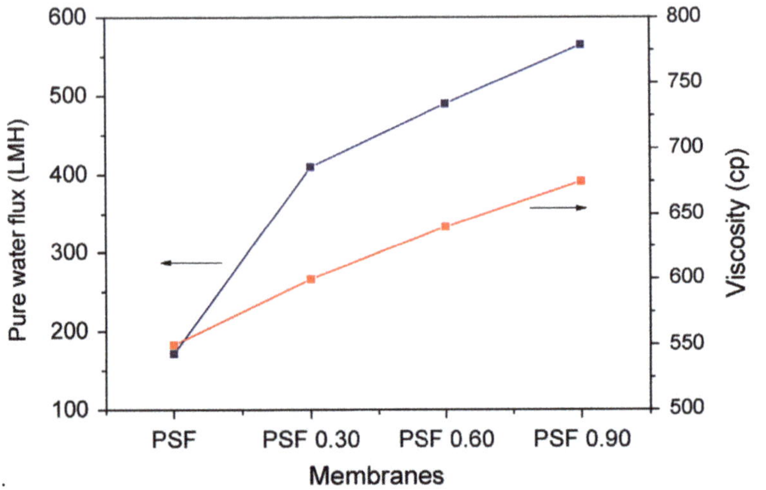

Fig 14: The different TiO_2 loaded membranes on the viscosity and the pure water flux (LMH=L/m²h)[89].

6.6 Effect of TiO2 on Fouling Resistance of Membrane

The hybrid nanocomposite membranes exhibit excellent fouling resistance on incorporation of TiO_2.The fouling mitigation effects of TiO_2 /polymer nanocomposite membrane was studied [91] quantitatively by measuring the filtration resistances of membranes Fig. 15.

The major fouling contributor i.e. cake layer resistance of the nanocomposite membrane was dramatically decreased as compared to that of the PES membrane due to the presence of nanoparticles, which increases the hydrophilicity of the polymeric membrane surfaces. The decrease in cake layer resistance strongly suggested that adsorbed foulants on the nanocomposite membranes could be more readily dislodged by shear force than those on a neat polymeric membrane and thereby the amount and the rate of membrane fouling decreased.

Similar results were observed for filtration resistances of neat and TiO_2 self-assembled nanocomposite membranes and their fouling mitigation effects during the filtration of MBR mixed liquor [91].The fouling mitigation effect increases with the amount of TiO_2 nanoparticles and the cake layer resistance was dramatically decreased[90]. TiO_2 nanoparticles can impart a fouling –resistant property to neat polymeric membranes used for filtration of membrane bioreactor (MBR) sludge [92].

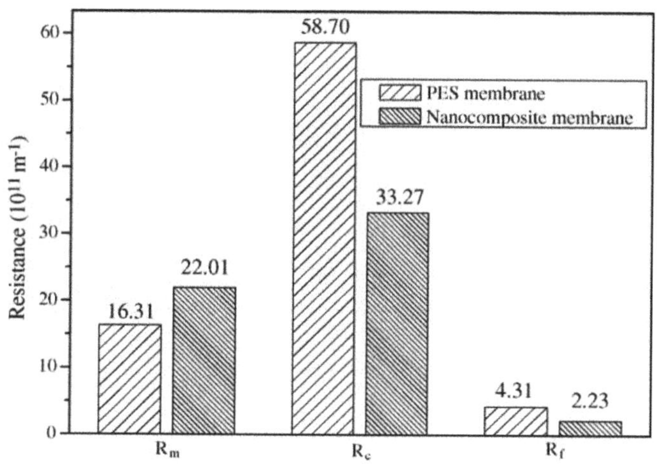

Fig. 15: Filtration resistances of PES and nanocomposite membranes [90].

The antifouling property of the membrane was improved by adding TiO_2 particles due to decrease in hydrophobic adsorption between milk protein and TiO2 entrapped PES membranes surface and deposited foultants were readily removed by cross flow filtration [43].

6.7 Effect on TiO2 on Salt Rejection

The effect of TiO2 concentration on the membrane performance in terms of flux and salt rejection was investigated and results are shown in Fig. 16. As evident from the figure, as TiO_2 concentration increased up to 5.0 wt%, the flux slightly decreased and the salt rejection ($MgSO_4$) increased. However, above 5.0 wt%, the flux abruptly increased and salt rejection decreased sharply. It has also been reported that the curing of nanocomposite membrane at 70° C for 5 min exhibited the high and stable rejection value of more than 95 % with respect to $MgSO_4$. Above this critical concentration, the mechanical strength of the membranes decreased significantly, producing easier peeling-off of the PA-TiO_2 layer from PES substrate after the filtration experiment. These results may be due to the fact that high TiO_2 concentrations interfere with the interfacial polymerization by TiO_2 leading to lower degree of polymerization of PA.

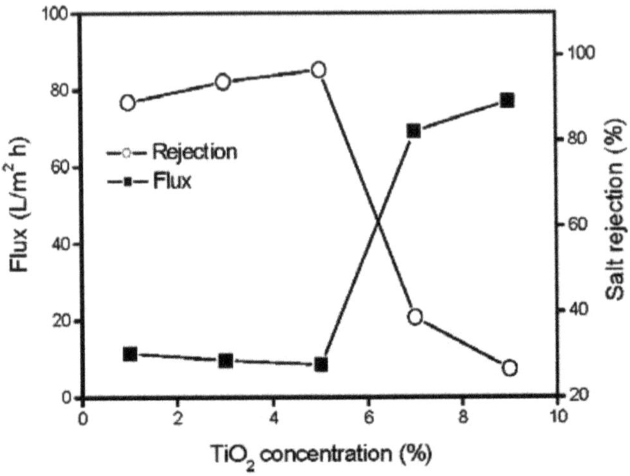

Fig. 16: Effect of TiO_2 concentration on the membrane performance of water flux and $MgSO_4$ rejection.

The improved hydrophilicity of the substrate together with the increased overall porosity upon addition of TiO_2 nanoparticles resulted in the increase of water flux and salt rejection. Similarly the TFC membrane exhibited 3.28 L/m^2 h water permeability and 97% salt rejection when it was tested with 20 mM NaCl aqueous solution at operating pressure of 2.5 bar. When 0.60 wt% TiO_2 nanoparticles were introduced into the PSf substrate, the water permeability was significantly increased and TFN 0.60 membrane achieved 6.57L/m^2 h, which was 100% higher than that of the control TFC membrane (3.27L/m^2h). However, the salt rejection of TFN 0.60 membrane was slightly decreased from 97.0 to 93.6% and further decreased drastically to 77.5% for TFN 0.90 membrane [93]. The lower degree of cross-linking formed in the PA selective layer due to the porous morphology and/ortheTiO$_2$ agglomeration on PSf substrate which led to salt rejection.

6.8 Effect of TiO_2 on Disinfection Performance and Biofouling

Polymeric –TiO2 nanocomposite membranes showed good disinfection performance and antibactericidal effect for E. coli and possibly can be used as anti-biofouling TFC membranes. Photo disinfection of the hybrid membrane occurs due to the generation of various active oxygen species such as hydroxyl radical, hydrogen peroxide, etc. formed by various oxidation and reductions reactions occurring on the surface. These active oxygen species act as anti-bacterial reagents and prevent biofouling of the membranes either by killing the bacteria or preventing their growth cycle [94-95]. A comparative study of the photocatalytic bactericidal efficiency of the hybrid TFC membrane was found to be remarkably higher for the hybrid TFC membrane under UV light illumination [34] than that without illumination and the neat TFC membranes for Escherichia coli *(E. coli).* Such photo disinfection property of hybrid membrane is expected to reduce the microbiological fouling of RO membrane.

6.9 Effect of TiO2 Loading for Degradation of Organic Pollutants and Self Cleaning of Membranes

A number of review papers and book chapters have been published on the photocatalytic behavior of TiO2 nanocomposite membranes [65,15]. The incorporation of TiO2 photocatalyst on the membranes incorporates the response of membrane towards the artificial source or solar light induced degradation of pollutants for water decontamination as well as for cleaning of the membrane. Synthetic compounds like dyes, phenols, pesticides, fertilizers, herbicides and surfactants, etc present in the aqueous environment can be successfully phased out using polymer-TiO2 nanocomposite membranes [96-102]. The degradation of the pollutants takes place due to the oxidation – reduction reactions occurring on the membrane surface.

In membrane separation technology, membrane cleaning is the most important aspect in terms of cost as well as performance which affect the membrane operations. Twin properties of TiO2 i.e. photocatalysis and super –hydrophilicity, paved the way for the development of self-cleaning membranes. The self-cleaning approach relies on photocatalytic activity which resulted in the partial or total decomposition of pollutants especially organic compounds present at the surface under UV light, while photo induced super hydrophilicity allowed the elimination of residual contaminants through a simple water rinsing operation [103]. The TFC -TiO2 coated membrane shows the self-cleaning property which is highly dependent on light irradiation. The property gets increased with a longer time of UV radiation until a specific time, afterwards no significant increase in self-cleaning property was observed. Increasing the ultra hydrophilicity of the membranes by immersing the coated membranes in water followed by UV light irradiation has been reported [104}. Thus the polymer-TiO2 nanocomposite membranes have shown great potential for the development of energy-efficient water purification and waste water treatment technology.

Conclusion

Different polymeric membranes are one of the most commonly employed membranes for water purification and treatment. In the present chapter, the advancements in synthetic and technical development of organic polymer based TiO_2 nanocomposite membranes for water treatment has been discussed. The improvement in membrane properties such as antimicrobial, super-hydrophilic and photocatalytic etc. on incorporation of metal oxide, TiO_2 are important attributes for finding energy and cost effective solutions for water purification and treatment.

References

[1] Horike, S.M, Tamaki, K., Long, J.R. et al. (2008). Size-selective lewis acid catalysis in a microporous metal-organic frame work with exposed Mn+2 coordination sites. J. Am.Chem.Soc.130:5854–5855. *http://dx.doi.org/10.1021/ja800669j*

[2] Xiao, G.A., Huang, H., Su, T. Tan. et al. (2013). The activity of acrylic–silicon/nano-TiO2 films for the visible-light degradation of formaldehyde and NO2. Build.Environ.65: 215–221. *http://dx.doi.org/10.1016/j.buildenv.2013.04.014*

[3] Stein, A. Melde., Schroden, R., Cet, B.J. et al. (2000). Hybrid inorganic–organic mesoporous silicates-nanoscopic reactors coming of age.Adv.Mater.12:1403–1419. *http://dx.doi.org/10.1002/1521-4095(200010)12:19<1403::AID-ADMA1403>3.0.CO;2-X*

[4] Yang, Z., Shen, J., Jayaprakash., N., Archer, LA et al. (2012). Synthesis of organic–inorganic hybrids by mini emulsion polymerization and their application for electrochemical energy storage. Energy Environ.Sci. 5:7025–7032. *http://dx.doi.org/10.1039/c2ee03230a*

[5] Corma, A., Díaz, U., Arrica, M., Fernández, E., Ortega, Í. et al. (2009). Organic–inorganic nano spheres with responsive molecular gates for drug storage and release. Angew. Chem.Int.Ed.121:6365–6368. *http://dx.doi.org/10.1002/ange.200902208*

[6] Yan, L., Li, Y.S., Xiang, C.B., Xianda, S. et al. (2006). Effect of nano-sized Al2O3-particle addition on PVD Filtration membrane performance. J.Membr.Sci.276:162–167. *http://dx.doi.org/10.1016/j.memsci.2005.09.044*

[7] Tripathi, B.P., Kumar, M., Saxena, A., Shahi, V.K. et al.(2010). Bifunctionalized organic– inorganic charged nano composite membrane for pervaporation dehydration of ethanol.J.Colloid Interface Sci.346:54–60. *http://dx.doi.org/10.1016/j.jcis.2010.02.022*

[8] Bosc, F., Lacroix-Desmazes, P., Ayral, A. et al. (2006).TiO2 anatase-based membranes with hierarchical porosity and photocatalytic properties. J.Colloid Interface Sci.304:545–548. *http://dx.doi.org/10.1016/j.jcis.2006.09.064*

[9] Pang, R., Li, X., Li, J., Lu, Z., Sun, X., Wang, L.et al. (2014). Preparation and characterization of ZrO2/PES hybrid ultra filtration membrane with uniform ZrO2 nanoparticles. Desalination. 332:60–66. *http://dx.doi.org/10.1016/j.desal.2013.10.024*

[10] Han, P., Yahui, H., Yang, W., Linlin, L. et al.(2006). Preparation of polysulfone–Fe3O4 composite ultrafiltration membrane and its behavior in magnetic field. J. Membr.Sci.284:9–16. *http://dx.doi.org/10.1016/j.memsci.2006.07.052*

[11] Li, X., Pang, R., Li, J., Sun, X., Shen, J., Han, W., Wang, L. et al. (2013). In situ formation of Ag nanoparticles in PVDF ultrafiltration membrane to mitigate organic and bacterial fouling. Desalination.324:48–56. *http://dx.doi.org/10.1016/j.desal.2013.05.021*

[12] Rahimpour, A., Jahanshahi, M., Rajaeian, B., Rahimnejad, M. et al. (2011). TiO2 entrapped nano-composite PVDF/SPES membranes: preparation, characterization, anti- fouling and antibacterial properties. Desalination.278:343–353. *http://dx.doi.org/10.1016/j.desal.2011.05.049*

[13] Ng, L.Y., Mohammad, A.W., Leo, C.P., Hilal, N. et al. (2013). Polymeric membranes incorporated with metal/metal oxide nanoparticles: a comprehensive review. Desalination.308: 15–33. *http://dx.doi.org/10.1016/j.desal.2010.11.033*

[14] Buonomenna, M.G. (2013). Membrane processes for a sustainable industrial growth. Rsc. Adv.3:5694–5740 *http://dx.doi.org/10.1039/c2ra22580h*

[15] Yin, Jun., Deng, Baolin.et al. (2015).Polymer-matrix nanocomposite membranes for water treatment. Journal of MembraneScience. 479:256–275 *http://dx.doi.org/10.1016/j.memsci.2014.11.019*

[16] Van der Bruggena, B., M¨antt¨ ari b, M., Nystr¨omb, M.et al. (2008).Drawbacks of applying nanofiltration and how to avoid them: Separation and Purification Technology. 63:251–263 *http://dx.doi.org/10.1016/j.seppur.2008.05.010*

[17] Alzahrania, Salem., Mohammad, Abdul Wahab.et al. (2014). Challenges and trends in membrane technology implementation for produced water treatment. Journal of Water Process Engineering. 4:107–133 *http://dx.doi.org/10.1016/j.jwpe.2014.09.007*

[18] Kiriakidou, F., Kondarides, D.I., Verykios, X.E. et al. (1999). The effect of operational parameters and TiO2-doping on the photocatalytic degradation of azo-dyes. Catal.Today. 54: 119-130. *http://dx.doi.org/10.1016/S0920-5861(99)00174-1*

[19] Sun, B. and Smirniotis, P.G.et al.(2003). Interaction of anatase and rutile TiO2 particles in aqueous photooxidation. Catal.Today. 88: 49-59. *http://dx.doi.org/10.1016/j.cattod.2003.08.006*

[20] Kumar A, Sharma G, NaushadM, Kalia S, Singh P (2014) A Polyacrylamide/Ni0.02Zn0.98O nanocomposite with high solar light photoacatalytic activity and efficient adsorption capacity for toxic dyes removal. J IndEngChem Res 53:15549-15560.

[21] Hu, Y., Tsai, H.L., Huang, C.L.et al.(2003). Effect of Brookite Phase on the Anatase-Rutile Transition in Titania Nanoparticles. Eur. Ceram. Soc. 23: 691-696. *http://dx.doi.org/10.1016/S0955-2219(02)00194-2*

[22] Nicholls, D. (1974). Complexes and First-Row Transition Elements. MacMillan Education: Hong Kong. *http://dx.doi.org/10.1007/978-1-349-02335-6*

[23] Shao, Y., Tang, D., Sun, J., Lee, Y., Xiong, W.et al. (2004). Lattice deformation and phase transformation from nano-scale anatase to nano-scale rutile TiO2 prepared by a sol-gel technique. China Particuology.2: 119. *http://dx.doi.org/10.1016/S1672-2515(07)60036-0*

[24] Rehman, S., Ullah, R., Butt, A.M., Gohar, N.D.et al. (2009). Strategies of making TiO2 and ZnO visible light active. J. Hazard. Mater. 170: 560–569. *http://dx.doi.org/10.1016/j.jhazmat.2009.05.064*

[25] Suppan, P. (1994). Chemistry and Light. Royal Society of Chemistry: Cambridge, 107.

[26] Hoffman, A., Carraway, E.R., Hoffman, M.et al. (1994). Photocatalytic production of H2O2 and organic peroxides on quantum-sized semiconductor colloids. Environ. Sci. Technol. 28: 776-785. *http://dx.doi.org/10.1021/es00054a006*

[27] Mahdavi, F., Burton, T.C., Li, Y.et al. (1993). Photoinduced reduction of nitro compounds on semiconductor particles. J. Org. Chem. 58: 744-746. *http://dx.doi.org/10.1021/jo00055a033*

[28] Choi, W., Termin, A., Hoffmann, M.R.et al. (1994). The role of metal ion dopants in quantum-sized TiO2: correlation between photoreactivity and charge carrier recombination dynamics. J. Phys. Chem. B. 98: 13669–13679. *http://dx.doi.org/10.1021/j100102a038*

[29] Vinodgopal, K. and Kamat, P.V.et al. (1995). Enhanced rates of photocatalytic degradation of an Azo Dye using SnO2/TiO2 coupled semiconductor thin films. Environ. Sci. Technol. 29: 841-845. *http://dx.doi.org/10.1021/es00003a037*

[30] Kaur, P. and Sud, D.et al. (2012). Photocatalytic degradation of quinalphos in aqueous TiO2 suspension: Reaction pathway and identification of intermediates by GC/MS.J.Mole.Catal.A:Chemical, *http://dx.doi.org/10.1016/j.molcata.2012.08.005.*

[31] Li, Y., Hwang, D.S., Lee, N.H., Kim, S.J.et al. (2005). Synthesis and characterization of carbon-doped titania as an artificial solar light sensitive photocatalyst. Chem. Phys. Lett. 404: 25-29. *http://dx.doi.org/10.1016/j.cplett.2005.01.062*

[32] Xu, Z., Shang, J., Liu, C., Kang, C., Guo, H., Du, Y.et al.(1999). The preparation and characterization of TiO2 ultrafine particles. Mater. Sci. Eng. B. 63: 211–214. *http://dx.doi.org/10.1016/S0921-5107(99)00084-7*

[33] Rahimpour, A., Jahanshahi, M., Rajaeian, B., Rahimnejad, M.et al. (2011). TiO2 entrapped nano- composite PVDF/SPES membranes: preparation, characterization, antifouling and antibacterial properties. Desalination, 278: 343–353. *http://dx.doi.org/10.1016/j.desal.2011.05.049*

[34] Kim, Sung Ho., Kwak , Seung-Yeop., Sohn, Byeong-Hyeok ., Park , Tai Hyun.et al.(2003). Design of TiO2 nanoparticle self-assembled aromatic polyamide thin-

film-composite (TFC) membrane as an approach to solve biofouling problem. J. Membr Sci 211:157–165
http://dx.doi.org/10.1016/s0376-7388(02)00418-0

[35] Madaeni, S.S., Mansourpanah Y. (2004).Chemical cleaning of reverse osmosis membranes fouled. Desalination. 161: 13-24
http://dx.doi.org/10.1016/S0011-9164(04)90036-7

[36] Razmjou, A., Mansouria, J., Chen, V., Lim, M., Amal, R. et al. (2011). Titania nanocomposite polyethersulfone ultrafiltration membranes fabricated using a low temperature hydrothermal coating process. J. Membr. Sci.380:98– 113.
http://dx.doi.org/10.1016/j.memsci.2011.06.035

[37] Wang, Q., Wang, X., Wang, Z., Huang, J., Wang, Y.et al. (2013). PVDF membranes with simultaneously enhanced permeability and selectivity by breaking the tradeoff effect via atomic layer deposition of TiO2. J. Membr. Sci. 442: 57–64.
http://dx.doi.org/10.1016/j.memsci.2013.04.026

[38] Meng, S., Mansouri, J., Ye, Y., Chen, V.et al. (2014). Effect of templating agents on the properties and membrane distillation performance of TiO2 -coated PVDF membranes. J. Membr. Sci. 450: 48–59.
http://dx.doi.org/10.1016/j.memsci.2013.08.036

[39] Luo, M.et al. (2003). Preparation and characterization of polyethersulfone/TiO2 nanoparticles composite ultrafilteration membrane with high hydrophilicity and its pollution mechanism. Ph.D. Thesis, xi'an Jiaotong university.

[40] Razmjou, A., Holmes, A.R.L., Li, H., Mansouri, J., Chen, V.et al. (2012). The effect of modified TiO2 nanoparticles on polyethersulfone ultrafiltration hollow fiber membranes, Desalination 287:271-280.
http://dx.doi.org/10.1016/j.desal.2011.11.025

[41] Luo, M., Tang, W., Zhao, J., Pu, C.et al. (2006). Hydrophilic modification of poly (ether sulfone) used TiO2 nanoparticles by a sol–gel process. J. Mater. Process. Technol, 172:431–436.
http://dx.doi.org/10.1016/j.jmatprotec.2005.11.004

[42] Li, J.H., Xu, Y.Y., Zhu, L.P., Wang, J.H., Du, C.H. et al.(2009).Fabrication and characterization of a novel TiO2 nanoparticle self-assembly membrane with improved fouling resistance. Journal of Membr. Sci. 326: 659-666
http://dx.doi.org/10.1016/j.memsci.2008.10.049

[43] Rahimpour, A., Madaeni, S.S., Taheri, A.H., Mansourpanah, Y.et al. (2008). Coupling TiO2 nanoparticles with UV irradiation for modification of polyethersulfone ultrafiltration membranes. J. Membr. Sci. 313:158– 169 *http://dx.doi.org/10.1016/j.memsci.2007.12.075*

[44] Lee, H.S., Im, S.J., Kim, J.H., Kim, H.J., Kim,J.P., Min,B.R. et al. (2008). Polyamide thin –film nanofiltration membranes containing TiO2 nanoparticles. Desalination.219:48-56. *http://dx.doi.org/10.1016/j.desal.2007.06.003*

[45] Rajesh, S., Senthilkumar, S., Jayalakshmi, A., Nirmala, M.T., Ismail, A.F., Mohan, D.et al. (2013). Preparation and performance evaluation of poly (amide–imide) and TiO2nanoparticles impregnated polysulfone nanofiltration membranes in the removal of humic substances. Coll. Surf. A Physicochem. Eng. Asp, 418:92–104. *http://dx.doi.org/10.1016/j.colsurfa.2012.11.029*

[46] Wanbin, Li., Zhihong, Yang., Qin, Meng., Chong, Shen., Guoliang, Zhang.et al. (2014). Thermally stable and solvent resistant self-crosslinked TiO2/PAN hybrid hollow fiber membrane fabricated by mutual supporting method. Journal of Membr.Sci. 467:253–261 *http://dx.doi.org/10.1016/j.memsci.2014.05.019*

[47] Cruz, N.K.O., Semblante, G.U., Senoro, D.B., You, S.J., Lu, S.C.et al. (2013). Dye degradation and antifouling properties of polyvinylidene fluoride/titanium oxide membrane prepared by sol–gel method. J. Chinese Inst. Chem. Eng.

[48] Madaeni, S.S., Zinadini, S., Vatanpour, V.et al. (2011). A new approach to improve antifouling property of PVDF membrane using in situ polymerization of PAA functionalized TiO2 nanoparticles. J. Membr. Sci. 380:155–162. *http://dx.doi.org/10.1016/j.memsci.2011.07.006*

[49] Mills, A., Hunte, S.L. (1997).Anoverviewof semiconductor photocatalysis. J. Photochem. Photobiol. A: Chem. 1:108.

[50] Ebert, K., Fritsch, D., Koll, J., Tjahjawiguna, C.et al. (2004).Influence of inorganic fillers on the compaction behaviour of porous polymer based membranes. J. Membr. Sci. 233:71–78 *http://dx.doi.org/10.1016/j.memsci.2003.12.012*

[51] Mansourpanah, Y., Madaeni, S.S., Rahimpour, A., Farhadian, A., Taheri, A.H. et al. (2009). Formation of appropriate sites on nanofiltration membrane surface for binding TiO2 photo-catalyst: performance, characterization and fouling resistant capability. J. Membr. Sci. 330: 297–306.

http://dx.doi.org/10.1016/j.memsci.2009.01.001

[52] Kwak, S.Y., Kim, S.H., Kim, S.S. et al. (2001). Hybrid organic/inorganic reverse osmosis (RO) membrane for bactericidal antifouling Part 1. Preparation and characterization of TiO2 nanoparticle self-assembled aromatic polyamide thin-filmcomposite (TFC) membrane. Environ. Sci. Technol. 35:2388 *http://dx.doi.org/10.1021/es0017099*

[53] Flemming, H.C.et al. (1997).Reverse osmosis membrane biofouling.Thermal Fluid Sci. 14:382. *http://dx.doi.org/10.1016/S0894-1777(96)00140-9*

[54] Kimura, K., Amy, G., Drewes, J., Watanabe, Y.et al. (2003).Adsorption of hydrophobic compounds onto NF/RO membranes: an artefact leading to overestimation of Rejection. J. Membr. Sci. 221: 89–101. *http://dx.doi.org/10.1016/S0376-7388(03)00248-5*

[55] M¨antt¨ ari, M., Puro, L., Nuortila-Jokinen, J., Nystr ¨om, M.et al. (2000). Fouling effects of polysaccharides and humic acid in nanofiltration. J. Membr. Sci. 165:1–17.

[56] Sch¨afer, A.I., Andritsos, N., Karabelas, A.J., Hoek, E.M.V., Schneider, R., Nystr ¨om, M.et al. (2005).Fouling in nanofiltration, in: A.I. Sch¨afer, A.G. Fane, T.D.Waite (Eds.), Nanofiltration: Principles and Applications, Elsevier, Oxford, UK.

[57] Van der Bruggen, B., Kim, J.H., DiGiano, F.A., Geens, J., Vandecasteele, C.et al.(2004).Influence of MFpretreatment on NF performance for aqueous solutions containing particles and an organic foulant. Sep. Purif. Technol. 36 (3):203–213. *http://dx.doi.org/10.1016/S1383-5866(03)00216-8*

[58] Zhang, M., Song, L.et al. (2000).Mechanisms and parameters affecting flux decline in crossflow microfiltration and ultrafiltration of colloids. Environ. Sci. Technol. 34: 3767–3773. *http://dx.doi.org/10.1021/es990475u*

[59] Costa, A.R., de Pinho, M.N.et al. (2005).Effect ofmembrane pore size and solution chemistry on the ultrafiltration of humic substances solutions. J. Membr. Sci. 255: 49–56. *http://dx.doi.org/10.1016/j.memsci.2005.01.016*

[60] Song, L., Chen, K.L., Ong, S.L., Ng, W.J. et al. (2004).A new normalization method for determination of colloidal fouling potential in membrane processes. J. Colloids Interface Sci. 271:426–433

http://dx.doi.org/10.1016/j.jcis.2003.12.016

[61] Ivnitsky, H., Katz, I., Minz, D., Volvovic, G., Shimoni, E., Kesselman, E., Semiat, R., Dosoretz, C.G. et al. (2007).Bacterial community composition and structure of biofilms developing on nanofiltration membranes applied to wastewater treatment. Water Res. 41 (17): 3924–3935. *http://dx.doi.org/10.1016/j.watres.2007.05.021*

[62] Li, J., Xu, Z., Yang, H., Yu, L., Liu, M.et al. (2009).Effect of TiO2 nanoparticles on the surface morphology and performance of microporous PES membrane. Appl. Surf. Sci. 255: 4725–4732. *http://dx.doi.org/10.1016/j.apsusc.2008.07.139*

[63] Potts, D.E., Ahlert, R.C., Wang, S.S. et al. (1981). A critical review of fouling of reverse osmosis membranes, Desalination. 36:235. *http://dx.doi.org/10.1016/S0011-9164(00)88642-7*

[64] Matthiasson, E., Sivik, B. et al. (1980).Concentration polarization and fouling. Desalination. 35: 59. *http://dx.doi.org/10.1016/S0011-9164(00)88604-X*

[65] Bet-moushoul E., Mansourpanah, Y., Farhadi, Kh., Tabatabaei, M. et al.(2016). TiO2 nanocomposite based polymeric membranes: A review on performance improvement for various applications in chemical engineering processes Chemical Engineering Journal, 283: 29–46 *http://dx.doi.org/10.1016/j.cej.2015.06.124*

[66] Shon, H.K., Vigneswaran, S., Ben Aim, R., Ngo, H.H., Kim, I.S., Cho, J. et al.(2005).Influence of flocculation and adsorption as pretreatment on the fouling of ultrafiltration and nanofiltration membranes: application with biologically treated sewage effluent. Environ. Sci. Technol. 39 (10):3864–3871. *http://dx.doi.org/10.1021/es040105s*

[67] Song, W., Ravindran, V., Koel, B.E., Pirbazari, M. et al. (2004). Nanofiltration of natural organic matter with H2O2/UV pretreatment: fouling mitigation and membrane surface Characterization. J. Membr. Sci. 241 (1): 143–160. *http://dx.doi.org/10.1016/j.memsci.2004.04.034*

[68] Hillis, P., Padley, M.B., Powell, N.I., Gallagher, P.M. et al. (1998).Effects of backwash conditions on out-to-in membrane microfiltration. Desalination. 118;197. *http://dx.doi.org/10.1016/S0011-9164(98)00128-3*

[69] Sagiv, A., Semiat, R.et al. (2005). Backwash of RO spiralwound membranes. Desalination. 179:1.

http://dx.doi.org/10.1016/j.desal.2004.11.050

[70] Tarazaga, C.C., Campderros, M.E., Padilla, P.E. et al. (2006). Physical cleaning by means of electric field in the ultrafiltration of a biological solution. J. Membr. Sci. 278:219. *http://dx.doi.org/10.1016/j.memsci.2005.11.004*

[71] Ulbricht, M., Ridel, M., Marx, U. et al. (1996). Novel photochemical surface functionalization of polysulfone ultrafiltration membranes for covalent immobilization of Biomolecules. J. Membr. Sci. 120:239. *http://dx.doi.org/10.1016/0376-7388(96)00148-2*

[72] Rahimpour, A., Madaeni, S.S. (2007).Polyethersulfone (PES)/cellulose acetate phthalate (CAP) blend ultrafiltration membranes: preparation, morphology, performance and anti-fouling properties. J. Membr. Sci. 305: 299. *http://dx.doi.org/10.1016/j.memsci.2007.08.030*

[73] Bequet, S., Remigy, J.-C., Rouch, J.C., Espenan, J.-M., Clifton, M., Aptel, P.et al. (2002).From ultrafiltration to nanofiltration hollow fiber membranes: a continuous UV photografting process. Desalination 144:9. *http://dx.doi.org/10.1016/S0011-9164(02)00281-3*

[74] Zhao, Z.P., Li, J., Zhang, D.X., Chen, C.X. et al. (2004).Nanofiltration membrane prepared from polyacrylonitrile ultrafiltration membrane by low-temperature plasma. I. Graft of acrylic acid in gas. J. Membr. Sci. 232:1. *http://dx.doi.org/10.1016/j.memsci.2003.11.009*

[75] Qiu, C., Xu, F., Nguyen, Q.T., Ping, Z.et al. (2005).Nanofiltrationmembrane prepared fromcardo polyetherketone ultrafiltration membrane by UV induced grafting method. J.Membr. Sci. 255:107. *http://dx.doi.org/10.1016/j.memsci.2005.01.027*

[76] Zhao, Y.H., Zhu, B.K., Kong, L., Xu, Y.Y.et al. (2007). Improving hydrophilicity and protein resistance of poly(vinylidene fluoride) membranes by blending with amphiphilic hyperbranched-star polymer. Langmuir. 23:5779–5786. *http://dx.doi.org/10.1021/la070139o*

[77] Mackay, M.E., Carmezini, G. (2002).Manipulation of hyperbranched polymers' conformation. Chem. Mater. 14: 819–825. *http://dx.doi.org/10.1021/cm010623r*

[78] Miyauchi, M., Tokudome, H. (2007).Super-hydrophilic and transparent thin films of TiO2 nanotube arrays by a hydrothermal reaction. J. Mater. Chem. 17: 2095–2100. *http://dx.doi.org/10.1039/b700387k*

[79] Song, S., Jing, L., Li, S., Fu, H., Luan, Y.et al. (2008).Superhydrophilic anatase TiO2 film with the micro- and nanometer-scale hierarchical surface structure. Mater. Lett. 62: 3503–3505. *http://dx.doi.org/10.1016/j.matlet.2008.03.005*

[80] Nakajima, A., Koizumi, S., Watanabe, T., Hashimoto, K.et al. (2000). Photoinduced amphiphilic surface on polycrystalline anatase TiO2 thin films. Langmuir. 16:7048–7050. *http://dx.doi.org/10.1021/la0004348*

[81] Yu, J., Zhao, X., Zhao, Q.et al. (2000).Effects of surface morphology of photocatalytic porous TiO2 thin films on hydrophilicity. J.Chin. Ceram. Soc. 3: 245.

[82] Garp, O., Huisman, C.L., Reller, A.et al. (2004). Photoinduced reactivity of titanium dioxide, Prog. Solid State Chem. 32: 33–177. *http://dx.doi.org/10.1016/j.progsolidstchem.2004.08.001*

[83] Langlet, M., Permpoona, S., Riassetto, D., Berthome, G., Pernot, E., Joud, J.C.et al. (2006).Photocatalytic activity and photo-induced superhydrophilicity of sol–gel derived TiO2 films. J. Photochem. Photobiol. A. 181: 203–214. *http://dx.doi.org/10.1016/j.jphotochem.2005.11.026*

[84] Fujishima, A., Rao, T.N., Tryk, D.A.et al. (2000).TiO2 photocatalysts and diamond electrodes. Electrochim. Acta 45:4683–4690. *http://dx.doi.org/10.1016/S0013-4686(00)00620-4*

[85] Wang, K., Zhang, J., Lou, L., Yang, S., Chen, Y.et al. (2004).UV or visible light induced photodegradation of AO7 on TiO2 particles: the influence of inorganic anions. J. Photochem. Photobiol. A 165 201–207. *http://dx.doi.org/10.1016/j.jphotochem.2004.03.025*

[86] Mills, A., Elliott, N., Parkin, I.P., O'Neill, S.A., Clark, R.J.et al. (2002).Novel TiO2 CVD films for semiconductor photocatalysis. J. Photochem. Photobiol. A 151:171–179. *http://dx.doi.org/10.1016/S1010-6030(02)00190-9*

[87] Guan, K. (2005).Relationship between photocatalytic activity, hydrophilicity and selfcleaning effect of TiO2/SiO2 films. Surf. Coat. Technol. 191:155–160. *http://dx.doi.org/10.1016/j.surfcoat.2004.02.022*

[88] Madaeni, S.S., Ghaemi, N., Alizadeh, A., Joshaghani, M.et al. (2011). Influence of photo-induced superhydrophilicity of titanium dioxide nanoparticles on the anti-

fouling performance of ultrafiltration membranes. Applied Surface Science 257:6175–6180.
http://dx.doi.org/10.1016/j.apsusc.2011.02.026

[89] Emadzadeh, D., Lau, W.J. ., Matsuura, T., Rahbari-Sisakht, M., Ismail, A.F.et al. (2014). A novel thin film composite forward osmosis membrane prepared from PSf–TiO2 nanocomposite substrate for water desalination. Chemical Engineering Journal 237: 70–80.
http://dx.doi.org/10.1016/j.cej.2013.09.081

[90] Bae, Tae-Hyun., Kim, In-Chul., Tak, Tae-Moon. et al. (2006). Preparation and characterization of fouling-resistant TiO2 self-assembled nanocomposite membranes. J. Membr. Sci. 275:1–5.
http://dx.doi.org/10.1016/j.memsci.2006.01.023

[91] Bae, T.H., Tak, T.M. (2005). Preparation of TiO2 self-assembled polymeric nanocomposite membranes and examination of their fouling mitigation effects in a membrane bioreactor system. J. Membr. Sci. 266:1–5.
http://dx.doi.org/10.1016/j.memsci.2005.08.014

[92] Bae, T.H., Tak, T.M. (2005). Effect of TiO2 nanoparticles on fouling mitigation of ltrafiltration membranes for activated sludge filtration. J. Membr. Sci. 249:1–8.
http://dx.doi.org/10.1016/j.memsci.2004.09.008

[93] Emadzadeh, D., Lau,W.J., Matsuura, T., Ismail, A.F., Rahbari-Sisakht, M.et al. (2014) Synthesis and characterization of thin film nanocomposite forward osmosis membrane with hydrophilic nanocomposite support to reduce internal concentration polarization. Journal of Membr. Sci. 44: 974–85.
http://dx.doi.org/10.1016/j.memsci.2013.08.014

[94] Sunada, K., Kikuchi, Y., Hashimoto, K., Fujishima, A. et al. (1998).Bactericidal and detoxification effects of TiO2 thin film Photocatalysts. Environ. Sci. Technol. 32:726.
http://dx.doi.org/10.1021/es970860o

[95] Brading, M.G., Jass, J., Lappin-Scott, H.L. et al. (1995).Dynamics of bacterial biofilm formation. in: M.L.-S. Hilary, J.W. Costerton (Eds.), Microbial Biofilms, Cambridge University Press, NewYork, NY, pp. 46–63.
http://dx.doi.org/10.1017/cbo9780511525353.004

[96] Damodar, R.A., You, S., Chou, H.et al. (2009).Study the self cleaning, antibacterial and photocatalytic properties of TiO2 entrapped PVDF membranes. J. Hazard. Mater. 172: 1321–1328.
http://dx.doi.org/10.1016/j.jhazmat.2009.07.139

[97] Alaoui, O. Tahiri., Nguyen, Q.T., Mbareck, C., Rhlalou, T.et al. (2009). Elaboration and study of poly (vinylidene fluoride) – anatase TiO2 composite membranes in photocatalytic degradation of dyes. Appl. Catal. A Gen. 358: 13–20. *http://dx.doi.org/10.1016/j.apcata.2009.01.032*

[98] Yang, S., Gu, J.S., Yu, V., Zhou, J., Li, S.F., Wu, X.M., Wang, L. et al. (2011).Polypropylene membrane surface modification by RAFT grafting polymerization and TiO2 photocatalysts immobilization for phenol decomposition in a photocatalytic membrane reactor. Sep. Purif. Technol. 83:157–165. *http://dx.doi.org/10.1016/j.seppur.2011.09.030*

[99] Dzinun, H., Othman, M.H.D., Ismail, A.F., Puteh, M.H.et al. (2015).Photocatalytic degradation of nonylphenol by immobilized TiO2 in dual layer hollow fibre membranes. Chem. Eng. J. 269: 255–261. *http://dx.doi.org/10.1016/j.cej.2015.01.114*

[100] Song, H., Shao, J., Hea, Y., Liu, B., Zhong, X.et al.(2012).Natural organic matter removal and flux decline with PEG–TiO2-doped PVDF membranes by integration of ultrafiltration with photocatalysis. J. Membr. Sci. 405–406:48–56. *http://dx.doi.org/10.1016/j.memsci.2012.02.063*

[101] Leong, Sookwan., Razmjou, Amir., Wang, Kun., Hapgood, Karen., Zhang, Xi wang.et al. (2014). TiO2 based photocatalytic membranes. Huanting Wang Journal of Membr. Sci. 472:167–184 16 17.

[102] You, S.J., Semblante, G.U., Lu, S.C., Damodar, R.A., Wei, T.C.et al. (2012). Evaluation of the antifouling and photocatalytic properties of poly(vinylidene fluoride) plasma-grafted poly(acrylic acid) membrane with self-assembled TiO2. J. Hazard. Mater. 237–238:10–19. *http://dx.doi.org/10.1016/j.jhazmat.2012.07.071*

[103] Garp, O., Huisman, C.L., Reller A. et al. (2004) Photoinduced reactivity of titanium dioxide, Prog. Solid State Chem. 32 : 33–177. *http://dx.doi.org/10.1016/j.progsolidstchem.2004.08.001*

[104] Madaeni, S.S., Ghaemi, N. (2007) Characterization of self-cleaning RO membranes coated with TiO2 particles under UV irradiation. J. of Membr. Sci. 303: 221–233 *http://dx.doi.org/10.1016/j.memsci.2007.07.017*

CHAPTER 9

Advances in Electrochemical Cholesterol Nanosensors

Mitali Saha
Department of Chemistry, National Institute of Technology, Agartala-799046, Tripura, India
mitalisah@gmail.com

Abstract

Electrochemical biosensors have benefited extensively from advances in nanomaterials (e.g., metal nanoparticles, carbon nanotubes and graphene). Among the various biomolecules, cholesterol has aroused considerable interest in recent years on account of its being an important marker in clinical diagnosis. Recent research has shown that cholesterol plays an important role in the brain synapses and also in the immune system including protection against cancer. As biosensors have the ability to detect and measure constantly the concentration of specific organic or inorganic substances in desired specimens, the thrust to monitor the presence of cholesterol in blood is driving the need for the development of efficient cholesterol biosensors. This chapter highlights the recent advances in the field of enzymatic and non enzymatic cholesterol sensors with the use of nanotechnology.

Keywords

Nanomaterials, Electrochemical, Cholesterol, Biosensor

Contents

1. Introduction

Nanotechnology is one of the most exciting forefront fields and impacts several areas of material science. It is also useful in solving challenging bioanalytical problems. The combined efforts of chemists, biochemists, physicists and engineers towards coupling of biological molecules with electronics have been the key to recent successful developments in the area of biosensors. The last decades have shown unprecedented interest in the development of analytical devices for the detection, quantification, and monitoring of different biological and chemical compounds.

Cholesterol and its fatty acid esters are the main constituents of mammalian cell membrane and are precursors of other biological materials, such as bile acid and steroid hormones. Accurate measurement of cholesterol level in blood has long been recognized as an important clinical test for diagnosing cardiovascular diseases. Currenlty, many patients are using disposable cholesterol strips for self monitoring of their blood cholesterol levels. Recent improvement in the biosensor technology has been made by using various types of materials for their fabrication such as conductive electroactive polymers, composite materials, composites of transition metal-based complexes and organometallic compounds, sol–gel and hydro–gel materials, nanomaterials, nano-metal composites, nano-metal oxides, etc. This chapter highlights the recent advances in the developments of nanomaterials based enzymatic and non enzymatic cholesterol biosensors.

2. Overview of biosensors

A biosensor is a self-contained device consisting of two functional parts. A bioselective element is in direct contact with a physical transducer, which transforms the information from biorecognition domain into an electrical or optical signal. The amplitude of such signal depends on the concentration of the analyzed compound (analyte) in the sample. For a biosensor to be pragmatic, efficient and capable it must inherently possess the following features and the biocatalyst must be highly specific for the analyte under investigation:

- It must not be labile when subjected to normal storage conditions
- It should exhibit good and uniform stability over a significant number of assays
- The reaction should be independent of various physical parameters
- The cofactors/coenzymes must be co-immobilized with the biocatalyst
- Accurate, precise and linear results must be obtained without the need to resort to dilution/concentration
- While using the biosensor for invasive monitoring, care must be taken to ensure that the probe is tiny and biocompatible and should eschew toxic or antigenic effects.

3. Electrochemical biosensors

Electrochemical biosensors provide an attractive solution for analyzing the content of a biological sample due to the direct conversion of a biological event to an electronic signal. However, converting biological information to an easily processed electronic signal is challenging due to the complexity of connecting an electronic device directly to a biological environment. An electrochemical sensor is a device that transforms electrochemical information into an analytically useful signal. Electrochemical sensors are usually composed of two basic components, a chemical (molecular) recognition system which is the most important part of a sensor and a physicochemical transducer which is a device that converts the chemical response into a signal that can be detected by modern electrical instrumentations. These two parts form a working (or sensing) electrode. A reference electrode and sometimes a counter electrode are also used in electrical measurements.

3.1 Potentiometric biosensors

Potentiometric biosensors are based on ion-selective electrodes (ISE) and ion-sensitive field effect transistors (ISFET). The primary outputting signal is possibly due to ions accumulated at the ion-selective membrane interface. Current flowing through the electrode is equal to or near zero. The electrode follows the presence of the monitored ion resulting from the enzyme reaction [1]. There are three types of ion-selective electrodes which are of use in biosensors: (a) Glass electrodes for cations (e.g. normal pH electrodes) in which the sensing element is a very thin hydrated glass membrane which generates a transverse electrical potential due to the concentration-dependent competition between the cations for specific binding sites. The selectivity of this membrane depends upon the composition of the glass. The sensitivity to H^+ is greater than that achievable for NH_4^+; (b) Glass pH electrodes coated with a gas-permeable membrane selective for CO_2, NH_3 or H_2S. The diffusion of the gas through this membrane causes a change in pH of a sensing solution between the membrane and the electrode which is then determined and (c) Solid-state electrodes where the glass membrane is replaced by a thin membrane of a specific ion conductor made from a mixture of silver sulphide and a silver halide.

3.2 Amperometric biosensors

Amperometric biosensors are quite sensitive and more suited for mass production than the potentiometric ones [2]. The working electrode of the amperometric biosensor is usually either a noble metal or a screen-printed layer covered by the biorecognition component [3]. Amperometric biosensors can work in two- or three-electrode configurations. The former case consists of reference and working (containing immobilized biorecognition component) electrodes. The main disadvantage of the two-electrode configuration is limited control of the potential on the working electrode surface with higher currents, and because of this, the linear range could be shortened. To solve this problem, a third auxiliary electrode is employed so that the voltage is applied between the reference and the working electrodes, and current flows between the working and the auxiliary electrodes. The amperometric biosensors are often used on a large scale for analytes such as glucose, lactate and sialic acid [4, 5]. Biological agents such as model Bacillus cereus and Mycobacterium smegmatis, the serological diagnosis of Francisella tularensis, a pharmacology study and the detection of pesticides and nerve agents have also been described [6-9].

4. Nanomaterials based cholesterol sensor

Importance of cholesterol biosensors has already been recognized in the clinical diagnosis of cardiac and brain vascular diseases. Nevertheless, the practical application of a majority of the fabricated cholesterol biosensors is ordinarily limited by their inadequate performance in terms of one or more analytical parameters including stability, sensitivity and detection limit. In recent years, with the development of nanotechnology, novel nanomaterials are being fabricated and their useful properties and applications in biosensors are being examined. For example, nanomaterials-based biosensors, represent integration of material science, molecular engineering, chemistry and biotechnology which can improve the sensitivity and specificity of biomolecule detection, hold the capability of detecting or manipulating atoms and molecules, biomolecular recognition, pathogenic diagnosis and environment monitoring. Nanoscale materials offer distinctive size tunable electronic, catalytic and optical properties which opened new opportunities for designing highly efficient biosensor devices. Incorporation of nanomaterials in biosensing devices has been found to improve the electroactive surface, electronic conductivity and biocompatibility of the electrode surfaces. All these parameter, ultimately result in enhancing the analytical performance of the biosensors.

4.1 Working principles of cholesterol oxidase and cholesterol esterase

The measurement of cholesterol level is usually performed using cholesterol esterase (CE), cholesterol oxidase (CO), and peroxidase (POD). In 1974, Roeschlau and Allain [10,11] described the first fully enzymatic method. Cholesterol esters were cleaved by the action of cholesterol esterase to yield free cholesterol and fatty acids (scheme-1). Cholesterol oxidase then catalyzes the oxidation of cholesterol to cholest-4-en-3-one and hydrogen peroxide. The hydrogen peroxide so formed can be used to oxidize an indicator in the presence of peroxidase. In this test, 4-aminoantipyrine and phenol are converted into quinoneimine dye in the presence of POD by the oxidizing action of H_2O_2. The color intensity of the resultant dye is directly proportional to the cholesterol concentration and thus the level of cholesterol is determined by measuring the increase in absorbance at 512 nm.

$$\text{Cholesterol Esters} \xrightarrow{\text{C.E}} \text{Cholesterol} + \text{Fatty Acids}$$

$$\text{Cholesterol} + O_2 \xrightarrow{\text{C.O}} \text{Cholesten-3-one} + H_2O_2$$

$$2\,H_2O_2 + \text{4-aminoantipyrine} + \text{Phenol} \xrightarrow{\text{Peroxidase}} \text{Quinoneimine} + 4\,H_2O$$

Scheme 1. Mechanism of cholesterol oxidase and cholesterol esterase

The enzymatic method is the principal method used for measuring the level of cholesterol in plasma, which consists of solubilizing the cholesterol in an aqueous solution through its incorporation in lipoproteins. In order to measure accurately the cholesterol content of cells or tissue by this method, identification of a suitable solvent should be done to allow good recovery of free and esterified cholesterol in solution and that is compatible with the substrates and enzymes. Previously, for cells and tissues, the use of the enzymatic cholesterol assay has been limited due to the inadequate quantitative recovery of free cholesterol and cholesterol esters in enzyme compatible solvents. In fact, it has been proposed that the enzymatic method cannot be used for this purpose due to poor extraction and solubilization, which results in a wrong estimation of the cholesterol mass compared with an analytical method [12].

4.2 Enzyme based cholesterol biosensor

Electrochemical biosensors based on immobilized enzymes have found wide application for detection of several biological substrates. In general, enzyme-immobilization methods can be divided into four groups; adsorption, entrapment, cross-linking and covalent attachment [13]. For the adsorption method, the enzyme is attached on the matrix either by hydrogen bonds or through various charge attractions. For example, a multilayer of poly(2,6-pyridinedicarboxylic acid) with negative charge in the first layer and poly(allylamine hydrochloride) with positive charge in the second layer immobilizes negatively charged horseradish peroxidase by adsorption [14]. The one step entrapment process by electro-polymerization, using conducting polymers, such as polypyrrole [15,16] and polyaniline [17], is a fast, convenient and popular method. During the entrapment process, the enzyme is entrapped inside the polymer and the thickness can be arbitrarily controlled by changing the polymerization time. In the cross-linking method, the use of a cross-linking agent such as glutaraldehyde [18] is necessary to link the enzyme and the substrate. However, in a covalent attachment, chemical bonds are formed between the enzyme and the matrix through functional groups, such as amino or carboxylic groups. For instance, urease can be covalently immobilized on a copolymer film of poly(N-3-aminopropyl pyrrole) and polypyrrole, to fabricate an urea biosensor [19]; cholesterol oxidase (ChO) can be bound on a polyaniline-carbon nanotube composite film for a cholesterol biosensor [20]. The most recent approaches do not make use of mediators, but monitor directly the redox behavior of the enzyme at the electrode surface. A redox reaction between an enzyme and an analyte can be transduced to an electrical signal via subsequent direct electron transfer between the enzyme and the electrode. For the majority of enzymes, the protein matrix of the enzyme insulates the active site and thus prevents the direct electron transfer. In such cases, chemical mediators are commonly employed as transducers to convert the enzyme activity into an electrical signal. Carbon nanotubes (CNT) have been recognized as one of the most promising electrode materials since the first electrode application in the oxidation of dopamine in 1996 [21]. With the discovery of carbon nanotubes by Iijima, the combination of CNT with polyaniline (the polymer with strong biomolecular interactions, excellent stability and very good electronic property) was found to improve significantly the electrical, mechanical properties and reversible control of conductivity both by charge-transfer doping and protonation.

Since then, the research has been focused on their electrocatalytic behaviors toward the oxidation of biomolecules and their performance has been found to be much superior to those of other carbon electrodes in terms of reaction rate, reversibility and detection limit. Cholesterol biosensor based on enzymatic action at carbon nanotube modified electrode has been studied by various groups [22-25]. Mechanism of enzymatic detection of cholesterol at CNT modified electrode proposed by Solanki et al. [24] and Yan et al. [25] is shown in scheme 2.

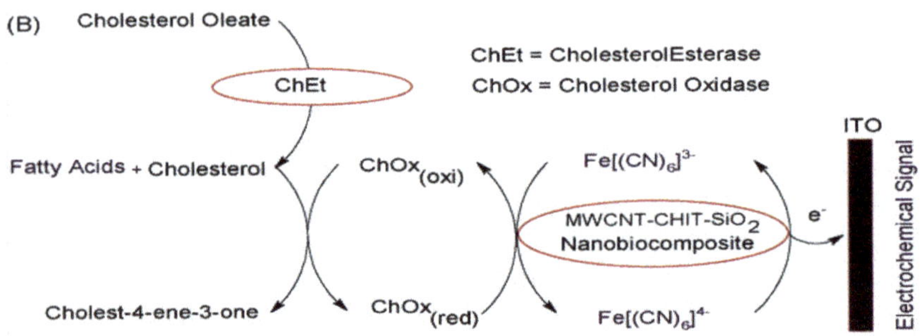

Scheme 2. Proposed biochemical reaction at ChEt–ChOx/MWCNT/SiO_2–CHIT/ITO bioelectrode.

Metallic nanowires, in particular Au nanowires are very promising as smart building blocks in emerging electronic and sensing applications [26-29]. As a bio-sensing material, metallic nanowires can be covalently modified with functional groups like –SH, –COOH and –NH_2 [30, 31], which can be linked further to a specific enzymes, such as cholesterol oxidase and cholesterol esterase. Sandro Carrara et al. [32] has shown that modification of screen-printed rhodium–graphite electrodes with gold nanoparticles integrated with cytochrome P450scc yields a bioelectrocatalytic system for cholesterol measurements. B.D. Malhotra et al. [33] prepared a cholesterol biosensor by immobilizing cholesterol oxidase onto electrophoretically deposited conducting polymer film on indium-tin-oxide glass plate derived from nano-structured polyaniline (PANI) colloidal suspension using N-ethyl-N′-(3-dimethylaminopropyl) corbodiimide and N-hydroxysuccinimide chemistry. These bio-electrodes exhibited linearity within he concentration range of 25 to 400 mg/dL of cholesterol, detection limit as 25mg/dL and sensitivity as 7.76×10^{-5} Abs $(mg/dL)^{-1}$. In the past few years, chitosan, which is a transformed polysaccharide, has been widely used as a biomaterial for immobilizing enzyme. It exhibits excellent membrane-forming ability, biocompatibility, nontoxicity, high mechanical strength, and hydrophilicity [34]. It has also been used as a medium for dispersion of CNTs in aqueous solution [35, 36].

Nanostructured oxides of metals such as tin, zirconium, titanium, zinc, iron, cerium and magnesium have been found to exhibit interesting non-toxic and catalytic properties, functional biocompatibility and nanomorphological surface activity. These materials also show enhanced electron-transfer kinetics and strong adsorption capability, providing suitable environments for the immobilization of biomolecules and resulting in enhanced electron transfer and improved biosensing characteristics. Scientists are now inclined to use biocompatible nanomaterials as suitable matrixes for the enzyme immobilizations for the effective detection of cholesterol [37–41]. Among nanomaterials, ZnO has attracted much attention due to its wide range of applications. ZnO as a wide band gap (3.37 eV) semiconductor plays an important role in sensors, actuators optics and optoelectronics, due to its semiconducting, pyroelectric and piezoelectric properties.

The discovery of graphene in 2004, added a new dimension to electrochemical biosensor research [42]. The unique thin structure of graphene nanosheets with an excellent capability for the rapid signal communication endorses their potential for such practical applications. Researchers have fabricated a sensitive sensor that can detect minute traces of cholesterol in human blood serum. They made the sensor by modifying multilayered graphene sheets with a compound derived from sugar molecules. This sensor is potentially useful for measuring the levels of serum cholesterol. The researchers found that the sensor was able to detect cholesterol even in the presence of interfering biochemicals such as salts, amino acids and lipids which are usually present in human blood serum. The sensor could detect cholesterol at micromolar concentrations, retaining stability and sensitivity even after 100 cycles of electrochemical studies [43]. The use of graphene can avoid the problems associated with metal alloy NP and CNT. The unique properties of graphene (fast electrontransportation, high thermal conductivity, excellent mechanical flexibility and good biocompatibility) justify its potential applicability as electrochemical biosensor [44-47].

4.3 Non- enzymatic cholesterol biosensor

The reproducibility of enzyme-based sensors is not very good and the activity of the enzyme is liable to be affected by temperature, pH, and toxic chemicals. Although, the enzyme-based sensors show high selectivity and excellent sensitivity, the activity of enzyme decreases with the use of the sensor and the enzyme is easily denatured during its immobilization due to the intrinsic nature of enzyme [48]. In order to overcome the disadvantages of enzymatic cholesterol sensors, nonenzymatic cholesterol sensors have been designed and fabricated, which have several attractive advantages, such as simple fabrication, high stability, reproducibility, low cost and freedom from oxygen limitations. Dai et al. synthesized porous tubular Ag nanoparticles by electrodeposition of Ag into a CdS modified porous anodic alumina template and subsequent removal of CdS [49], the modified electrode has a wide linear range but the sensitivity is relatively low.

It is the first example of a nonenzymatic cholesterol biosensor based on Ag nanoparticles. This biosensor possessed an acceptable reproducibility, good stability and low interferences. Park and co-workers fabricated a Pt nanoparticle modified macroporous gold electrode [50], while Lee used nanoporous Pt electrode for the nonenzymatic determination of cholesterol [51]. These two Pt-based sensors have good sensitivity but the detection limit was not enough. These sensors are highly effective at detecting low levels of cholesterol. Lee et al. [52] prepared a layer-by-layer assembled CNT network as a matrix for loading platinum nanoparticles in order to accomplish the non-enzymatic determination of cholesterol. The sensor has good anti-interferent properties in the presence of glucose, ascorbic acid, and dopamine. Hai Binh Nguyen et al. [53] reported a novel cholesterol biosensor based on electrochemical microelectrode with graphene films coated on PANi/Fe_3O_4 films which showed high sensitivity (74μAmM^{-1} cm^{-2}), fast response time (<5 s) and correlation coefficient square (R^2) of 0.9986. According to these authors, by taking the layer-by-layer fabricated electrode and advantage of graphene like pattern, excellent analytical quantification of cholesterol sensor like fast response time, high selectivity and reliability, high sensitivity can be obtained. Singh et al. [54] reported a new non enzymatic sensor using functionalized nanographite as a working electrode, Ag/AgCl as a reference and platinum as a counter electrode. Under optimum detection conditions, the constructed sensor had a linear range of 50 to 500 mg/dl for cholesterol with a correlation co-efficient of 0.99784. Sensitivity of the developed cholesterol sensor was 1.0587 μA/mg. This biosensor showed good reproducibility, stability and low interferences.

5. Conclusion

During the last 40 years, various biosensors have been developed encompassing a wide range of applications but the number of commercially available biosensors is limited. Different strategies have been explored for the construction of cholesterol biosensors that employed nanomaterials along with other materials as sensing elements. Yet there is much scope for further research. Among the various detection techniques of cholesterol, voltammogramic biosensing method has been recently developed as an extremely significant technique. Unfortunately, the biosensor market has not expanded as predicted. But, the combination of several new techniques, derived from chemistry, biochemistry, materials science and electronics will lead to the development of accurate and low cost commercial cholesterol biosensor which will surely boost the market of biosensors for health care. In fact, miniaturization will play an important role in the trend of biosensor development in the near future.

Acknowledgement

Author is grateful to the Director, NIT Agartala to allow the publication of these results.

References

[1] J.M Kauffmann, G.G. Guilbault, Potentiometric enzyme electrodes, Bioprocess. Technol. 15 (1991) 63–82.

[2] A.L. Ghindilis, P. Atanasov, M. Wilkins, E. Wilkins, Immunosensors: electrochemical sensing and other engineering approaches, Biosens. Bioelectron. 13 (1998) 113–131. *http://dx.doi.org/10.1016/S0956-5663(97)00031-6*

[3] J. Wang, Amperometric biosensors for clinical and therapeutic drug monitoring: a review, J. Pharm. Biomed. Anal. 19 (1999) 47–53. *http://dx.doi.org/10.1016/S0731-7085(98)00056-9*

[4] H. Ohnuki, T. Saiki, A. Kusakari, H. Endo, M. Ichihara, M. Izumi, Incorporation of glucose oxidase into languir-blodgett films based on Prussian blue applied to amperometric glucose biosensor, Langmuir. 23 (2007) 4675–4681. *http://dx.doi.org/10.1021/la063175g*

[5] S.A.Marzouk, S.S. Ashraf, K.A.Tayyari, Prototype amperometric biosensor for sialic acid determination, Anal. Chem. 79 (2007) 1668–1674. *http://dx.doi.org/10.1021/ac061886d*

[6] M. Yemini, Y. Levi, E. Yagil, J. Rishpon, Specific electrochemical phage sensing for Bacillus cereus and Mycobacterium smegmatis, Bioelectrochemistry. 70 (2007) 180–184.
http://dx.doi.org/10.1016/j.bioelechem.2006.03.014

[7] M. Pohanka, P. Skládal, M. Kroča, Biosensors for biological warfare agent detection, Def. Sci. J. 57 (2007) 185–193.
http://dx.doi.org/10.14429/dsj.57.1760

[8] M. Pohanka, D. Jun, K. Kuča, Amperometric biosensor for evaluation of competitive cholinesterase inhibition by the reactivator HI-6, Anal. Lett. 40 (2007) 2351–2359.
http://dx.doi.org/10.1080/00032710701576007

[9] G. Liu, Y. Lin, Biosensor based on self-assembling acetylcholinesterase on carbon nanotubes for flow injection/amperometric detection of organophosphate pesticides and nerve agents, Anal. Chem. 78 (2006) 835–843.
http://dx.doi.org/10.1021/ac051559q

[10] C.C. Allain, L. Poon, S.G. Chan, W. Richmond, P. Fu, Enzymatic determination of total serum cholesterol, Clin. Chem. 20 (1974) 470-473.

[11] P. Roschlau, E. Bernt, W. Gruber, Enzymatishe Bestimmung des Cesamt-Cholesterins in Serum, Z. Klin. chem. Klin. Biochem. 12 (1974) 226-229.

[12] P. Cullen, K. Tegelkamp, M. Fobker, F. Kannenberg, G. Assmann, Measuring cholesterol in macrophages: comparison of high-performance liquid chromatography and gas-liquid chromatography with enzymatic fluorometry, Anal. Biochem. 251 (1997) 39 – 44.
http://dx.doi.org/10.1006/abio.1997.2227

[13] A.J.Cunningham, Introduction to Bioanalytical Sensors; John Wiley & Sons Inc.: New York, (1998) 167-192.

[14] M.H.Yang, C.X.Li, Y.H.Yang, G.L.Shen, R.Q.Yu, Hydrogen peroxide biosensor based on enzyme multilayers through electrostatic adsorption. Acta Chim. Sin. 62 (2004) 502-507.

[15] S. Singh, A. Chaubey, B.D. Malhotra, Amperometric cholesterol biosensor based on immobilized cholesterol esterase and cholesterol oxidase on conducting polypyrrole films, Anal. Chim. Acta. 502 (2004) 229-234.
http://dx.doi.org/10.1016/j.aca.2003.09.064

[16] S. Brahim, D. Narinesingh, A. Guiseppi-Elie, Amperometric determination of cholesterol in serum using a biosensor of cholesterol oxidase contained within a polypyrrole-hydrogel membrane, Anal. Chim. Acta. 448 (2001) 27-36. *http://dx.doi.org/10.1016/S0003-2670(01)01321-6*

[17] H.Y. Wang, S.L. Mu, Bioelectrochemical characteristics of cholesterol oxidase immobilized in a polyaniline film. Sensor. Actuator. B-Chem. 56 (1999) 22-30.

[18] E. Tamiya, Y. Sugiura, A. Akiyama, I. Karube, Ultramicro-H2O2 electrode for fabrication of the in-vivo biosensor. Ann. N.Y. Acad. Sci. 613 (1990) 396-400. *http://dx.doi.org/10.1111/j.1749-6632.1990.tb18185.x*

[19] V. Rajesh Bisht, W. Takashima, K. Kaneto, An amperometric urea biosensor based on covalent immobilization of urease onto an electrochemically prepared copolymer poly (N-3-aminopropyl pyrrole-co-pyrrole) film. Biomaterials. 26 (2005) 3683-3690. *http://dx.doi.org/10.1016/j.biomaterials.2004.09.024*

[20] C. Dhand, S.K. Arya, M. Datta, B.D. Malhotra, Polyaniline-carbon nanotube composite film for cholesterol biosensor. Anal. Biochem. 383 (2008) 194-199. *http://dx.doi.org/10.1016/j.ab.2008.08.039*

[21] P.J. Britto, K.S.V. Santhanam, P.M. Ajayan, Carbon nanotube electrode for oxidation of dopamine, Bioelectrochem. Bioenerg. 41 (1996) 121–125. *http://dx.doi.org/10.1016/0302-4598(96)05078-7*

[22] G. Li, J.M. Liao, G.Q. Hu, N.Z. Ma, P.J. Wu, Study of carbon nanotube modified biosensor for monitoring total cholesterol in blood, Biosens. Bioelectron. 20 (2005) 2140–2144. *http://dx.doi.org/10.1016/j.bios.2004.09.005*

[23] A.L. Gopalan, K.P. Lee, D. Ragupathy, Development of a stable cholesterol biosensor based on multi-walled carbon nanotubes–gold nanoparticles composite covered with a layer of chitosan–room-temperature ionic liquid network, Biosens. Bioelectron. 24 (2009) 2211–2217. *http://dx.doi.org/10.1016/j.bios.2008.11.034*

[24] P.R. Solanki, A. Kaushik, A. A. Ansari, A.Tiwari, B. D.Malhotra, Multi-walled carbon nanotubes/sol–gel-derived silica/chitosan nanobiocomposite for total cholesterol sensor, Sens. Actuators B. 137 (2009) 727–735. *http://dx.doi.org/10.1016/j.snb.2008.12.044*

[25] J.Y. Yang, Y. Li, S.M. Chen, M. Lin, Fabrication of a Cholesterol Biosensor Based on Cholesterol Oxidase and Multiwall Carbon Nanotube Hybrid Composites, Int. J. Electrochem. Sci. 6 (2011) 2223–2234.

[26] J.K.N. Mbindyo, B.D. Reiss, B.R. Martin, C.D. Keating, M.J. Natan, T.E. Mallouk, DNA-directed assembly of gold nanowires on complementary surfaces, Adv. Mater. 13 (2001) 249–254.
http://dx.doi.org/10.1002/1521-4095(200102)13:4<249::AID-ADMA249>3.0.CO;2-9

[27] S. Keebaugh, A.K. Kalkan, W.J. Nam, S.J. Fonash, Gold nanowires for the detection of elemental and ionic mercury, Electrochem. Solid-State Lett. 9 (2006) H88–H91.
http://dx.doi.org/10.1149/1.2217130

[28] N.I. Kovtyukhova, T.E.Mallouk, Nanowires as building blocks for self assembling logic and memory circuits, Chem. Eur. J. 8 (2002) 4355–4363.
http://dx.doi.org/10.1002/1521-3765(20021004)8:19<4354::AID-CHEM4354>3.0.CO;2-1

[29] C.D. Keating, M.J. Natan, Striped metal nanowires as building blocks and optical tags, Adv. Mater. 15 (2003) 451–454.
http://dx.doi.org/10.1002/adma.200390105

[30] S. Mann, W. Shenton, M. Li , S. Connolly , D. Fitzmaurice , Carbon nanotube-templated self-assembly and thermal processing of gold nanowires, Adv. Mater. 12 (2000) 1430–1432.
http://dx.doi.org/10.1002/(SICI)1521-4095(200001)12:2<147::AID-ADMA147>3.0.CO;2-U

[31] S. Connolly, S. Cobbe, D. Fitzmaurice, Effects of ligand-receptor geometry and stoichiometry on protein-induced aggregation of biotin-modified colloidal gold, J. Phys. Chem. B. 105 (2001) 2222–2226.
http://dx.doi.org/10.1021/jp001948i

[32] V. Victoria, S.C. Shumyantseva, B. Valter, D.J. Rileyd, V. B. Tatiana, G.S. Konstantin, I.A. Alexander, N. Claudio, Direct electron transfer between cytochrome P450scc and gold nanoparticles on screen-printed rhodium–graphite electrodes, Biosens. Bioelectron. 21 (2005) 217–222.
http://dx.doi.org/10.1016/j.bios.2004.10.008

[33] C. Dhand, S.P. Singh, S.K. Arya, M. Datta, B. D. Malhotra , Cholesterol biosensor based on electrophoretically deposited conducting polymer film derived from

nano-structured polyaniline colloidal suspension, Anal. Chim. Acta . 602 (2007) 244–251.
http://dx.doi.org/10.1016/j.aca.2007.09.028

[34] B. Krajewska, Application of chitin- and chitosan-based materials for enzyme immobilizations: a review, Enzyme Microb. Technol. 35 (2004) 126–139.
http://dx.doi.org/10.1016/j.enzmictec.2003.12.013

[35] Y.C. Tsai, S.Y. Chen, H.W. Liaw, Immobilization of lactate dehydrogenase within multiwalled carbon nanotube-chitosan nanocomposite for application to lactate biosensors, Sens. Actuators B: Chem. 125 (2007) 474–481.
http://dx.doi.org/10.1016/j.snb.2007.02.052

[36] D. Du, X. Huang, J. Cai, A. Zhang, Amperometric detection of triazophos pesticide using acetylcholinesterase biosensor based on multiwall carbon nanotube–chitosan matrix, Sens. Actuators B: Chem. 127 (2007) 531–535.
http://dx.doi.org/10.1016/j.snb.2007.05.006

[37] S. Singh, A. Chaubey, B.D. Malhotra, Amperometric cholesterol biosensor based on immobilized cholesterol esterase and cholesterol oxidase on conducting polypyrrole films. Anal. Chim. Acta. 502 (2004) 229-324.
http://dx.doi.org/10.1016/j.aca.2003.09.064

[38] A. Ansari, A. Kaushik, P. Solanki, B. D. Malhotra, Sol–gel derived nanoporous cerium oxide film for application to cholesterol biosensor. Electrochem. Commun. 10 (2008) 1246-1249.
http://dx.doi.org/10.1016/j.elecom.2008.06.003

[39] R. Khan, A. Kaushik, P. Solanki, A. Ansari, M. Pandey, B. Malhotra, Zinc oxide nanoparticles-chitosan composite film for cholesterol biosensor, Anal. Chim. Acta. 616 (2008) 207-213.
http://dx.doi.org/10.1016/j.aca.2008.04.010

[40] S. Singh, S. Arya, P. Pandey, B. Malhotra, S. Saha, K. Sreenivas, V. Gupta, Ultrasensitive cholesterol biosensor based on low-temperature grown ZnO nanoparticles, App. Phys. Lett. 91 (2007) 63901-63904.
http://dx.doi.org/10.1063/1.2768302

[41] C. Bongiovanni, T. Ferri, A. Poscia, M. Viralli, R. Santucci, A. Desideri, An electrochemical multienzymatic biosensor for determination of cholesterol. Bioelectrochem. 54 (2001) 17-22.
http://dx.doi.org/10.1016/S0302-4598(01)00105-2

[42] K.S. Novoselov, A.K. Geim, S.V. Morozov, Y . Zhang, S.V. Dubonos, I.V. Grigorieva, A.A. Firsov, Electric Field Effect in Atomically Thin Carbon Films, Science 306 (2004) 666–669. *http://dx.doi.org/10.1126/science.1102896*

[43] N. Agnihotri, et al. Non-enzymatic electrochemical detection of cholesterol using β-cyclodextrin functionalized graphene. Biosens. Bioelectron. 63 (2014) 212–217. *http://dx.doi.org/10.1016/j.bios.2014.07.037*

[44] M. Pumera, Electrochemistry of graphene: new horizons for sensing and energy storage. Chem. Rec. 9 (2009) 211–223. *http://dx.doi.org/10.1002/tcr.200900008*

[45] M.J. Allen, V.C. Tung, R.B. Kaner, Honeycomb carbon: a review of grapheme, Chem. Rev. 110 (2010) 132–145. *http://dx.doi.org/10.1021/cr900070d*

[46] M. Pumera, Graphene-based nanomaterials and their electrochemistry, Chem. Soc. Rev. 39 (2010) 4146–4157. *http://dx.doi.org/10.1039/c002690p*

[47] D.A.C. Brownson, C.E. Banks, Graphene electrochemistry: an overview of potential applications. Analyst. 135 (2010) 2768–2778. *http://dx.doi.org/10.1039/c0an00590h*

[48] P. Sejin, B. Hankil, C. T. Dong, Electrochemical non-enzymatic glucose sensors, review. Anal. Chim. Acta. 556 (2006) 46–57. *http://dx.doi.org/10.1016/j.aca.2005.05.080*

[49] Y. Li, H. Bai, Q. Liu, J. Bao, M. Han, Z. Dai, A non enzymatic cholesterol sensor constructed by using porous tubular silver nanoparticles. Biosens. Bioelectron. 25 (2010) 2356–2360. *http://dx.doi.org/10.1016/j.bios.2010.03.036*

[50] Y.J. Lee, J.Y.Park, Non enzymatic free-cholesterol detection via a modified highly sensitive macroporous gold electrode with platinum nanoparticles, Biosens. Bioelectron. 26 (2010) 1353–1358. *http://dx.doi.org/10.1016/j.bios.2010.07.048*

[51] Y.J. Lee, J.D. Kim, J.Y. Park, CMOS-integrable enzyme-free amperometric cholesterol nano-biosensor for U-Health and POC applications, Journal of the Korean Physical Society. 54 (2009) 1769–1773. *http://dx.doi.org/10.3938/jkps.54.1769*

[52] Y. Jiao, L. Hyuck, C. Misuk, N. Jeodo, L. Youngkwan, Non enzymatic cholesterol sensor based on spontaneous deposition of platinum nanoparticles on layer-by-layer assembled CNT thin film. Sensors and Actuators B. 171– 172 (2012) 374–379.

[53] N. H. Binh, N. V. Chuc, N.V. Tu, L. H. Doan, N. V. Quynh, T. N. T. Thanh, D. Q. Phuc, N. X. Nghia, P. N. Minh, T. D. Lam, Development of the layer-by-layer biosensor using graphene films: application for cholesterol determination. Adv. Nat. Sci. Nanosci. Nanotechnol. 4 (2013) 015013-015016. *http://dx.doi.org/10.1088/2043-6262/4/1/015013*

[54] B. Singh, B. Nitin, V. K. Jain, B. Vasuda, A novel nanographite based non-enzymatic cholesterol sensor", In: Jain. V. K., Verma. A. (eds.), Physics of semiconductor devices, Environmental science and engineering , Springer International Publishing , Switzerland, (2014) 531-534. *http://dx.doi.org/10.1007/978-3-319-03002-9_133*

CHAPTER 10

Retention Behaviour of Synthetic Dyes on Plain and Tri-n-butylamine Impregnated Silica Gel-G and Starch Layers in Aqueous Alcohol Media

S.D. Sharma[a*] and Charu Sharma[b]

[a] School of Sciences, IFTM University, Delhi Road, Moradabad-244102, India

[b] Manav Rachna College of Education, Faridabad (India)

charu15sharma@yahoo.com, dsdsharma144@rediffmail.com

Abstract

Retention behaviour of twenty four synthetic dyes on plain and tri-n-butylamine (TBA) impregnated silica gel-G and starch layers in aqueous alcohol media have been studied. Compact and sharp spots led to some useful separations of dyes, having similar chromophore groups on the basis of their differential solubilities and preferential adsorption on these layers. TBA impregnation has significant effect on the resolution of dyes. Diffusion ratios for some dyes were determined. Plain starch layers were found to have the least compactness and, therefore, not much useful for the separation of dyes. The effect of water content in the mobile phase on the development time has been discussed. The easy and inexpensive method can be applied for the trace analysis of synthetic dyes in some natural and synthetic samples.

Keywords

Retention Behaviour, Synthetic Dyes, TBA Impregnation, Silica Gel and Starch Layers

Contents

1. Introduction

The identification and separation of synthetic dyes is of great importance in analytical chemistry. Most of these dyes have diazo and amino groups and may be carcinogenic in nature. This is concerning as these dyes are extensively used in various foods and other products such as clothes, healthcare products, medicines, leather etc. The advantages of thin layer chromatography (TLC) in this field are now well known. A survey of the literature reveals that plain and impregnated silica gel-G, starch and their admixture layers have been extensively used for the purpose. The impregnation of the adsorbent layers has resulted into more compact and sharper spots leading to better and effective resolution of a variety of dyes. The TLC of synthetic dyes up to 2007 has been reviewed by Mohammad *et al.* [1].

However, little work has been done on the reverse phase thin layer chromatography (RPTLC) of synthetic dyes. Davidek and Janicek[2] were probably the first who carried out the TLC of fat soluble dyes on layers prepared by spreading on glass plates a suspension of starch in a 10% solution of paraffin oil in petrol ether. Later, RPTLC of synthetic dyes was also performed on C_{18} modified silica gel [3-5] layers. A mixture of dyes has been separated by electro chromatography on RP-18F_{254}s layers [6]. Normal and RPTLC of fluorescent dyes [7] and coal tar dyes in foods [8, 9] were also carried out. Planar chromatography with electro-osmotic flow was used to separate a mixture of dyes using 80% aqueous ethanol as a mobile phase [10]. Plain and ion pair TLC of synthetic dyes was performed on an admixture of silica gel and barium sulphate in an aqueous ethanol media [11].

In our laboratory RPTLC of metal ions was successfully carried out using tri-n-butyl phosphate (TBP) and tri-n-butylamine (TBA) impregnated silica gel – G [12-14] and talc [15] layers. We have also separated a number of amino acids on TBA impregnated silica gel-G layers [16-17]. Also, the sorption behaviour of few water soluble dyes using inorganic material [18] and activated carbon [19] as sorbents has been extensively studied. Normal phase soil TLC of synthetic dyes has been carried out and their impact on translocation of trace metals was also studied [20]. Adsorption of crystal violet dye on Pinous bark powder is carried out [21] and rice starch has been used as an unconventional TLC support for the separation of 3, 5-dinitrobenzoic acid esters of a homologous series of aliphatic C_1-C_{20} linear alcohols [22]. Mishra has carried out RPTLC of synthetic dyes on coconut oil impregnated egg shell layers using alcohol and alcohol-water systems [23].

Recently, a number of adsorption studies of synthetic dyes on waste materials have been carried out [24-26]. Also, ultra performance liquid chromatography (UPLC) with electrospray ionization tendem mass spectroscopy have been performed for simultaneous determination of various dyes in aquatic products [27]. A review on sequestration of dye from textile waste water using agricultural waste products as adsorbents signifying their potential has recently been published [28]. Recently development and validation of a TLC method for the analysis of synthetic food-stuff dyes has appeared [29]. Casoni et al has developed a TLC-Densitometric method for the determination of food colorants from pharmaceutical powders [30]. Very recently, determination of synthetic food dyes in commercial soft drinks by using TLC and ion pair HPLC, and also a rapid and green limit test method for some synthetic colorants in foods using polyamide TLC, have been reported [31-32].

However, all these studies, in no way, decrease the need of an easy and inexpensive technique of separations such as normal and RPTLC. Dyes are normally classified according to either their chemical constitution or their application to textile fibres and for other coloring purposes. These are also classified according to the chromophores or essential color producing groups that are present. However, a classification based primarily on characteristic structural units is more satisfactory. As far as we are aware, TBA impregnated silica gel G and starch layers have not been used for the separation of dyes. We have, therefore, decided to study the chromatographic behaviour of some closely related dyes on plain and TBA impregnated silica gel-G and starch layers in an alcohol media. Aqueous alcohol system is chosen as it gives very effective separation of synthetic dyes [10, 11, 23] and has also been used in structure driven retention optimization model for RPTLC [33]. The diffusion of the spots and the effect of water content of the mobile phase on the retention behaviour of dyes have also been discussed.

2. Materials and methods

All the chemicals and solvents used were of chromatographic grade. Glass plates of 15 × 2.5 cm and chromatographic clambers of 20 × 4 cm ID were used.

The thin layer chromatographic plates (thickness 0.5 mm) were prepared by spreading with an applicator a slurry of silica gel-G or starch on clean glass plates. The silica gel-G plates were activated at 100° ±5° C for 1 h in a temperature controlled electric oven. For the reverse phase thin layer chromatography, these layers were impregnated with tri-n-butyl amine (20% solution in benzene). 2-3µl of 0.1% solution of the dye was applied to the layers with the help of a micropipette. TLC plates were developed by the ascending technique. The solvent ascent was fixed to 11 cm in all cases in saturated glass jars at room temperature and under similar experimental conditions. All experiments were run in triplicates. After development, the dyes were self detected as their original color. The reproducibility of chromatograms on the same coatings was good and for a given dye the variation in hR_f does not exceed 5% of the average value.

3. Results and discussion

A cursory inspection of the hR_f data (Tables 1 and 2) reveals that some dyes such as methyl violet, rhodamine-B, tartrazine, methyl red and eriochrome black T show tailing on plain silica gel-G and starch layers. Contrary to it, very sharp and compact spots were observed for dyes on TBA impregnated layers. However, there is no significant change in hR_f values of the individual dyes even when their mixtures are applied for chromatography.

The effect of nature of adsorbent on hR_f values is quite revealing. In most cases, the hR_f values on starch were lower than on silica gel-G layers. This may be due to the strong adsorption of the dyes on starch layers. Further, for most of the dyes, the hR_f values on plain silica gel-G layers were higher than on TBA impregnated ones e.g. Acridine yellow, acridine orange, auramine, fast red, thymol blue and phenolphthalein in methanol; metanil yellow, acridine yellow and auramine in ethanol; thymol blue, metanil yellow, auramine, rhodamine-B eosin, neutral red, pheono-red, thymolphthalein & tartrazine in propanol and Bengal rosa, neutral red, acridine orange and eosin in butanol media. Thus, impregnation leads to higher adsorption on these layers.

The addition of water to alcohol in the mobile phase generally increases the hR_f values probably due to the increase in the solubility of dyes in the given media. In all cases, the difference of $\pm$ 20 or more in hR_f values has been taken as the criterion for satisfactory resolution of dyes. The separations are a direct consequence of the preferential adsorption of the dyes on the coatings and their differential solubilities in the given mobile phase. Other possible factors are hydrogen bonding, colloidal nature of some of the dyes and steric effects. On the basis of hR_f data (Tables 1 and 2), the separation of dyes having similar chromophore groups have actually been realized (Table 3).

The spot diffusion ratio [34] is a parameter which would indicate the diffusion after development, of the spots on the stationary phase. This parameter, which is the ratio of the diameter of the spot after development divided by the diameter of the applied spot, would indicate: (a) how well the adsorbent layer (particle size and distribution, density and compactness) is made; and (b) the degree of the resolution. The larger the ratio, the inferior is the layer and vice versa. Table 4 shows the diffusion ratios of the dyes spotted on silica gel-G, starch, TBA impregnated silica gel-G as well as starch layers which were developed for 11 cm in all cases. The data reported for each dye is the mean of the three different values. In general, diffusion ratios are lower for plain silica gel-G layers than for impregnated ones implying that impregnation leads to the decrease in the compactness of the adsorbent layer. However, reverse is true in case of starch layers where the diffusion ratios are much higher on plain layers than on the impregnated ones. The plain silica gel-G retains a small water content (0.18 mole H_2O/SiO_2) [35] and as such the *spot diffusion ratio* is much lower. However, starch contains many hydroxyl groups and is consequently strongly hydrophilic resulting in much higher diffusion ratios (table 4).

A plot of average diffusion ratio v/s adsorbent layers in different alcohol media reveals that the value for silica gel-G, impregnated silica gel-G and impregnated starch layers are quite different having maximum value as 4.66. However, on plain starch layers, the values are much higher (up to 8.33) thereby showing the less compactness of the starch layer and as such, these layers are not good for separation of synthetic dyes (Fig. 1). The TBA impregnation makes the starch surface more compact resulting in lower diffusion ratios. Further, it is observed that there is no correlation between the diffusion ratio and the alcohol chain length of the mobile phase.

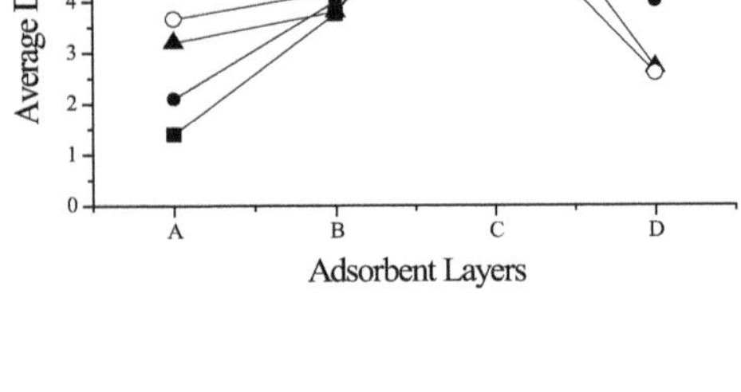

Fig. 1 Plot of average diffusion ratio vs adsorbent layers in different alcohol media

A. Silica gel-G B. TBA impregnated silica gel-G
C. Starch D. TBA impregnated starch

Table 5 shows the effect of water content in the mobile phase and that of impregnation on the development time. The salient features are:

(a) As the water content in the mobile phase increases, the development time also increases.

(b) In all cases, there is a sharp increase in the development time as one proceeds from pure alcohol to H_2O: alcohol mixture (2:8). On further addition of water, there is much less increase till the composition becomes (5:5). However, beyond this, no significant increase in the development time is noticed.

(c) In all cases, the development time for impregnated layers is always higher than for unimpregnated ones probably due to the less polarity of TBA impregnated layers. Further, the difference between the two is more and more pronounced as one proceeds from methanol to butanol.

(d) Alcohol chain-length also affects the development time. In case of pure alcohol as the mobile phase, the development time gradually increases with the alcohol chain length. This may be due to a decrease in the dielectric constant of the media.

Generally, the dyes contain one or more functional groups in their molecular structure. Some common groups are –OH, -COOH, -SO_3H, -NO_2, –N=N– etc. During chromatography, the interaction of the above functional groups with the matrix of adsorbent material leads to dye separations. These interactions may be anywhere from covalent to coulombic, H-bonding or the weak Van der Walls forces. The adsorption of a particular dye on the adsorbent layer depends on the strength and type of such interactions.

4. Conclusion

The normal and reversed phase TLC of synthetic dyes on different adsorbents show that TBA impregnation in silica gel-G and starch layers leads to stronger adsorption thereby increasing the possibility of separations. The average *spot diffusion ratio* revealed that, in general, impregnation results a decrease in the compactness of the adsorbent layer. However, for starch layers reverse is the trend and as such plain starch layers are not very useful for separations. The development time increases gradually when in aqueous alcohol media the water content increases up to 20 % of the mixture. Beyond that, no significant increase in development time is noticed. Also, the development time increases gradually with increase in alcohol chain length. Some useful separations of dyes having similar chromophore groups were achieved.

Table 1 hR_f values of synthetic dyes on various coatings in alcohol media

Name of the Dye	hR_f Values															
	CH_3OH				C_2H_5OH				C_3H_7OH				C_4H_9OH			
	A	B	C	D	A	B	C	D	A	B	C	D	A	B	C	D
Eriochrome Black T	T	75	50	75	05	75	00	70	T	70	00	00	05	20	00	00
Methyl red	90	95	T	100	50	75	T	95	50	95	00	00	40	70	00	00
Fast red	80	00	80	100	80	00	00	100	20	95	00	100	95	90	00	100
Thymol blue	90	75	50	100	50	85	90	55	100	95	00	00	95	80	30	30
Bromocresol	90	90	82	T	T	75	00	T	90	90	00	T	T	80	T	T
Phenolphthalein	95	80	80	100	20	70	90	95	75	90	00	95	85	30	95	90
Chromotropic acid	00	00	00	95	00	65	00	100	00	00	00	00	00	00	90	00
Metanil yellow	75	65	T	65	74	40	00	10	75	70	00	00	80	80	20	T
Acridine yellow	10	00	20	10	20	00	30	10	05	05	00	00	00	05	T	00
Methyl violet	T	00	T	00	T	10	T	05	T	10	00	00	T	15	T	00
Acridine orange	10	05	40	05	00	00	00	00	00	05	00	00	10	00	00	00
Methyl green	00	00	00	00	00	00	00	00	00	00	00	00	00	00	00	00
Auramine	40	20	20	00	10	05	00	05	25	00	00	00	00	10	20	00
Rhodamine B	T	70	T	00	00	35	T	05	15	05	00	T	05	10	85	T
Eosin	75	80	00	90	60	40	T	100	85	60	00	T	85	70	95	T
Bengal rosa	90	80	15	90	85	70	00	95	90	90	00	00	90	75	00	00
Alizarin	15	00	05	95	05	00	05	95	00	00	00	00	00	05	00	00
Alizarin red S	10	80	72	00	T	10	00	05	00	10	00	00	00	05	85	00
Neutral red	65	90	20	00	80	50	25	00	85	70	00	00	78	60	20	00
Tartrazine	T	75	00	90	05	85	00	00	20	05	00	00	00	15	00	00
Anthrone	80	00	75	00	15	90	95	95	90	95	00	10	95	30	15	00
Thymolphthalein	100	00	90	80	15	25	80	70	100	00	00	70	100	95	90	75
Phenol red	10	30	00	00	00	00	00	00	20	00	00	00	40	00	00	00
Rhodizonic acid	00	80	00	00	00	00	00	55	00	90	95	00	00	95	90	00

T – Tailing

A – Silica gel – G
B – TBA impregnated silica gel –G
C – Starch
D – TBA impregnated starch

Table 2 hR_f values of synthetic dyes on various coatings in alcohol – Water (1 : 1) mobile phases

Name of the Dye	hRf Values															
	$CH_3OH : H_2O$				$C_2H_5OH : H_2O$				$C_3H_7OH : H_2O$				$C_4H_9OH : H_2O$			
	A	**B**	**C**	**D**	**A**	**B**	**C**	**D**	**A**	**B**	**C**	**D**	**A**	**B**	**C**	**D**
Eriochrome Black T	85	75	45	40	95	95	00	70	05	95	85	45	25	75	10	00
Methyl red	85	75	T	90	90	95	T	85	90	90	90	50	50	90	90	85
Fast red	95	00	00	85	90	30	00	90	95	95	95	100	90	95	00	95
Thymol blue	80	85	90	55	95	75	90	55	85	60	85	65	60	75	95	60
Bromocresol	90	80	80	90	95	95	95	90	90	90	T	60	50	65	T	70
Phenolphthalein	90	85	50	90	45	75	55	00	95	60	95	90	00	80	00	95
Chromotropic acid	100	00	00	100	90	85	00	90	00	00	95	40	30	00	95	20
Metanil yellow	95	65	30	55	90	30	85	75	95	80	T	65	60	80	90	55
Acridine yellow	25	05	20	00	00	00	30	00	20	15	90	00	50	05	25	00
Methyl violet	T	05	25	05	10	05	T	05	35	45	T	10	50	20	10	00
Acridine orange	20	05	20	00	10	05	T	90	10	10	90	50	10	10	06	00
Methyl green	10	35	00	00	00	30	00	00	00	00	10	00	00	00	05	50
Auramine	30	25	15	05	20	15	20	00	30	35	T	35	20	30	T	00
Rhodamine B	35	50	T	90	T	40	T	85	85	70	80	50	45	50	T	T
Eosin	70	85	30	90	95	95	80	75	90	75	100	70	50	80	75	65
Bengal rosa	90	85	20	80	95	90	T	90	95	65	100	55	50	90	05	25
Alizarin	20	00	00	95	10	00	10	95	T	T	T	75	T	T	00	00
Alizarin red S	30	20	50	10	T	T	75	80	T	T	90	70	T	T	95	60
Neutral red	T	25	00	90	60	65	05	70	90	75	80	60	45	75	20	00
Tartrazine	90	95	T	90	90	85	10	85	90	90	88	55	55	75	30	T
Anthrone	100	00	00	90	85	00	75	10	95	80	95	25	95	75	00	30
Thymolphthalein	90	90	00	95	90	50	50	10	95	90	95	75	00	80	95	90
Phenol red	00	00	T	90	00	00	00	85	95	95	75	45	40	00	25	00
Rhodizonic acid	00	00	05	60	00	00	05	80	95	90	50	45	50	95	05	50

Table 3 Separation of dyes having same chromophore groups achieved on various coatings

Coatings	**Mobile phase**	**Class (Chromophore group)**	**Separations (hR_f)**
B	CH_3OH	Azo syes	Fast red (00) – Methyl red (95)
B	CH_3OH	Azo dyes	Chromotropic acid (00) – Metanil yellow (65)– Methyl red (95)
B	C_2H_5OH	Phthalein dyes	Phenol red (00) – Thymolphthalein (25) – Thymol blue (85)
B	C_4H_9OH	Phthalein dyes	Phenolphthalein (30) – Neutral red (60)
A	C_2H_5OH	Acridine dyes	Acridine yellow (20) – Acridine orange (00)
A	C_4H_9OH-H_2O (1:1)	Anthraquinone dyes	Alizarin (95) – Alizarin red S (00)
A	CH_3OH	Anthraquinone dyes	Alizarin or Alizarin red S (15) – Anthrone (80)
A	C_4H_9OH – H_2O (1:1)	Triphenyl methane dyes	Methyl green (00) – Methyl violet (50)
B	C_3H_7OH	Xanthene dyes	Rhodamine B (05) – Eosin (60) – Bengal rosa (90)
D	C_2H_5OH	Xanthene dyes	Eosin (95) – Neutral red (05)
A	C_2H_5OH	Xanthene dyes	Rhodamine B (05) – Neutral red (75)
D	CH_3OH	Xanthene dyes	Bengal rosa (85) – Neutral red (05)

Table 4 Diffusion ratios of some dyes on various coatings

Name of the Dye	Diffusion Ratios															
	CH_3OH				C_2H_5OH				C_3H_7OH				C_4H_9OH			
	A	B	C	D	A	B	C	D	A	B	C	D	A	B	C	D
Alizarin	1.0	2.0	9.0	1.0	1.5	5.0	12.5	3.7	4.0	2.0	5.9	2.5	2.3	4.0	9.0	3.3
Rhodomine B	0.8	10.0	12.5	7.5	1.0	5.0	7.5	8.0	5.0	8.6	15.0	10.0	4.0	5.0	12.5	5.0
Phenol red	1.0	1.3	1.0	3.0	2.6	5.0	1.0	3.0	1.0	3.0	1.0	1.0	4.0	5.0	1.0	1.0
Auramine	1.0	3.0	4.0	3.0	0.3	2.5	1.0	2.0	6.0	3.0	1.0	1.0	2.2	2.5	1.0	1.0
Methyl red	2.5	2.5	17.5	5.0	5.0	5.0	20.0	3.3	3.0	5.0	15.0	3.0	5.0	2.0	12.0	1.0
Metanil yellow	3.3	12.0	15.0	7.5	0.8	3.0	9.0	5.0	2.0	4.0	8.0	2.0	4.4	4.0	10.0	5.0
Eosin	1.0	1.0	10.0	9.0	2.5	2.6	10.0	6.0	6.0	5.0	11.0	2.0	4.0	6.0	5.0	5.0
Anthrone	1.0	1.0	5.0	2.5	3.3	4.5	1.0	2.6	1.0	3.0	1.0	1.0	2.6	5.0	1.0	1.0
Thymolphthalein	1.0	1.0	1.0	3.3	2.0	3.3	4.0	2.0	1.0	1.0	5.0	2.0	4.5	4.0	1.0	1.0

Table 5 Effect of water content in the mobile phase on development time (Mobile phase ascent being 11 cm)

Mobile phase composition		Development Time (in minutes)															
		CH_3OH				C_2H_5OH				C_3H_7OH				C_4H_9OH			
		A	B	C	D	A	B	C	D	A	B	C	D	A	B	C	D
Pure alcohol		45	50	30	40	50	70	35	45	50	70	75	85	45	64	60	90
H_2O	Alcohol																
20	80	60	75	50	65	70	85	50	55	60	80	105	115	60	75	100	135
30	70	70	78	52	70	80	92	52	60	67	84	115	125	75	77	102	145
40	60	75	82	60	77	82	96	57	66	70	90	117	128	80	81	110	150
50	50	78	95	70	85	90	98	65	75	75	92	120	135	90	83	115	175
80	20	84	98	75	89	93	103	68	80	81	98	125	138	94	86	119	180

References

[1] A. Mohammad and N. Ershad, TLC of synthetic dyes: A review, chem. environ. Res, 16 (3 & 4) (2007) 165-179.

[2] J. Davidek and U. G. Janicek, Chromatography of fat soluble food dyes on thin starch layers with stationary non-polar phases, J. Chromatog. 15 (1964) 542-545. *http://dx.doi.org/10.1016/S0021-9673(01)82816-8*

[3] H. Oka, Y. Ikai, N. Kawamura, M. Yamada, H. Inoue, T. Ohno, K. Inagaki, A. Kuno and N. Yamamoto, Simple method for the analysis of food dyes on reversed phase thin layer plates, J. Chromatog., 411 (1987) 437-444. *http://dx.doi.org/10.1016/S0021-9673(00)93996-7*

[4] A Kör ner, Ion pair reversed phase thin layer chromatography of sulfonated dye: A theoretical approach towards optimization J. Planar Chromatog – Mod. TLC, 6 (1993) 138-143.

[5] H. Oka, Y. Ikai, T. Ohno, N. Kawamura, J. Hayakawa, Ken-ichi Harada and M. Suzuki, Identification of unlawful food dyes by thin layer chromatography – Fast atom bombardment mass spectroscopy, J. Chromatog., 674 (1994), 301-307. *http://dx.doi.org/10.1016/0021-9673(94)85235-9*

[6] D. Nurok, M. C. Frost, C. L. Pritchard and D. M. Chenoweth, The performance of planar chromatography using electroosmotic flow, J. Planar Chromatog – Mod. TLC, 11 (1998) 244-246

[7] S. N. Shtykov, E. G. Sumina, E. V. Smushkina and N.V. Tyurina, Normal and Reversed phase thin layer chromatography of Fluoroscein dyes with aqueous micellar solutions as mobile phases, J. Planar Chromatog – Mod TLC, 12 (2) (1991) 129-134.

[8] T. Hayashi, N. Ozeki, H. Oka, Y. Ito and T. Kagami, Applicability of Aluminium-backed reversed phase TLC plate to the analysis of coal tar dyes in foods, Nippon Shokuhin Kagaku Gakkaishi, 5 (2) (1998) 106-110. C.A. 130 (24) 1999.130; 324428e.

[9] E. Ueno, T. Ohno, H. Oshima, I. Saito, Y. Ito, H. Oka, T. Kagami, H. Kijima and K. Okazaki, Identification of small amounts of coal tar dyes in foods by reversed phase TLC/Scanning densitometery with sample concentration technique, Shokuhin Eiseigaku Zasshi, 39 (4) (1998) 286-291, C.A. 129 (17) (1998) 129 : 215896 q.

[10] D. Nurok, M.C. Frost and D. M. Chenoweth, Separation of dyes using planar chromatography with electro-osmotic flow, J. Chromatog. A, 903 (2000) 211-217. *http://dx.doi.org/10.1016/S0021-9673(00)00889-X*

[11] A. K. Misra and U. Gupta, Plain and ion pair thin layer chromatography of synthetic dyes on admixture of silica gel-G and Barium sulphate in an aqueous ethanol system, J. Chromatog. Sc., 40 (2002) 297-303. *http://dx.doi.org/10.1093/chromsci/40.5.297*

[12] S. D. Sharma, S. Misra and S.C. Sharma, Reversed phase thin layer chromatographic behaviour of some toxic metal ions an tri-butyl phosphate modified silica gel-G layers in Acetone-nitric acid-water systems: Quantitative separations of Pb2+ and Cr3+ from numerous metal ions, J. Ind. Chem. Soc., 75 (1998) 81-83.

[13] S. D. Sharma and S. C. Sharma, Retention behaviour of metal ions on tri-butylphosphate and tri-n-butyl amine impregnated silica gel-G layers in Oxalic acid-oxalate mobile phases, J. Chromatog., 841 (1999) 263-271. *http://dx.doi.org/10.1016/S0021-9673(99)00293-9*

[14] S. D. Sharma and C. Sharma, Retention behaviour of some d- and f- block metal ions on untreated and tri-n-butylamine impregnated silica gel-G layers – Quantitative separation of Zr4+ and W6+ from other metal ions and alloy components, J. Planar Chromatog – Mod. TLC, 13 (2000) 93-100.

[15] S. D. Sharma, S. C. Sharma and C. Sharma, Reversed phase thin layer chromatography of some d-block metal ions on unconventional supports in complexing acid media, J. Planar chromatog.- Mod. TLC, 12 (1999) 440-445.

[16] S. D. Sharma, H. Sharma and S. C. Sharma, Retention behaviour of amino acids on plain, tri-n-butyl phosphate and tri-n-butyl amine impregnated silica gel-G layers in mixed mobile phases, J. Ind. Chem. Sec., 79 (2002) 743-746.

[17] S. D. Sharma, H. Sharma and C. Sharma, Reversed phase thin layer chromatography of amino acids on silicone fluid DC 200, triamyl phosphate and tri-n-butylamine impregnated silica gel-G layers. J. Ind. Chem. Soc., 80 (2003) 926-929.

[18] A. Jayaswal and U. Chudasama, Sorption of water soluble dyes using inorganic materials as sorbents, Ind. J. Chem. Technol., 13 (2006) 539-543.

[19] R. Kant and V. K. Rattan, Adsorption of dye green B from a textile industry effluent using two different samples of activated carbon by static batch method and continuous process, Ind. J. Chem. Technol., 16 (2009) 240-244.

[20] A Mohammad and N. Ershad, Normal phase soil thin layer chromatography of synthetic dyes and their impact on translocation of trace metals, Anal Agrar. Sc. (Georgia), 6 (1) (2008) 7-17.

[21] R. Ahmad, Studies on adsorption of crystal violet dye from aqueous solution onto coniferous pinus bark powder (CPBP), J. Hazard. materials, 171 (1-3) (2009) 767-773. *http://dx.doi.org/10.1016/j.jhazmat.2009.06.060*

[22] T. D. Sekulic and N. P. Janjic, Study of the characteristics and separating power of unconventional TLC support II. Principal-Components analysis, J. Plannar Chromatog – Mod. TLC, 20 (2007) 7-11.

[23] A.K. Mishra, Retention behaviour of synthetic dyes on coconut oil impregnated egg shell layers in pure and mixed solvent systems, Ind. J. Chem. Technol., 144 (2007) 407–411.

[24] R. Srivastava and D. C. Rupainwar, Removal of hazardous triphenyl methane dye having adsorption over waste material – Mango bark powder, Ind. J. Chem. Technol., 18 (6) (2011) 469-474.

[25] P. N. Dave, S. Kaur and Ekta Khosla, Removal of Eriochrome Block – T by adsorption onto Eucalyptus bark using green technology, Ind. J. Chem. Technol., 18 (1) (2011) 53-60.

[26] R. Srivastava and D. C. Rupainwar, A Comparative evaluation for adsorption of dye on neem bark and Mango bark powder, Ind. J. Chem. Technol., 18 (1) (2011) 67-75.

[27] Xing –Jiang XU, XIU – RUIJION, Xiu – Zhen Zhang, Hui Huang and Li- min Jhang, Simultaneous determination of malachite green, crystal violet, methylene blue and the metabolite residues in aquatic products by ultraperformance liquid chromatography with electrospray ionization tandem mass spectrometry, J. Chromatog. Sc., 50 (7) (2013) 591-597.

[28] S. Rangabhashiyam, N. Anu and N. Selvaraju, Sequestration of dye from textile industry waste water using agricultural waste products as adsorbents, J. Environ. Chem. Engg., 1 (2013) 629-641. *http://dx.doi.org/10.1016/j.jece.2013.07.014*

[29] Jelena Vias Kovic, Filip Andric, Petar Ristivogevic, Aleksandra Radoicic, Zivoslav Tesic and Dusarka Milojkovic – Opsenica, Development and validation of a TLC method for the analysis of synthetic food-stuff dyes, Journal of liquid chromatography and related techniques, 36 (2013) 2476-2488.

[30] Casoni, Dorina, Boldan, Anamaria, Cobzac, Simona Codruta, TLC- Densitometric determination of synthetic food colorants from pharmaceutical powders, Studia Universitatis Babas – Bolyai, Chemia, 57 (2012) 83-92.

[31] Francisca Ivani de Androde, Maria Izabel Florindo Guedes, Icaro Gusmao Pinto Vieira, Francisca Noelia Pereira Mendes, Paula Alves Salmito Rodrigues, Carla Soraya Costa Maia, Maria Marlene Marques Avila, Luzara de Matos Ribeiro, Determination of synthetic food dyes in commercial soft drinks by TLC and ion pair HPLC, Food Chemistry, 157 (2014) 193-198. *http://dx.doi.org/10.1016/j.foodchem.2014.01.100*

[32] Tie-Xin Tang, Xin-Junxu, Dong-Mei Wang, Zhi-Min Zhao, Long- Ping – Zhu and De-Po Yang, A rapid and green limit test method for five synthetic colorants in foods using polyamide thin layer chromatography, Food Anal. Methods, 8 (2015) 459-466.
http://dx.doi.org/10.1007/s12161-014-9907-6

[33] W. Kiridena and C. F. Poole, Structure-driven Retention optimization model for reversed phase thin layer chromatography, Anal. communications, 34 (1997) 195-198.
http://dx.doi.org/10.1039/a702566a

[34] H. J. Issaq, K. E. Seburn and J. R. Hightower, Thin layer chromatographic separations by conventional glass backed plates and empore sheets : A comparative study, J. Liq. Chromatog., 14 (8) (1991) 1511-1517.
http://dx.doi.org/10.1080/01483919108049632

[35] E. Stahl, Thin layer chromatography, second ed., Springer–Verlag, New York, 1969, Page 20.
http://dx.doi.org/10.1007/978-3-642-88488-7

CHAPTER 11

An Overview of Preparation, Properties and Applications of Ionic Polymer Composite Actuators

Inamuddin and Ajahar Khan

Department of Applied Chemistry, Aligarh Muslim University, Aligarh, India

*Corresponding author E-mail address: inamuddin@rediffmail.com

Abstract

It deals with the basic information regarding to the materials and their types used for manufacturing IPMC actuators using various fabrication processes under optimum experimental parameters and possible applications. Among the variety of electroactive polymers recently developed IPMCs are good candidates for use in bio-related, robotics and aerospace applications because of their biocompatibility. Several fabrication processes, their performance and mechanical characteristics, a number of recent applications of IPMCs have been reported in this chapter. The control and various factors affecting performance of IPMC have also been reported.

Keywords

Ionic Polymer Metal Composites, Bending Actuator, Electroless Plating, Structure of IPMC

Contents

1. Introduction

The development of mankind and human civilization can be correlated to the invention and manufacture of new materials. In present times, new materials are discovered and manufactured at a very high frequency. We could christen this time to be the decade of smart materials. Composite materials that respond to an external stimulus in real or near real time are known to be as "smart" materials. Smart materials are thus defined as materials that exhibit a fast, repeatable, reversible and significant change in at least one of their physical properties in response to an external stimulus[1]. Smart materials are present in all the basic forms (solid, compound, composite and solution). Smart materials on the basis of external stimulus can be classified as 'property changing' or 'energy exchanging' [2]. 'Property changing' materials are those in which the input external energy causes a mechanical, chemical, thermal or electrical change in material properties. Energy exchanging materials involve the changes in their internal energy level while the materials properties remain the same. The response of the materials in either category is reversible and direct. Smart materials in these two categories along with their characteristic properties are listed in **Table 1**. Among the different variety of materials listed in **Table 1**, electroactive polymers (EAPs) have been extensively studied and received significant attention due to their inherent attractive features such as mechanical flexibility, light weight, easy processing, flexibility and fracture tolerance. These polymers exhibit large property changes in response (change shape or size) to electrical stimulation which adds significant benefits to their potential applications. In addition to these properties, these polymers respond to an electric field or mechanical deformation [3–11]. Since the invention in the mid 1970's, EAP materials have been the subject of extensive studies by several research groups. EAPs are considered to be alternatives to predictable sensor and actuators [12–14] because of their low electric voltage dependence, flexibility, light weight, bio-compatibility, compliant properties, ability to perform in air and aquatic media and easy fabrication processes.

EAP materials respond mechanically to external stimulation with considerable change in their shape and size. Generally, electroactive polymers are classified into two major categories [2,15,16] **Table 2**. The classification is based on their activation mechanism responsible for actuation as electronic EAPs (which are driven by an electric field or coulomb forces) and ionic EAPs (which change shape by mobility or diffusion of ions and their conjugated substances). The electronic EAPs such as electrostrictive, electrostatic, piezoelectric and ferroelectric generally require high activation fields (>150

V μm^{-1}) which are close to the breakdown level of the material. Beside requiring high activation electric potential, electronic EAPs can be controlled in a better manner as they can hold the induced displacement on the application of dc voltage. The property of these materials to hold the induced displacement, when a dc voltage is applied, makes them potential materials in robotic applications and these materials can be operated in air without major constraints. The electronic EAPs also have high energy density as well as a rapid response time in the range of milliseconds. In general, these materials have a glass transition temperature inadequate for low temperature actuation applications [1]. On the other hand, ionic EAPs are materials that involve mobility or diffusion of ions and they consist of two electrodes and an electrolyte. The activation of ionic EAP can be achieved by relatively low voltage and generally induce bending displacement. Ionic EAP materials such as gels, ionic polymer-metal composites, conducting polymers and carbon nanotubes require low driving voltages, nearly equal to 1-6 V. One of the constraints of these materials is that they must be operated in a wet state or in solid electrolytes. Ionic EAPs predominantly produce bending actuation that induces relatively lower actuation forces than electronic EAPs. Often, operation in aqueous systems is plagued by the hydrolysis of water. Moreover, ionic EAPs have slow response characteristics compared to electronic EAPs. The amount of deformation of these materials is usually much more than electronic EAP materials and the deformation mechanism bears more resemblance to a biological muscle deformation. The induced strain of both the electronic and ionic EAPs can be designed geometrically to bend, stretch or contract [16]. But it is difficult to maintain their position constant under dc activation [1]. They can be used as actuators or sensors. As actuators, they are characterized by being able to undergo a large amount of deformation when they are stimulated. Similarly, when they exhibit the inverse effect, they can be used as sensors. **Table 3** shows comparison of electronic and ionic EAPs characteristics [10].

Table 1: Classification of smart material [2].

	Property changing			Energy changing		
S.No.	**Material**	**Stimuli**	**Change**	**Material**	**Stimuli**	**Effect**
1.	Magnetorheological	Magnetic field	Viscosity	Thermoelectric	Electrical current	Temperature differential
2.	Thermochromic	Thermal energy	Spectral reflectivity	Photovoltaic	Radiation energy	Electrical current
3.	Shape memory	Thermal energy	Crystalline phase change	Photoluminescent	Radiation energy in the ultraviolet spectrum	Radiation energy in the visible spectrum
4.	Thermotropic	Thermal energy	Phase change (conductivity, transmissivity, solubility and volumetric expansion)	Electroactive polymer	Electrical Current	Strain

Table 2: List of leading electroactive polymers.

S.No.	Ionic EAP	Electronic EAP
1	Ionic polymer gels (IPG)	Electrostrictive paper
2	Conducting polymers (CP)	Electrostatic
3	Carbon nanotubes (CNT)	Dielectric EAP
4	Ionic polymer metal composite (IPMC)	Piezoelectric
5		Electro-viscoelastic elastomers
6		Ferroelectric polymers
7		Liquid crystal elastomers (LCE)

Table 3: Properties of electronic and ionic EAPs [10].

S.No.	Properties	Ionic EAPs	Electronic EAPs
1	Controllability	Difficult	Easy
2	Activation voltage	Small	Large
3	Energy density	Poor	Good
4	Electrolyte	Present (wet)	Absent (dry)
5	Mechanism	Mobility or diffusion of ions	Coulomb forces
6	Displacement	Bending	Changes shape or dimensions
7	Examples	Ionic polymer gels (IPG), Ionic polymer metal composites (IPMC), Conductive polymers (CP), Carbon nanotubes (CNT)	Electrostrictive, Elastomers, Electrostatic, Piezoelectric, Ferroelectric, Polymers

EAP materials have been termed as "artificial muscles" because of their electromechanical response under electrical stimulation, exhibiting large strain. For this capability, EAPs are particularly attractive in biomimetic, since they can be used to mimic the movement of animals, humans and insects for making biologically inspired mechanisms. These materials also have the potential for application in the field of robotics, where large linear movement is often needed. The general characteristic properties of EAPs and their significance [10] are summarized in **Table 4**.

Electroactive polymers have great potential as polymer-based actuators and sensors. The actuation in electronic EAPs (e-EAPs) is driven by the electrostatic mechanism due to the presence of dipoles. On the other hand, ionic EAPs (i-EAPs) actuate due to the diffusion or transport of ions within the polymer membrane. While e-EAPs can hold induced displacement under dc activation, i-EAPs require an electrolyte or solvent for actuation. However, due to their bi-directional actuation and low voltage requirements, i-EAPs are being considered for biomedical, space exploration, robotics and naval applications. Among the EAPs, dielectric elastomers, conducting polymers, ionic gels, carbon nanotubes based actuators and ionic polymer metal composites (IPMCs) actuators are considered to be the most promising and attractive materials to researchers because of their light weight, large bending displacement under low applied voltage and small force generation capability with dexterous behaviour [17–26].

Typically, an IPMC consists of a semipermeable polymer membrane sandwiched between Pt, Au or Ag metal electrodes and isneutralized with certain counter ions that balance the electricalcharge of the anions covalently fixed to the back-bone membrane or in simple words an IPMC comprises of a neutralized semi permeable ion-exchange polymer membrane coated with metal Pt or Au as electrode at both surfaces of membrane and water as inner medium for the dissociation for metal cations. Ionic polymer metal composites (IPMCs) are capable of electromechanical and mechanoelectrical response to applied voltage and mechanical deformation respectively [27–30]. **Figure 1** depicts the outline of general introduction of IPMC.

Table 4: The general characteristic properties of EAPs and their significance [10].

S.No.	Measurement		Properties	Significance
1	Mechanical		Tensile strength (Pa)	Mechanical strength of the actuator material
			Stiffness (Pa)	Required to calculate blocking stress, mechanical energy density and mechanical loss factor/ band width.
			Thermal expansion coefficient (ppm/C)	Affects the thermal compatibility and residual stress.
2	Electrical		Maximum voltage (V)	Necessary to determine limits of safe operation.
			Impedance spectra (Ohms and phase angle)	Provides both resistance and capacitance data. Used to calculate the electrical energy density; electrical relaxation/dissipation and equivalent circuit.
			Nonlinear current (A)	Used in the calculation of electrical energy density, quantify nonlinear responses/driving limitations.
			Sheet resistance (Ohms per square)	Used for quality assurance.
3	Electro active properties	Strain	Electrically induced strain (%) or displacement (cm)	Used in calculation of blocking stress and mechanical energy density.
		Stress	Electrically induced force (N) or mechanically induced charge (C)	Electrically induced force/torque or stress induced charge.
		Stiffness	Stress/strain curve.	Voltage controlled stiffness.
4	Microstructure		Thickness (electrode & EAP), internal structure, uniformity, anisotropy and hosted defects	These are features that will require establishing standards to assure the quality of the material.
5	Environmental behavior		Operation at various temperatures, humidity and pressure conditions	Determine material limitations at various conditions.

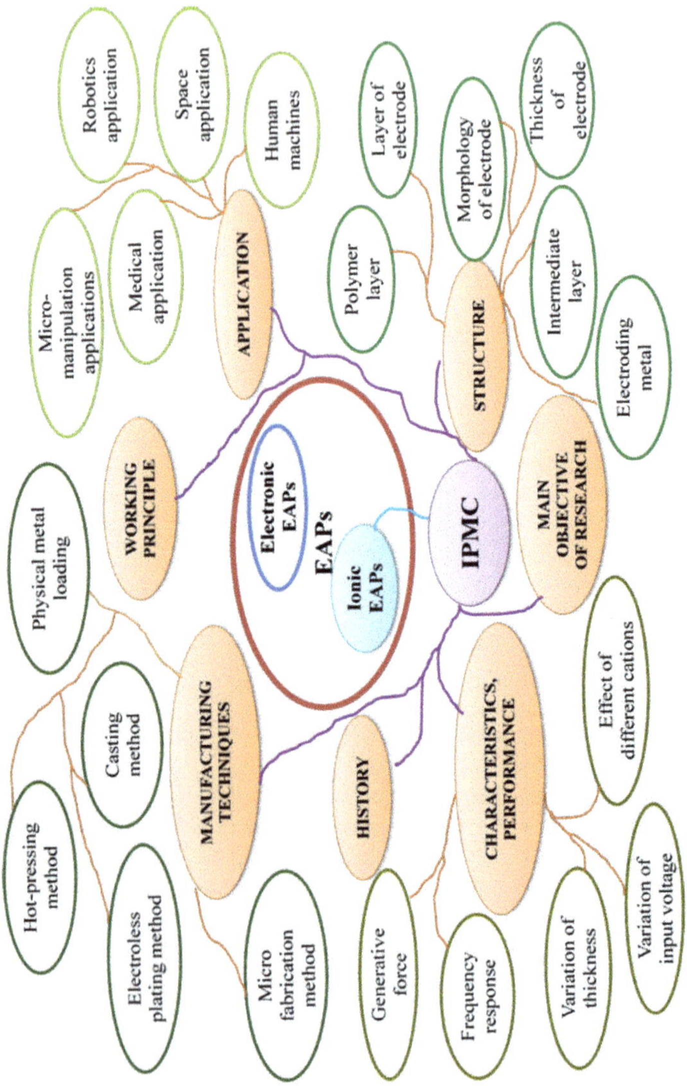

Figure 1: Outline of general introduction of IPMC.

2. History of IPMC

As mentioned earlier, IPMC is a type of i-EAP which gives a mechanical response upon applied voltage for the first time in 1992, was discovered by Shahinpoor [31], Oguru *et al.* [32] and Shadeghipour *et al.* [3]. Shadeghipour et al. developed a novel accelerometer cell based on nafion coated with platinum and also calibrated it. Shahinpoor prepared swimming robotic structures using ionic polymer gel muscles. In this conceptual design, all the onboard electronics was to be placed in a head-like structure. A flexible membrane was attached to this head and was filled with an aqueous ionic gel. It was possible to electrically control the expansion and contraction of this polymeric fiber filled gel. Oguru *et al.* showed that upon given voltage, mechanical deformation is produced across the thickness of a nafion based membrane. Scanning electron microscope (SEM) image of a nafion based IPMC is shown in **Figures 2. Figure 3** shows the chemical structure of nafion based IPMC [28,33].

3. Working principle of IPMC

The actuation response of IPMCs were discovered in 1992 by Shahinpoor, Sadeghipour *et al.* and Oguro *et al.*[3,31,32]. When an adequate potential (~1-5 V) is applied across the thickness of an IPMC actuator, the hydrated counter cations are able to break away from the fixed anionic groups and move toward the negatively charged surface. The extent of bending is related to the magnitude of the applied potential and the direction of the bending motion is reversed by reversing the polarity of this potential. The polymer membrane allows selective ions to pass through the membrane while blocking others ions in the hydrated state. The water present on the surface of IPMC serves as a medium for ion migration, so the quantity of water molecules present on the surface of the IPMC is a very important factor which influences the performance of the IPMC. During hydration, water molecules surround the cations and make the entire hydrated entity mobile. There are two possible physical forces responsible for actuation of IPMC. One is the fluid-induced swelling force [34]. In this case when an electrical potential is applied across the IPMC actuator, cations are redistributed and move towards the cathode side (cation rich region) bringing with their water molecules resulting in a locally imbalanced net charge density which produces internal stresses acting on the backbone of polymer causes differential swelling across the polymer thickness. In the anion-rich region, the polymer chains relax, while in the cation rich region, they further extend, consequently, the IPMC will bend towards the anode side (cation poor region). The force is the electrostatic force that arised due to the imbalanced net charges in the polymer membrane [35]. On the other hand, if a mechanical force is applied to the IPMC and makes it bend, the ion movement will generate sensing charges on the electrodes. So an IPMC can be used both for sensing and actuation [35,36]. The pressure from the strained polymer matrix causes water to diffuse out of the cation rich areas, resulting in a slow relaxation toward the cathode. Under dry conditions, the cross-linked cations are not free to move. However, in wet condition (on hydration) cations are surrounded by water molecules to make the whole film mobile. Thus, transduction in the IPMCs film is because of the movement of cations with the water molecules [37]. They operate best in a humid environment and can be made as self-contained encapsulated actuators to operate in dry environments as well. They have been modelled as both capacitive and resistive element actuators that behave like biological muscle and provide an attractive means of actuation as artificial muscles for biomechanics and biomimetics applications [38]. Transport of hydrated cations within an IPMC beam under the applied voltage and associated electrostatic interactions lead to the bending of an IPMC beam. The responsible factors are:

a) Swelling and contraction caused by the water content.

b) Electrostatic force generated by deviation of fixed charges of sulphonic acid groups.

c) Momentum conservation effect concerning the ion migration and water travel.

d) Conformation change of the polymer structure according to the ionic migration.

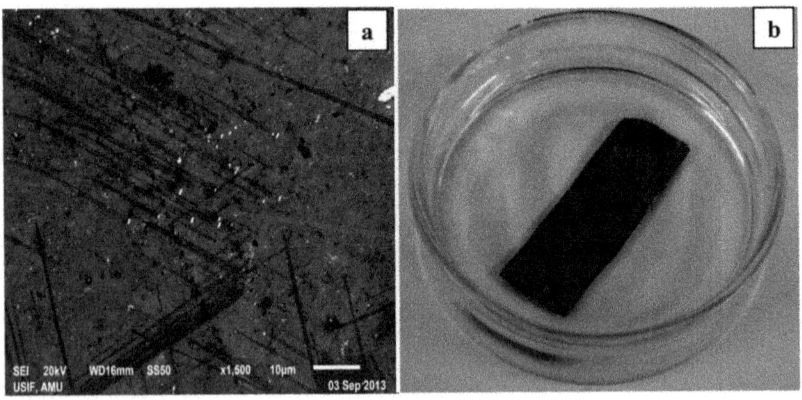

Figure 2: SEM image showing surface morphology of a nafion based IPMC membrane (a) and actual view of fabricated nafion based IPMCmembrane actuator (b).

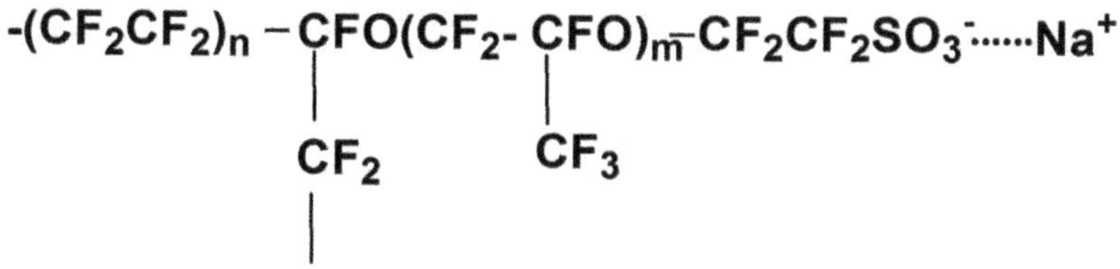

Figure 3: Chemical structure of nafion (perfluorinated sulfonic acid ionic polymer).

This conformation change actually represents bending of the polymer. The transduction principles of an IPMC are illustrated in **Figure 4**.

A bending response under the influence of a low voltage of about 1-5 V of an IPMC was first reported by Oguro *et al.*[32] and Shahinpoor *et al.* [31] also reported similar findings at the same time. Use of mechanoelectric transduction for the development of a pressure sensor has been reported by Sadeghipour *et al.*[3]. Relatively high bandwidth (several hundred Hertz), flexibility, operational feasibility in aqueous environment and chemical stability are some of the advantages inherited by IPMCs. A lifetime of up to 250000 cycles has been recorded for an ionic liquid coated IPMC operating in air [39].Hunter *et al.* [40] did an extended comparison of the performances of artificial actuators. Comparison of some of the smart materials with strength [10] is shown in **Table5**. Thus, from the **Table 5**, it can be concluded that IPMC has potential as an actuator in low-mass, large bending displacement, sensors and other applications. From static test results such as those blocking force tests and displacement, an IPMC actuator can be corresponding to a composite beam [41]. Coulomb forces and diffusion are associated with reverse effects and considered as bending [35].

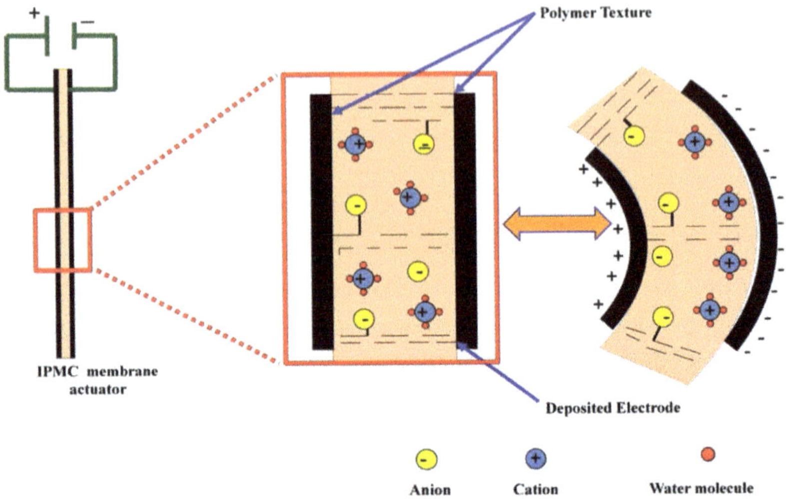

Figure 4: Graphical representation of bending mechanism of IPMC membrane actuator.

Table 5: Comparison of an IPMC with other smart materials [10].

S.No.	Smart Materials	Strain (%)	Stress (MPa)	Efficiency (%)
1	Magnetostrictive	0.2	-	< 50
2	Piezoelectric	0.1	35	> 75
3	SMA	> 4	> 300	> 3.8
4	IPMC	> 40	0.3	> 30

Generally, IPMC is capable of actuating and sensing in harsh conditions. It can responds at low temperature as -100°C where an increase in voltage allows compensation for the loss in efficiency. Furthermore, IPMC based artificial muscle becomes less conductive because decrease in temperature increases electrical resistance. The results come out to resist the generally accepted fact that resistance of metallic conductors with increasing/decreasing temperature increases/decreases respectively. It has been reported that the IPMC displays less ionic current activities and becomes less active at lower temperature [28]. Upon varying the applied frequency signal the point where large deformations are observed at a critical frequency called resonant frequency, beyond which the actuator response is diminished. Mechanical hysteresis and time varying performance is the main drawbacks of IPMC materials. With the research over the past decade, the materials have become more stable but these problems are not completely resolved. In the recent years, research pertaining to IPMC materials has prolonged more in terms of better adaptability to the environment applicability and getting higher efficiency [42,43].

4. Structure of IPMC

As previously mentioned, an IPMC actuator imparting a layered structure consists of a polymer or composite material membrane coated with novel metal electrodes. Scanning electron microscopes of an IPMC membrane actuator, shows a maximum of three layered (**Figure 5**). The intermediate polymer layer is followed by outer most layers of metal electrodes on both the surfaces. The intermediate polymer layer comprises of Pt, Au or Ag metal particles dispersed or deeply embedded inside the polymer matrix. **Figure 5** represents the cross sectional view of three layers in an electroless plated IPMC membrane. In a sputter-coated sample the number of layers is limited to two: the polymer layer and the electrode. Each layer contributes to the overall electromechanical and mechanoelectric effects in the IPMC. Research has been conducted to understand the electrode layer comprising of porous electrodes [44–48], the intermediate layer [49–51] and the polymer layer [52–56]. The role of these different layers in the transduction of IPMC actuator can be better explained in the following section.

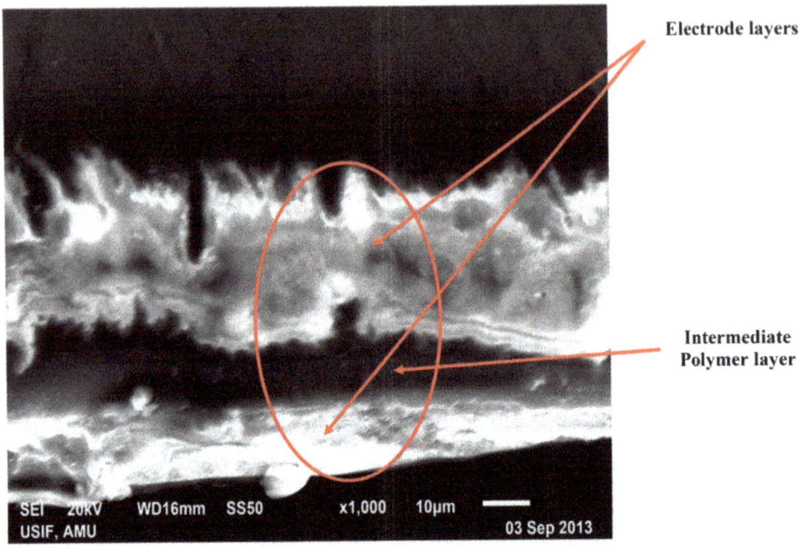

Figure 5: Scanning electron micrograph shows the three layers in an electrolessplated nafion based IPMC membrane actuator [34].

4.1 Polymer layer

Typically, an ion-exchange composite and polymer membrane with selective ion diffusivity and ion conductivity are used for fabricating an IPMC actuator. The ion-exchange membrane acts as a pathway for selective ions to move. The transduction in the IPMC actuator depends on the counter ions, ionomer and the state of hydration of the polymer membrane. Two ionomers with the name of nafion (perfluorosulfonate) by DuPontor flemion (perfluoro carboxylate) by Asahi Glass arecommonly available. The major differences between these two ionomers are in the ion-exchange capacities and functional groups (nafion having sulfonate and the flemion having carboxylateg roups). The carboxylate group has better mechanical strength and higher ion-exchange capacity corresponding to the sulfonate group. Larger ion-exchange sites cause a higher interfacial area between the metal electrode and the polymer in flemion based samples [57]. A flemion based IPMC has higher water uptake, larger bending, higher stiffness, no back relaxation, higher dielectric strength, better mechanical strength and higher capacitance than nafion based one [58,59]. However, due to commercial availability, nafion is used as a base polymer for fabricating IPMC because of several advantages such as higher working temperature (up to 120°C), chemical resistance and high ion conductivity of around 0.1 S cm^{-1} in 1 M H_2SO_4 at 20°C measured using a dc current pulse as compared to other polymers [60,61]. Nafion is a perfluorosulfonate ionomer that contains a poly(tetrafluoroethylene) backbone and perfluoroether side chains that terminate in a sulfonate group. The anions are attached to the polymer chain while the cations are free to move upon hydration [62]. Its derivatives can be synthesized normally by the copolymerization of a derivative of a perfluoro (methyl vinyl ether) with sulfonyl acid fluoride and tetrafluoroethylene. Beside the various advantages, perfluorinated polymers have some limitations including low blocking force, back relaxation, high cost and environmental unfriendliness. Therefore, to replace the perfluorinated polymers and improve the performance of IPMCs actuators, non-perfluorinated EAPs are considered to be important alternatives [63]. It has also been shown that the IPMC deflection and force are positively affected by an increase in ionic conductivity. Since the key role of the ionomers in the transduction of IPMC is charge transport, hence an attempt to improve the ionic conductivity has been made. Other properties that affect the transduction capabilities are the ion-exchange capacity, water uptake and elastic modulus.

4.2 Layer of electrode

The electrode layer plays an important role in IPMC transduction. Different electrode properties like electrode thickness, morphology and the types of metal used influence the performance of IPMC actuators. It is also much easier to modify the electrode layer to change the IPMC properties than the ionomer or intermediate layers.

4.3 Morphology of electrode

The morphology of electrode affects the transduction property of an IPMC actuator [64]. The overall capacitance of the IPMC increases as the surface area of the electrode increased. Porosity and thickness of the novel metal electrode play a very important role in controlling the resistance of the electrode [45,65]. On the other hand, the particle gap, thickness and the distance between the surfaces of particles in the intermediate layer directly affect the capacitance of the electrode. In addition, the transduction of IPMC actuator is affected by the surface area of the electrode, which in turn is controlled by the morphology. IPMC transduction can also be improved by increasing the surface area which improves the capacitance of an IPMC. The effect of surface roughening on IPMC properties was studied by Jin *et al.*[66] and Tiwari [67]. Both studies confirmed that transduction in IPMC is improved by roughening. An improvement in tensile modulus and capacitance due to better electrode adhesion and electrode penetration was observed. The electrode morphology is also affected by the type of metal used for electroding. Gold and palladium electrodes form smoother electrode surfaces thus show better transduction [68,69].

4.4 Thickness of electrode

Increase or decrease of the electrode thickness andconductivity positively affects the transductionin an IPMC. This is because the resistivity, ρ, of the electrode is inversely proportional to the charge transfer:

$$\frac{dQ}{dT} = \frac{VL_tW}{2h\rho} \tag{1}$$

Where$Q$ is the charge, V is the voltage across the thicknessof the electrode ($2h$),length(Lt) and width(W) of the IPMC.

Table 6. Effect of plating cycle and reduction temperature on the particle size and capacitance of an actuator [2,49].

S.No.	Studies	No. of plating cycles			Reduction temperature (°C)		
		1	2	3	30	40	50
1	Capacitance @10 Hz (mF cm^{-2})	1	1.9	2.48	0.2	0.5	1
2	Surface electrode thickness (μm)	1.5	2.5	3	0.9	1.1	1.5
3	Particle size (nm)	60 (±10)	85 (±15)	110 (±20)	40 (±20)	75 (±15)	90 (±10)
4	Sheet resistance (Ω/sq)	9.8	4.95	4.3	13	11	6
5	Intermediate layer thickness (nm)	300	200	<200	1500	400	200
6	Particle gap in intermediate layer (nm)	100	50	<50	<200	<100	<50

Lee and Yoo [70] studied the effect of alcohols (methanol, n-propanol and ethanol) on the thickness of the electrode layer in the first deposition step. It was found that ethanol produced more uniform and thicker layer whereas in cases of propanol and methanol a thinner electrode were produced compared to water based samples. An alcohol based samples was found to have higher surface resistance than water based sample (methanol: 9 kΩ cm−1; ethanol: 3.8 kΩ cm−1; 1-propanol: 7 kΩ cm−1). It was observed that the electrode thickness increases with the number of plating cycles used for electroding and the reduction in temperature employed for the fabrication of IPMC actuator. The first reduction temperature mainly affects the electrode thickness while the second reduction temperature has more effect on the electrode density [71]. The results of Kim et al. [45] study summarized in **Table 6** shows the effect of temperature reduction and the plating cycle on electrode properties. An increase in capacitance and particle size was observed with the reduction temperature and plating cycle.

4.5 Electroding metal

The hydration is required for IPMCs operation hence care must be taken with the type of metal used for electroding. Platinum [72,73] and gold remain stable under hydration, resist corrosion and have high conductivity, hence are commonly used for electroding. Platinum has good chemical stability and availability in different cationic forms. The properties of platinum electrodes could be modified by using palladium [68] or silver nano powders [48]. Gold electroding on IPMCs actuator also offer several advantages such as high electric conductivity, softness, non-toxicity and high conductivity (454-500 S cm^{-1}) but a low surface area of 0.4-0.75 m^2 g^{-1}. Three different techniques have been used for electroless plating of Au electrode: galvanic displacement, autocatalytic and substrate-catalyzed. The impregnation reduction (IR) technique for gold electroding was patented by Fujiwara *et al.*[74,75]. Gold flakes have also been used for electroding [76,77]. Despite their favorable properties both gold and platinum are expensive.

Due to high availability (in the form of a powder, woven cloths, belts, fibers [78] or tapes), low price, specific capacitance (10 μF cm^{-2}) with a density of around 0.6 g cm^{-3} and large surface area of 1900 m^2 g^{-1}, carbon black has been widely used for electroding[76,79]. However, electroding using hot pressing causes the carbon black electrode to delaminate. Carbon nanotube (CNT) has demonstrated large force generation and small strain of around 0.1-1 % [80]. These advancements have led to the manufacture of IPMCs with CNTs as electrodes [71,80–85]. Single wall nanotubes (SWNTs) have a high surface area of 1000 m^2 g^{-1}. However, the poor dispersion inside the membrane and high cost limit their application [86]. Iridium [87] and ruthenium [88] due to their specific capacitances and large surface areas of 80-1000 μF cm^{-2} and 50-70 m^2 g^{-1} respectively, are preferred over Au and Pt. Other metals such as silver, lead [89], copper [90,91], palladium [72] and nickel [69,89,92] have also been used for electroding because of their similar properties to Pt and Au electroded IPMCs actuators. However, nickel, silver and copper are corrosive and have limited applications in aqueous media. Organic conductors such as polyaniline (PANI), polypyrrole and Poly(3,4-ethylenedioxythiophene) poly(styrenesulfonate) (PEDOT:PSS) have also been used for the development of all organic conductors with a typical conductivity of 100 S cm^{-1}[93].

4.6 Intermediate layer

The research on IPMCs actuators has been focused mostly on the polymer layer and the electrode layer. The intermediate layer plays an important role in the transduction of IPMC by affecting the interfacial area between the polymer electrode and the IPMC capacitance. The role of the intermediate layer in IPMC transduction has been overlooked [67,94]. It has been observed that with an increase in the interfacial area, bending responses as well as the charge output of IPMC have improved. Sunghee *et al.*[95] demonstrated the importance of optimizing the interfacial area. They observed that penetration of platinum particles inside the polymer membrane increases after soaking and reaches to saturation after 3 h of immersion. They also concluded that after fourth plating cycle when saturation in the interfacial area was occurred further plating leads to an increase in the electrode stiffness, hence reduce the displacement of the IPMC membrane tip under applied electric potential. Mechanical deformation in the IPMC membrane occurs due to charge accumulation at the interface between the polymer and the electrode because of a stronger correlation between capacitance and strain. The existence of this relation between the actuation of an IPMC and its capacitance was experimentally demonstrated by Akle *et al.*[96].

5. Manufacturing techniques

Multiple studies and various efforts have been performed to manufacture IPMC actuators, but the variability from sample to sample created difficulty to draw definite conclusions and repeatable results that may be readily reproduced [48,97–100]. Fabrication of IPMC starts by selecting the base ion-exchange polymeric materials, typically manufactured from organic or organic-inorganic composite polymers with covalently bonded fixed ionic groups [46,101,102].

The most commonly used ion-exchange materials for the fabrication of IPMCs are perfluorinated hydrocarbon: with short sidechains terminated by ionic groups (*i.e.* nafion from DuPont™) [34], typically sulfonate or carboxylate (-SO^{3-} or -COO^{-}) for cation exchange or ammonium cations for anion exchange [103–108]. Perfluorinated ion exchange materials are organofluorine compounds with all hydrogen replaced by fluorine on a carbon chain having at least one functional group or different atoms. The chemical formula of a typical perfluorinated compound of nafion is shown in **Figure 3**.

The mechanical strength of IPMC materials is determined by the large polymer backbones, while the short side-chains provide ionic groups that interact with the inner solvent and the passage of selective ions. These features are different in other ionic polymers that are primarily limited by cross linking such as styrene/divinylbenzene families, the ability of the ionic polymers to expand (due to their hydrophilic nature). Fabrication of IPMC actuator is a lab-intensive work hence safety measures should be taken in consideration during chemical processing. The initial compositing process and surface electroding process are the two distinct IPMC fabrication processes. Based on the preparation processes, the morphologies of precipitated metals significantly differ.

The purpose of the compositing process is to metalize the polymer membrane by immersing polymer membrane in a salt solution of platinum-containing cations via the ion-exchange process. A proper reducing agent ($LiBH_4$ or $NaBH_4$) is used to metalize the polymer. However, the distribution of metallic particles across the polymer membrane is not homogeneous but concentrations of metal particles are nearer to the interface boundaries. Kim and Shahinpoor [109]have presented a detailed description of various techniques and procedures used in manufacturing of IPMCs including their performance and comparison.

The surface electroding process is carried out by using proper reducing agents to carry out reduction, predominantly on the initial platinum electrode layer on the membrane surface formed by the initial compositing process leading to a rather smooth metallic surface. The mechanical and electrical behavior of IPMC is also influenced by the variation of the deposition of electrode material. The process of deposition of metal electrodes on the membrane surface is known as electroding. Irrespective of the method used, electroding is aimed at:

a) Deposition of metal particles on both the surface of the membrane.

b) Formation of a uniform metal layer that is chemically and mechanically stable.

c) Maximization of interfacial area between the electrode and membrane.

d) Minimization of electrical resistance of the electrode.

The first step is to roughen the material surface where it will serve as an effective electrode. This involves sand blasting or sand papering the surface of the polymer in order to increase the surface area density where platinum salt penetration and reduction occur, as well as ultrasonic cleaning and chemical cleaning by acid boiling (HCl or HNO_3, low concentrates). The second step is to incorporate the ion exchanging process using a metalcomplex solution such as tetraamine platinum chloride hydrate as an aqueous platinum complex ($[Pt(NH_3)_4]Cl_2$ or $[Pt(NH_3)_6]Cl_4$) solution. To carry out the ion-exchange process, immerse the membrane in the Pt salt solution, adding ammonium hydroxide solution (5%) to neutralize the solution and leave the membrane at room temperature for a few hours (preferably overnight) [10,63]. Although, the equilibrium condition depends on the type of charge of the metal complex, such complexes were found to provide good electrodes. The third step (initial platinum compositing process) is to reduce the platinum complex cations to the metallic state in the form of nanoparticles by using effective reducing agents such as an aqueous solution of sodium or lithium borohydride (5%) at favourable temperature (*i.e.* 60°C).

Before the reduction is carried out, rinse the polymer membrane with DMW with constant stirring at 40 °C, add 2 mL of the aqueous sodium borohydride solution (5 wt% $NaBH_4$ aqueous) every 30 minutes for several times while gradually raising the temperature up to 60 °C. Add 20 mL of prepared aqueous solution of $NaBH_4$ and stirred for 1.5 h at 60 °C. This gives a black layer of Pt particles deposited on the surface of the polymer membrane. To terminate the reduction process, rinse the membrane with DMW followed by immersing in 0.1 N HCl solution for 1 h.

A small amount of sample is warmed with $NaBH_4$ to check the end point. Precautions should be taken because adding reducing agent ($NaBH_4$ powder) in a hot solution can cause a gas explosion. If there is any trace of Pt ion remains in the plating solution, the colour of the solution turns to black. In such case, continue deposition of Pt particles with addition of NH_2OH-HCl and NH_2NH_2 solution, until no Pt ion remains.

Finally after washing, treat the membrane with dilute 0.1 N HCl. H^+ in the composite membrane can be simply exchanged for any cation by immersing it in a solution of the chloride salt of the cation, e.g., LiCl in order to exchange it into the lithium cation form. Palladium, silver, gold, and copper have also been successfully used as electrodes. The primary reaction for platinum composites is:

$$LiBH_4 + 4[Pt(NH_3)_4]^{2+} + 8OH^- \longrightarrow 4Pt + 16NH_3 + LiBO_2 + 6H_2O \qquad (2)$$

Other IPMC fabrication processes are described below.

5.1 Physical metal loading

Fabrication of IPMCs by physically metal loading method is a novel technique which is faster than conventional methods [47]. The principle of this technique involves two simple steps. First of all, a conductive primary powder is physically loaded into the ionic polymer, forming a dispersed particulate layer. In the second step, the chemical plating method is used to secure smaller secondary particles using reducing agents, similar to the conventional methods. Furthermore, an electroplating process can be applied to integrate the entire conductive phase intact, which serves as an effective electrode.

5.2 Casting method

The performance of the IPMC membrane varies with its thickness. Hence to achieve a desired thickness of IPMC film other than commercialized thickness, the casting method is widely used. However, the casting method has a reproducibility problem because it requires great attention to process variables such as temperature and concentration of specific solvent [110]. According to the technical information concerning nafion [34],the volume of nafion dispersion is approximately ten times the volume of cast nafion. The overall casting process consists of three steps (a) proper mixing, (b) constant stirring and sonicating and (c) thermal treatment. The mechanical stiffness of the cast polymer film is enhanced by thermal treatment by increasing strength of the molecule's bonding structure. Finally, the polymer film has to be treated with the H_2O_2 solution at temperature ranging between 75-100°C for 1 h followed by boiling in demineralised water for 1 h. This method has been used to achieve IPMC actuators of desired thickness [111].

5.3 Hot-pressing method

The hot-pressing method commonly uses a hot pressing machine to make several thin polymer films to adhere together, which enhances the force performance, reproducibility and bending stiffness (**Figure 6**). Easy to control the thickness, repeatability and simplicity are the main advantages of this method [110]. Preparation of IPMC actuators by this method is described below.

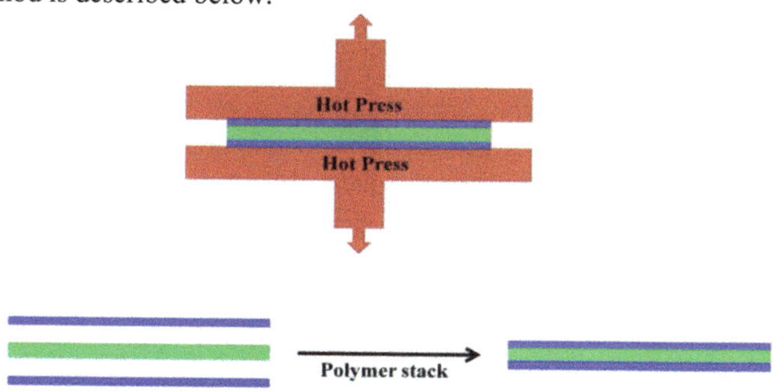

Figure 6: Fabrication of IPMC by stacking and hot press techniques.

Take the polymer films or membranes with proper dimensions, rinse with acetone followed by stacking in a mold with polyimide film. The mold is then pressesed at 180°C without pressure for 20 min. in preheated presses, followed by pressing for 10 min. at 50MPa. Cooling the polymer films to ambient temperature in the air then treated with 3wt% sulfuric acid at 70°C for 1 h. Finally treat the film with 10wt% H_2O_2 at 70°C for 1 h followed by rinsing with DMW for 0.5 h. Ion-exchange process is carried out by immersing the stack membrane in an aqueous solution of platinum ammine complex ($[Pt(NH_3)_4Cl_2]$) for 1-2 days. After rinsing the film with DMW, place it in 500 mL of stirring water at 40°C and add 5 mL of 5wt% aqueous solution of $NaBH_4$ after every 30 min. at 60°C.

5.4 Electroless plating method

There are several fabrication methods including sputtering, taping with conductive film, or chemical reduction among others. However, electroless plating is the best process to fabricate IPMCs, using conductive metal ions (platinum or gold).Generally, IPMC is fabricated by the chemical reduction method due to the low cost and uniform surface coating with metal by using this method. The adhesion of metal electrode on the surface of the polymer membrane is one of the main problems during the IPMC fabrication. The electroless plating technique can also be used with other methods like electroplating, polymer coating and hot-embossing to modify the adhesion between electrode and polymer, decreasing fabrication time and lowering the cost [21].Chung *et al.*[112] prepare the EAPs membrane by using a solution casting method and electroless plating for platinum electroding for the study of electrodes behaviours, membrane properties and processing conditions. Their study clearly shows that the smoothness of the surface increase by decreasing the surface roughness to 3 nm from the typical 52 nm but the surface resistance was too high to activate the bending motion. They coated the electrodes again in order to reduce the surface resistance, the fabricated IPMC showed longer bending life time than typical IPMC. Zhang et al. [113] fabricated the IPMC actuator based on Ag-nafion using penetration reduction and chemical plating methods for electroding palladium, platinum and gold due to their high conductivity and corrosion resistance. Metals such as copper have also been used as electrodes [114]. Electrode deposition using sputter coating and electrode deposition on dry and solvated membranes, using a direct assembly process, have also been considered [96,115]. However, the most commonly used manufacturing technique is electrode deposition because by using this technique the mechanical and chemical stabilities of electrodes are enhanced [72]. Kim *et al.* [111] have reported that the performance of IPMCs is directly affected by the concentration and size distribution of platinum particles on the membrane surface. The performance of IPMC can increase by using some of these techniques.

a) The multiple platinum reduction process has been effective to produce IPMCs exhibiting larger force characteristics.

b) During the platinum platinization process, use of dispersing agents gives significantly improved force characteristics showing a quick response upon applied electric potential with increased force generation.

The base material stretching process results in the higher permeability compared to the unstretched base material platinum composition process. It was clear from these techniques that the particle penetration within the material is much more effective, and forms a much denser platinum particle phase and distribution.

5.5 Micro fabrication method

Chung *et al.* [48] proposed micro-fabrication technologies which integrates the physical and chemical methods for the fabrication of IPMC actuators free from surface roughening pretreatment for good adhesion reducing both cost and fabrication time. The basicprinciple of this method is the casting of silver nano-powders with polymer composite followed by plating silver metal electrode by electroless plating method.

6. IPMC actuator: characteristics, performance and measurements

6.1 Generative force

A larger actuation displacement and force with reasonable actuation frequency is necessary for the successful application of IPMCs membrane actuators. The blocking force is mainly dependent on the length of IPMC, thickness and width. According to Kim *et al.*[41], the blocking force is proportional to the square of the thickness of IPMC and linearly proportional to the width and inversely proportional to the length of the IPMC.

With thick IPMC membrane, a high tip force can be generated for working relatively longer duration while its tip displacement and response speed are decreased. A long IPMC membrane cannot produce a high tip force but can create a large displacement. Tip displacement along with the tip force increases with increase in the applied voltage. In this case, electrolysis is accelerated which increase the evaporation of solvent by the heat generated from the IPMC actuator resulting in poor repeatability.

Therefore, a trade off in the actuation displacement, response speed, actuation force and endurance must be considered during the development of IPMCs actuator. Reduced surface resistance of metal electrodes in an IPMC also contributes to improvement of actuation performance and repeatability at the same time.

6.2 Frequency response test

While the deformation and blocking force characteristics of IPMC might be changed due to the micro cracks of the surface electrode that can increase the surface resistance, the natural frequencies are little affected by the surface electrode condition[41].Therefore, the results of a frequency response test are used to establish the equivalent Young's modulus of IPMC. Frequency response of IPMC contains not only the structural characteristics of IPMC but also electrical characteristics. It was seen that the electrical characteristics change the IPMC system, while the 1st and 2ndnatural frequencies are quite similar confirming that the electrical characteristics do not have an effect on the 1st and 2nd natural frequencies. Mechanical characteristics of IPMC also depend on the frequency, generally as the frequency increases there is corresponding decrease of displacement. At low frequencies (0.1~1.0Hz) and an imposed voltage, the effective elastic modulus of the IPMC cantilever strip is small, while at higher frequency (5~20Hz), the elastic modulus is high and the displacement is smaller. It happens due to the simple fact that during low frequency the hydrated ions have time to gush out of the surface electrode while in the high frequency they are rather contained inside the polymer.

6.3 Characteristics of IPMC under the variation of thickness

Nakamura *et al.*[116]established the relationship between the structure of IPMC and force generation as well as process of fabrication by investigating electromechanical and electrochemical characteristics of IPMC using impedance analysis. This study reveals that the generated force is dependent on the repeated plating, preheating and thickness of IPMC. The generated force almost linearly increases for the current range of 0-300 mA. It is assumed that the movement of counter ions followed by trapped water molecules being driven by the applied voltage on both sides of IPMC actuator. It was also seen that the generated force and capacitance of IPMC increased with the increase of the thickness but show non linear behaviour. A similar behaviour was also observed with resistance, an increase in resistance resulting from increasing thickness but the trend of resistance was almost linear to the thickness [117].

6.4 Characteristics of IPMC under the variation of input voltage

As the applying voltage increases, both bending displacement and the generative tip force increase. The linearity of the bending displacement and generative tip force is improved, as the thickness gets thinner. This is because as the thickness of the actuator gets thin, the uniformity of the electrodes is improved [10].

6.5 Effect of different cations

Electromechanical and chemical responses of IPMCs actuator depend on the nature of ion-exchange polymer, the conductivity and morphology of the metal electrodes, cations, and solvent uptake [58,118,119].The performance and bending displacement towards the positive electrode depend on the type of solvent used. Different alkali metal cations acquire different speeds and total tip displacements for the same IPMC membrane.

After performing the experiments, Kim and Shahinpoor [109] realised that the properties of IPMC differ depending upon different cations. For this nine samples were used and each of which was soaking in an appropriate salt solution (1.5N of NaCl, LiCl, KCl, HCl, $CaCl_2$, $MgCl_2$, $BaCl_2$, $CH_3[Ch_2]_3NBr$, $[CH_3]_4NBr$), respectively at modest temperature (30°C), for 3 days. **Figure 7** shows the blocking force of different IPMC containing various cations relative to the Na^+ ions containing IPMC at zero displacement[109]. From this figure it is clear that Li^+ containing IPMC is superior. It clearly shows that hydration processes with respect to mobile cations play an important role in the actuation behaviour.

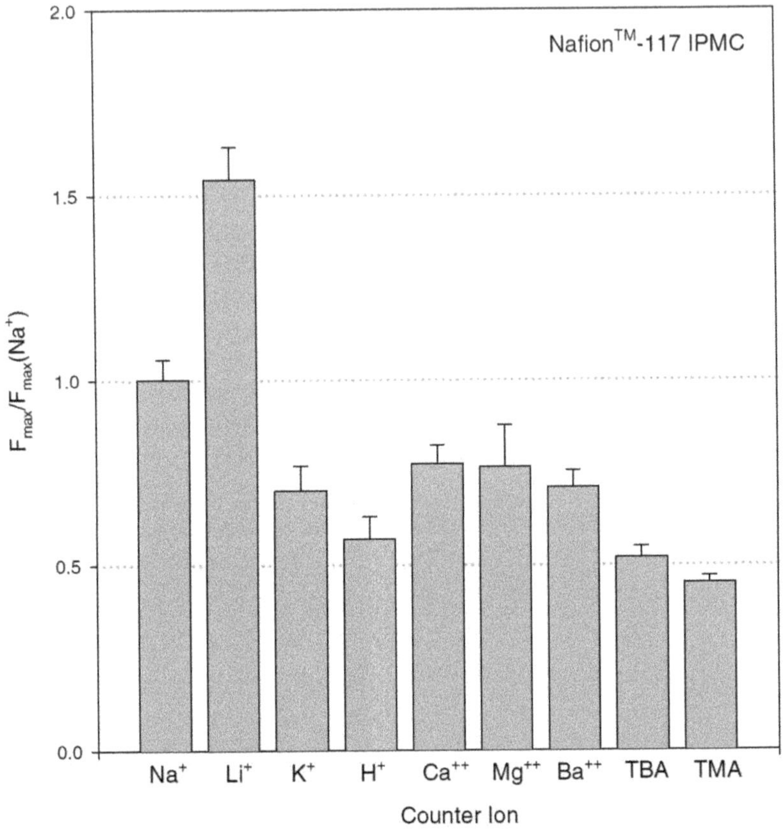

Figure 7: Effects of various cations on the actuation of the IPMC muscle [119]

7. Application

IPMCs have many advantages such as (1) produce high displacement, (2) require low drive voltage, (3) can operate very well in wet environment and (4) can be cut into small strips or any desired form. Therefore, IPMC shave significant potential in low mass, high displacement actuation and other applications. **Figure 8** demonstrates the applications of IPMC in different fields, a few are discussed below.

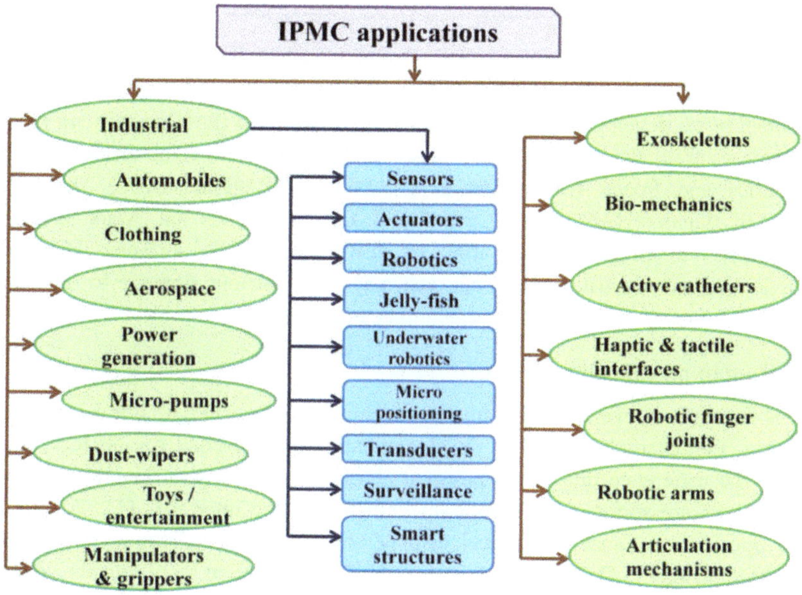

Figure 8: Applications of IPMCs in different areas.

7.1 Robotics application

For years efforts have been made in the direction of creating humanoids or robots, which resemble man. With the development of IPMC the goal of creating robots which can walk, talk and more importantly, show emotions like man appears to be closer. Hence, IPMC shows a potential to act like human muscles, thus we might be able to build robots which are closer to a human. Fabrication of biomimetic robots requires (a) development of specific circuitry,(b) power and regulators integrated into the mobility elements to allow a limited degree of self-powered maneuvering [1]. A conceptual design of a swimming robot structure using IPMC was presented by Shahinpoor [31]. Such a swimming robotic structure in the form of a submarine structure partially encapsulated in a flexible membrane filled with an electrolyte like water. Incidentally, the first commercial application of IPMC was a swimming fish robot developed by Eamex, Japan which was exhibited recently. Bar-Cohen *et al.* designed and fabricated a four finger based gripper with hooks miniature robotic arm at Jet Propulsion Laboratory (JPL) [120].

7.2 Space application

To meet the most challenging and technologically demanding space applications new technologies and materials find immediate application in space need to be developed. Bar-Cohen *et al.* developed a planetary dust wiper based on IPMC to remove planetary dust particles from the surface of a Nanorover [30]. It was a joint effort of National Aeronautics and Space Administration (NASA) and National Space Development Agency of Japan (NASDA). NASA found that operations on Mars involve an environment that causes accumulation of dust on hardware surfaces. Hence, they proposed a planetary dust wiper. This dust may cause serious damage to delicate and precious instruments such as a high-resolution camera, etc.

7.3 Human-machine interface

There is demand to develope better human-machine interfaces to help our sensory system efficiently. For example, a highly developed interface is needed for doctors to avoid realizing the difference due to his physical absence in telesurgery. Haptic devices and tactile sensors are some of the important human-machine interfaces [1]. To provide a human operator, artificial tactile feel display with the required stimuli was produced using IPMC actuators [121]. The idea of using EAPs in active tactile display device to present textual and graphical information to a blind person was presented by Bar-Cohen. The display medium can be constructed as a planar array of small cones called 'reading pins' [1].

7.4 Medical applications

Ease of operation, flexibility, softness and large displacement similar to biological muscles make the IPMC good candidates to operate as substitutes for human muscles [1] Shahinpoor [122] used IPMC as artificial muscles. A human body contains large percentage of water and IPMC operates very well in presence of a solvent medium like water which makes it a very good material for making artificial muscles. They can be inserted into the body to perform surgeries, a closed loop real-time control system can control it and a haptic interface like joystick can be used in this closed loop. IPMC can also be used in developing micro-robots, which can guide catheters in the blood stream [1].

7.5 Micro-manipulation applications

Since the last decade there has been tremendous escalation in the field of micro-electro mechanical systems (MEMS). Presently MEMS devices such as micro-pumps fail to achieve very high displacements or forces because of many limitations in terms of applied voltage or energy consumption. EAPs in general and IPMCs in particular because of their large displacements and low power consumption, can be a suitable material in MEMS [1]. IPMC actuators are well suited in micromanipulation devices due to easily micromanipulation in micro-assembly of small parts, in micro-fabrication, etc. Lumia and Shahinpoor gave a design of a micro-gripper, which used both the actuation and sensing capabilities of IPMC [123]. Kim and Bhat initiated an idea of using IPMC strips as fingers for a micro-gripper system [124].

8. Research dimensions

Typically, perfluorinated polymer-basedionomers membranes with the trade name of nafion are being used as actuators and dynamic sensors are commercially available for IPMC actuator applications because of their high proton exchange capacity, thermal, mechanical and chemical stabilities [8,96,104,105,125–129]. However, these polymers have certain drawbacks such as loss of proton conductivity at temperatures higher than 80 °C, high cost of fabrication, hazardous nature, small actuation bandwidth, low blocking force, little humidity due to drops in proton conductivity, as well as membrane dehydration. Therefore, the main objective of this work was to find suitable non-perfluorinated ionic polymers/composite materials as alternative to the perfluorinated ionic polymers for IPMC actuators. Non-perfluorinated polymers have various advantages such as good film forming capability, short processing time, low cost of fabrication, good flexibility, high thermo-mechanical stabilities and easy modification of chemical properties such as ion-exchange and water holding capacities and proton conductivity [8,33]. Therefore, non-perfluorinated polymers based IPMC actuators can providean easy and reliable solution for the realization of novel actuators as well as the possibility of industrial applications. It is stimulating to build smart material actuators, which can accomplish bio/micro manipulation, bio-sensing, and micro robotic tasks.

The goals of this work are as follows:

Design an IPMC actuator with a low cost and high reliability manufacturing process modified from other previous protocol.

a) To verify the actuator by a series of dynamic testing in an aqueous environment.

b) To control the bending motion of the actuator by implementing a PID controller.

9. Literature review

A vast variety of bending actuators based on different types of EAPs have been developed during last decade. It is very difficult to compile all the work carried out in this field. Thus, an attempt has been made to arrange the work carried out during last 15 years. The work published on perluorinated and non-perfluorinated polymer/composite based different types of actuators during the last fifteen years has been presented briefly in **Table 7.**

Table 7: Perfluorinated and non-perfluorinated polymer or composite based IPMC actuators.

S.No.	Title	Remarks	Ref.
Perfluorinated polymer or composite based actuators			
1	Bending of polyelectrolyte membrane platinum composites by electric stimuli Part II. Response kinetics	A solid composite polymer electrolyte membrane-platinum (SPM-Pt) showing bending response with electric stimuli. A model was developed in which mechanical deformationwas applied to the kinetics of the bending response. In the solid polymer electrolyte membrane(SPM) under electric fields via electro-kinetically induced pressure gradients.	[8]
2	Electric field sensitive neutral polymer gels	In this paper, the effects on themotility and shape of polymer gelsinduced by electric field and new driving mechanism to induce bending deformation of a neutral polymer gels was described.	[130]
3	Morphology of electrodes and bending response of the polymer electrolyte actuator	In this study, the bending behavior of a perfluorocarboxylic acid polymer film plated with Au metal electrode was measured by applied potential in water or saline solution.	[64]
4	Bending deformation of neutral polymer gels induced by electric fields	This study showed when finely distributed colloidal particles were incorporated into a swollen neutral polymer gel bending responseunder electric field occurs.	[131]
5	State of water and ionic conductivity of solid polymer electrolyte membranes in relation to polymer actuators	Examination of state of the water and the ionic conductivity of various ionic forms of solid polymer electrolyte membranes upon polymer actuators was discussed.	[132]
6	The effects of counter ions on characterization and performance of a	For perfluorinated polymer membrane when exchanged with alkali metal ions as counter ions, the displacement was rapid *i.e.* fast responseunder electric	[133]

	solid polymer electrolyte actuator	stimuli.	
7	Novel ionic polymer metal composites equipped with physically loaded particulate electrodes as biomimetic sensors, actuators and artificial muscles	The novel fabrication method of manufacturing IPMCs actuator based on perfluorinated polymer equipped with physically loaded electrodes as artificial muscles and biomimetic actuators, sensors was proposed.	[47]
8	A novel method of manufacturing three dimensional ionic polymer metal composites (IPMCs) biomimetic sensors, actuators and artificial muscles	Commercially available perfluorinated ion exchange polymer films of proper thickness (100-300 μm) permits fast mass transfer in various chemical processes. Hence can be used as artificial muscles.	[97]
9	Design of high authority flexural actuator using an electro-strictive polymer	In this study,a poly(vinylidine fluoride–trifluoroethylene) polymer based actuator faces laminated with internal electrode has been devloped that achieves maximum bending response subjected to an electric field.	[134]
10	Ionic polymer metal composites: II. Manufacturing techniques	Presents a detailed description of various techniques andexperimental procedures in manufacturing IPMCs.	[109]
11	Electrochemical characterization of polymer actuator with large interfacial area	Electrochemical and actuating behaviors were strongly dependent on the interfacial area between the polymer electrolyte and electrode. A large surface area composite actuator was fabricated using replication method.	[94]
12	Investigations on actuation characteristics of IPMC artificial muscle actuator	Electromechanical response characteristics of IPMC, having potential for artificial muscle actuators, were investigated using driving methods.	[135]
13	Novel electroactive, silicate nanocomposites prepared to be used as actuators and artificial muscles	A newly developed nanocompositing technique was proposed to optimize the desirable properties of the base ion exchange polymeric materials used for fabrication of EAP actuators and artificial muscles.	[136]

14	Low temperature characteristics of ionic polymer–metal composite actuators	This study describes IPMCs actuators as an attractive solution for cold operation, because of their actuation capability with relatively low electric fields.	[137]
15	A novel fabrication of ionic polymer-metal composites (IPMC) actuator with silver nano-powders	Silver nano-powders applied for the fabrication of IPMC actuator shows good adhesion between polymer membrane and the metal electrodes without surface roughening, high repeatability, low cost and short processing time.	[48]
16	A multi-wall carbon nanotube tower electrochemical actuator	A patterned multiwalled carbon nanotube arrays were synthesized successfully and electrochemical actuation was demonstrated.	[138]
17	A self-sensing ion conducting polymer metal composite (IPMC) actuator	Development of self sensing IPMC actuator as a position sensor.	[139]
18	Characteristics and performance of ionic polymer metal composite actuators based on nafion/layered silicate and nafion/silica nanocomposites	This study shows that the electrical and mechanical properties of the nafion polymer may greatly alter by incorporation of a small amount (3-10 wt.%) of nanoparticulates such as layered silicate (montmorillonite) or fumed silica. These nanocomposites were used for fabrication of IPMCs as polyelectrolytes.	[140]
19	Effects of humidity on the performance of ionic polymer metal composite actuators: experimental study of the back-relaxation of actuators	Study on the effect of water uptake on the displacement, mechanical force, velocity and charging profiles of perfluorinated ionomer-platinum/Li^+ based actuators.	[141]
20	An electrode model for ionic polymer metal composites	An IPMC was fabricated by electroless plating of platinum and parametric studies were carried out to estimate the electrical properties (resistance and capacitance) of the electrode.	[45]
21	The cause of nanohole and nano particle	Nanoparticles and nanoholes on the surfaces of gold electrode after actuation	[142]

	formation on Au electrode after actuation of electro active paper actuator	strongly associated with the degradation of the actuator performance.	
22	Time resolved mapping of water diffusion coefficients in a working soft actuator device	In this study, diffusion weighted imaging was employed to map the distribution of the diffusion coefficient of water in bare, water-soaked, Li^+ exchanged, cast nafion and in an nafion based IPMC actuator.	[65]
23	Influence of ordered morphology on the anisotropic actuation in uniaxially oriented electroactive polymer systems	A new anisotropic actuation response was yield by perfluorinated polymer nanostructure when altered by uniaxial orientation, mimicks the structure of biological muscle tissue. IPMCs cut from films oriented perpendicular to the draw direction yield tip-displacement response that is higher than those of unoriented IPMCs actuator.	[143]
24	The role of water in transport of ionic liquids in polymeric artificial muscle actuators	Here in the present study quantifying the delicate interplay between ionic liquid and water absorbed in an ion-exchange polymer membrane as used in ionomeric polymer actuators.	[144]
25	Influence of the conductor network composites on the electromechanical performance of ionic polymer conductor network composite actuators	Investigations regarding the electromechanical performance of the ionic polymer conductor network composite actuatorsinfluence by the conductor network composites (CNCs),manufactured by the direct assembly method with ionic liquids as the solvent was presented.	[145]
26	A highly porous nafion membrane templated from polyoxometalates based supramolecule composite for ion exchange polymer-metal composite actuator	Developments of a new technique for manufacturing a highly porous hybrid SiO_2/nafion ionomeric polymer membrane based IPMC actuator, attainingsignificantly high electromechanical performance.	[146]

27	Electrically stimulated gradients in water and counterion concentrations within electroactive polymer actuators	Applications of applied potential atomic force microscopy, neutron imaging, and current sensing atomic force microscopy techniques for investigating actuation mechanism of electroactive polymer actuator.	[147]
28	Enhanced electromechanical performance of graphite oxide-nafion nanocompositeactuator	Graphene oxide/nafion based actuators were tested in terms of conductivity, tip displacement, bulk and surface morphology and blocking force.	[148]
29	Ionic polymer metal composite material as a diaphragm for micropump devices	Two different IPMC diaphragms was prepared and experimentally tested. The stroke volume of each IPMCs diaphragm was examined with different values of applied electrical potential.	[149]
30	Electro-chemical operation of ionic polymermetal composites	The potential origin of the relaxation phenomena of IPMC actuator related to electro-chemically induced surface reactions with electrodes were discussed.	[107]
31	Preparation and performance of IPMC actuators with electrospun Nafion®–MWNT composite electrodes	A new IPMC actuator was prepared with nafion-117 ionomer membrane and electrodes made of an electrospun nafion/multiwalled carbon nanotube web were tested. To reduce the surface resistance the composite electrodes were coated with gold layers of 2-3 μm thickness.	[150]
32	Synthesis and performance evaluation of thin film PPy-PVDF multilayer electroactive polymer actuators	A spin coating technique was used to fabricate a porous polyvinylidene fluoride thin membrane with a thickness of 32 μm with electrochemically deposited polypyrrole layers on both sides to make ultra thin membrane polymer actuators.	[151]
33	Equivalent circuit modeling of ionomer and ionic polymer conductive network composite actuators containing ionic	This study described the electrical equivalent circuits that model the complex frequency dependent impedance of 1-ethyl-3-methylimidazolium trifluoromethanesulfonate containing ionic polymer conductor network	[152]

	liquids	composite devices and electroactive polymer membranes.	
34	High actuation response of PVDF/ PVP/ PSSA based ionic polymer metal composites actuator	The tip displacements of polyvinylidene fluoride (PVDF)/polyvinyl pyrrolidone (PVP)/polystyrene sulfonic acid (PSSA) based IPMCs actuators were measured at low AC and DC voltages. The results were compared with commercially obtain nafion polymer based IPMC actuator.	[105]
35	Enhanced electromechanical response of ionic polymer actuators by improving mechanical coupling between ions and polymer matrix	Poly[(vinylidene difluoride)-co-(chlorotrifluoroethylene)] (P(VDF−CTFE)) and P(VDF−CTFE)/poly(methyl methacrylate) (PMMA) cross-linked blends were used as i-EAP actuators with electrolyte ionic liquid 1-ethyl-3-methylimidazolium trifluoromethane sulfonate [C_2mim][TfO] for the first time. The results were compared with perfluorinated nafion or aquivion ionomers.	[153]
36	Enhanced and fast actuation of fullerenol/PVDF/PVP/PSSA based ionic polymer metal composite actuators	This study reports the enhanced and fast actuation of the fullerenol (Ful)/polyvinylidene fluoride (PVDF)/polyvinyl pyrrolidone (PVP)/polystyrene sulfonic acid (PSSA) composite polymer membrane based IPMC actuator, at low AC and DC voltages as compare to the PVDF/PVP/PSSA membrane based IPMC actuator.	[154]
37	Effect of nucleation time on bending response of ionic polymer metal composite actuators	An autocatalytic electroless plating of nickel metal as electrode is attempted to substitute electroless impregnation reduction technique in IPMC actuators to reduce processing time and cost.	[155]
38	Electrochemical and electromechanical properties of high-performance polymer actuators containing vapor grown carbon	The effects ofionic liquid and metal oxide on the electromechanical and electrochemical properties of poly(vinylidene fluoride-co-hexafluoropropylene) polymer based actuators were investigated using a metal	[156]

	nanofiber and metal oxide	oxide containing a vapour grown carbon nanofiber (VGCF)-IL gel electrode.	
39	Ionic polymer metal composite actuators based on triple layered polyelectrolytes composed of individually functionalized layers	In this paper an IPMC actuators were fabricated based on two types of triple-layered nafion polymer membranes via consecutive solution recasting and electroless plating methods.	[157]
40	Repetitive control of a trilayer conjugated polymer actuator	This study demonstrate a polypyrrole based conducting electroactive polymer actuator desired to track repeating commands for position control applications.	[158]
41	Durable and water floatable ionic polymer actuator with hydrophobic and asymmetrically laser scribed reduced graphene oxide paper electrodes	An asymmetrically and hydrophobic laser-scribed reduced graphene oxide paper electrodes were employed to developed a highly durable and water floatable ionic polymer artificial muscle.	[159]
42	Effect of dehydration on the mechanical and physicochemical properties of old- and palladium -ionomeric polymer metal composite (IPMC) actuators.	This paper describe the dehydration effect on physicochemical and mechanical properties of Au (gold) and Pd (palladium) metal electrode based IPMC using water as solvent.	[160]
43	Single active finger IPMC microgripper	A new designcomprising of one stationary and one actuating finger for a single active finger IPMC microgripper was proposed.	[61]
Non-perfluorinated polymer or composite based IPMC actuators			
44	A tunable dendritic molecular actuator	In this study an electroresponsive molecular actuator was devloped based upon a diblock copolymer of a negatively charged linear chain and a positively charged dendrimer.	[161]
45	Actuator based on MWNT/PVA	This study reveals the development of first actuator based on multiwalled	[162]

	hydrogels	carbon nanotube (MWNT) hydrogels with poly(vinyl alcohol) through simple hydrogelation.	
46	Electric field frequency and strength effects on Au electrode damage for an electroactive paper actuator coated with polypyrrole	The causes of electrode damage to an electroactive paper actuator coated with Au and polypyrrole were investigated at various electrical potential frequencies and strengths.	[163]
47	Polypyrrole actuators with inverse opal structures	Paper polypyrrole (PPy) films synthesized by electrochemical polymerization exhibited higher actuation rates than those of the actuators made from flat PPy films.	[164]
48	A mechanical actuator driven electrochemically by artificial molecular muscles	A microcantilever, coated with redox-controllable, bistable rotaxane molecules shows reversible deflections by alternating oxidizing and reducing electrochemical potentials. One surface of micro-cantilever devices were prepared by precoating with a gold film.	[165]
49	Polymer actuator based on PVA/PAMPS ionic membrane: Optimization of ionic transport properties	In this paper, water soluble poly (vinyl alcohol) and highly ionic conductive poly 2-acrylamido-2-methyl-1-propanesulfonic acid (PAMPS) composite was blend to synthesized and develop electroactive polymer as artificial muscle.	[166]
50	Polypyrrole actuator with a bioadhesive surface for accumulating bacteria from physiological media	A gold/polypyrrole based bilayer actuator was prepared by electrochemical deposition, with surface modification with a bioadhesive polymer, polydopamine. The actuator exhibited high actuation performances in physiological media.	[167]
51	Poly(ethylene oxide)-poly(butadiene) interpenetrated networks as electroactive	Molecular dynamics (MD) techniques have been used to study coordination stability and ionic transport in an interpenetrating polymer (IPN) network used	[168]

	polymers for actuators: A molecular dynamics study	as electrolyte for actuator devices.	
52	Electroactive artificial muscle based on crosslinked PVA/SPTES	Based on a crosslinked ionic networking membrane of sulfonated poly(arylenethioethersulfone) copolymer (SPTES) and poly(vinyl alcohol) (PVA), a novel biomimetic artificial muscle with electrically driven bending deformation was developed.	[169]
53	Electric-stimuli-responsive bending actuator based on sulfonated polyetherimide	In this study, biocompatible sulfonated polyetherimide (SPEI) membrane based bending actuator was fabricated by a solution casting method. It shows good electro-chemical properties, IEC, water uptake and proton conductivity required for a fast and high bending response of ionic polymer actuator.	[170]
54	Electroactive polymer actuators with carbon aerogel electrodes	Carbon aerogel was used for fabrication of nanoporous electrodes for EAP actuators. Their electromechanical and electrochemical characteristics were compared with actuators based on carbide-derived carbon and activated carbon electrodes.	[171]
55	High-performance, low-voltage and easy-operable bending actuator based on aligned carbon nanotube/polymer composites	The coefficient of thermal expansion of the polymer matrix can remarkably reduce by incorporating super-aligned carbon nanotube sheets into a polymer matrix (polydimethylsiloxane). The fabricated actuator was easily operable and capable to generate large bending actuation with controllable actuation.	[172]
56	Sulfonated polystyrene-based ionic polymer metal composite (IPMC) actuator	Acomparative study was carried out between sulfonated polystyrene (sPS)and perfluorinated nafion based IPMC. The tip displacement, the current density, the response rate and the blocking force were significantly higher for the sPS-IPMC than that of the nafion IPMC actuator.	[173]

57	Nanoporous carbide-derived carbon based actuators modified with gold foil: Prospect for fast response and low voltage applications	Carbide-derived nanoporous carbon (CDC) based ionomeric actuators were developed in combination with gold foil. The gold-foil modified actuators required remarkably low operating voltage and exhibited high frequency response.	[174]
58	Novel sulfonated styrenic pentablock copolymer/silicate nanocomposite membranes with controlled ion channels and their IPMC transducers	A novelelectrically driven IPMC actuator has been design and fabricated by ionic thermoplastic elastomer, poly((t-butyl-styrene)-b-(ethylene-r propylene)-b-(styrene-rstyrene sulfonate)-b-(ethylene-r-propylene)-b-(t-butyl-styrene)) (tBS-EP-SS-EP-tBS; SSPB) pentablock copolymer and its nanocomposites with sulfonated montmorillonite (s-MMT).	[175]
59	Electric-field-actuation of in situ composites that contain silver-coated carbon fibers in sodium sulfonate ionomers	Anelectric-field-stimulated actuation system was developed by impregnating sodium-sulfonated ionomers, poly(styrene-co-butyl acrylate-co sodium allyl sulfonate, PSBS) with *in-situ* deposited silver-modified carbon fibers (SCCF).	[176]
60	Imidazolium sulfonate containing pentablock copolymer ionic liquid membranes for electroactive actuators	Thermally stable electroactive actuators have been fabricated by imidazolium sulfonate containing pentablock copolymer ionic liquid composite. The fabricated actuator exhibit effective actuation response under a low potential	[177]
61	Comparative displacement study of bilayer actuators comprising of conducting polymers, fabricated from polypyrrole, poly(3,4 ethylene dioxythiophene) or poly(3,4-propylene dioxythiophene)	A report on comparative study of the displacement of bilayer actuators manufactured using different conducting polymers was presented.	[178]

62	Enhancement of actuation ability of ionic-type conducting polymer actuators using metal ion implantation	The effect of gold-ion implantation on the actuation performance of ionic type conducting polymer actuators has been examined.	[179]
63	Fabrication of ionic polymer actuator with graphene nanocomposite electrodes and its characterization	The polystyrene based ionic polymer actuator was fabricated by conventional solvent mixing using graphene nanocomposite (GN) as its electrodes.	[180]
64	Novel sulfonated polysulfone ion exchange membranes for ionic polymer metal composite actuators	A series of sulfonated polysulfone (SPSU) membranes with different degrees of sulfonation were prepared for fabricating cost effective and high performance IPMC actuators.	[181]
65	Kraton based ionic polymer metal composite (IPMC) actuator	The tip displacement, proton conductivity, current density, water uptake and ion-exchange capacity of kraton non-perfluorinated IPMCs were examined and compared with nafion based IPMCs membranes.	[63]
66	Ionic polymer actuator based on anion conducting methylated ether linked polybenzimidazole	A new type of ionic polymer actuator based on anion conducting methylated poly[(1-(4,4-diphenylether)-5-oxybenzimidazole)-benzimidazole] (PBI–OO) membranes and a carbon composite electrode material was successfully developed.	[108]
67	Low-voltage and high-performance electrothermal actuator based on multi walled carbon nanotube/ polymer composites	Environment friendly electrothermal bimetallic actuators was fabricated by multi-walled carbon nanotubes with the matrix of water borne polyurethane or silicone rubber.	[182]
68	Development of sulfonated poly(vinyl alcohol)/polpyrrole based ionic polymer	A novel sulfonated poly(vinyl alcohol)/polypyrrole polymer membrane sandwiched between platinum (SPVA-Py-Pt) was fabricated for a bending	[33]

	metal composite (IPMC) actuator and its characterization	actuator which can be used in microrobotic applications.	
69	Study and preparation of highly water stable polyacrylonitrile-kraton-graphene composite membrane for bending actuator toward robotic application	Polyacrylonitrile-kraton-graphene (PAN-KR-GR) ionomeric polymer membrane sandwiched between Pt electrode based IPMC actuator is developed. The study opened a route to design and prepare multifunctional IPMC membrane for robotic application.	[42]
70	Fabrication of silver nano powder embedded kraton polymer actuator and its characterization	A novel silver nano powder (Ag Pw) embedded kraton (KR) ionic polymer actuator was fabricated. The KR-AgPw ionic polymer metal composite (IPMC) membrane was prepared by a solution casting method.	[29]

Acknowledgements

The authors are thankful to Department of Applied Chemistry, Aligarh Muslim University, India for providing facilities. Dr. Inamuddin is also thankful to the Department of Science and Technology for awarding the Young Scientist award under Project no. SR/FT/CS-159/2011.

References

[1] Y. Bar-Cohen, No TitleElectroactive polymer [EAP] actuators as artificial muscles, 1st ed, SPIE Press, Bellingham, Washington, 2001.

[2] R. Tiwari, E. Garcia, The state of understanding of ionic polymer metal composite architecture: a review, Smart Mater. Struct. 20 (2011) 83001.

[3] K. Sadeghipour, R. Salomon, S. Neogi, Development of a novel electrochemically active membrane and "smart" material based vibration sensor/damper, Smart Mater. Struct. 1 (1992) 172.

[4] K. Oguro, H. Takenaka, Y. Kawami, Actuator element, (1993).

[5] K. Asaka, K. Oguro, Y. Nishimura, M. Mizuhata, H. Takenaka, Bending of Polyelectrolyte Membrane-Platinum Composites by Electric Stimuli I. Response Characteristics to Various Waveforms., Polym. J. 27 (1995) 436–440. doi:10.1295/polymj.27.436.

[6] M.S. M. Mojarrad, No Title, in: Proc. SPIE Smart Struct. Mater. Symp., 1997: p. 294.

[7] J.S. Y. Bar-Cohen, M. Shahinpoor, J.O. Harrison, No Title, in: Proc. SPIE Smart. Struct. Mater. Symp., 1999: pp. 3669, 51.

[8] K. Asaka, K. Oguro, Bending of polyelectrolyte membrane platinum composites by electric stimuli: Part II. Response kinetics, J. Electroanal. Chem. 480 (2000) 186–198. doi:http://dx.doi.org/10.1016/S0022-0728(99)00458-1.

[9] P.G. de Gennes, K. Okumura, M. Shahinpoor, K.J. Kim, Mechanoelectric effects in ionic gels, EPL (Europhysics Lett. 50 (2000) 513.

[10] B. Bhandari, G.-Y. Lee, S.-H. Ahn, A review on IPMC material as actuators and sensors: Fabrications, characteristics and applications, Int. J. Precis. Eng. Manuf. 13 (2012) 141–163. doi:10.1007/s12541-012-0020-8.

[11] Inamuddin, R.K. Jain, S. Hussain, M. Naushad, Poly (3,4-ethylenedioxythiophene): polystyrene sulfonate zirconium(IV) phosphate compsiteionomeric membrane for artifical muscle application. RSC Adv 5 (2015) 84526-84534.

[12] G. Alici, G. Spinks, N.N. Huynh, L. Sarmadi, R. Minato, Establishment of a biomimetic device based on tri-layer polymer actuators—propulsion fins, Bioinspir. Biomim. 2 (2007) S18.

[13] S. McGovern, G. Alici, V.-T. Truong, G. Spinks, Finding NEMO (novel electromaterial muscle oscillator): a polypyrrole powered robotic fish with real-time wireless speed and directional control, Smart Mater. Struct. 18 (2009) 95009.

[14] G. Alici, M.J. Higgins, Normal stiffness calibration of microfabricated tri-layer conducting polymer actuators, Smart Mater. Struct. 18 (2009) 65013.

[15] Y. Bar-Cohen, Q. Zhang, Electroactive Polymer Actuators and Sensors, MRS Bull. 33 (2008) 173–181. doi:10.1557/mrs2008.42.

[16] Y. Bar-Cohen, No Title Electroactive polymer (EAP) actuators as artificial muscles: reality, potential, and challenges, 2nd ed, SPIE Publications, Bellingham, Washington, 2004.

[17] J.-Y. Jung, I.-K. Oh, Novel Nanocomposite Actuator Based on Sulfonated Poly(styrene-b-ethylene-co-butylene-b-styrene) Polymer, J. Nanosci. Nanotechnol. 7 (2007) 3740–3743. doi:10.1166/jnn.2007.004.

[18] J. Lu, S.-G. Kim, S. Lee, I.-K. Oh, Fabrication and actuation of electro-active polymer actuator based on PSMI-incorporated PVDF, Smart Mater. Struct. 17 (2008) 45002.

[19] K. Ikeda, M. Sasaki, H. Tamagawa, {IPMC} bending predicted by the circuit and viscoelastic models considering individual influence of Faradaic and non-Faradaic currents on the bending, Sensors Actuators B Chem. 190 (2014) 954–967. doi:http://dx.doi.org/10.1016/j.snb.2013.09.016.

[20] G. M'boungui, B. Semail, F. Giraud, A.A. Jimoh, Development of a novel plane piezoelectric actuator using Hamilton's principle based model and Hertz contact theory, Sensors Actuators A Phys. 217 (2014) 116–123. doi:http://dx.doi.org/10.1016/j.sna.2014.06.026.

[21] M. Shahinpoor, K.J. Kim, Ionic polymer–metal composites: IV. Industrial and medical applications, Smart Mater. Struct. 14 (2005) 197–214. doi:10.1088/0964-1726/14/1/020.

[22] M.J. Han, J.H. Park, J.Y. Lee, J.Y. Jho, Ionic Polymer-Metal Composite Actuators Employing Radiation-Grafted Fluoropolymers as Ion-Exchange Membranes, (2006) 219–222. doi:10.1002/marc.200500694.

[23] J.-H. Jeon, S.-P. Kang, S. Lee, I.-K. Oh, Novel biomimetic actuator based on {SPEEK} and {PVDF}, Sensors Actuators B Chem. 143 (2009) 357–364. doi:http://dx.doi.org/10.1016/j.snb.2009.09.020.

[24] G.-H. Feng, J.-W. Tsai, Investigation of electrical to mechanical energy conversion of a three-dimensional four-electrode multidirectional-controllable IPMC transducer with/without an optical fiber enclosed, Smart Mater. Struct. 20 (2011) 15027.

[25] Y. Fang, X. Tan, A novel diaphragm micropump actuated by conjugated polymer petals: Fabrication, modeling, and experimental results, Sensors Actuators A Phys. 158 (2010) 121–131. doi:http://dx.doi.org/10.1016/j.sna.2009.12.012.

[26] H. Lei, W. Li, X. Tan, Encapsulation of ionic polymer-metal composite (IPMC) sensors with thick parylene: Fabrication process and characterization results, Sensors Actuators A Phys. 217 (2014) 1–12. doi:http://dx.doi.org/10.1016/j.sna.2014.05.013.

[27] M.M. M. Shahinpoor, K.J. Kim, Artificial muscles: applications of advanced polymeric nanocomposites, Taylor and Francis, New York, 2007.

[28] M. Shahinpoor, Y. Bar-Cohen, J.O. Simpson, J. Smith, Ionic polymer-metal composites (IPMCs) as biomimetic sensors, actuators and artificial muscles - a review, Smart Mater. Struct. 7 (1998) R15.

[29] A. Khan, Inamuddin, R.K. Jain, M. Naushad, Fabrication of a silver nano powder embedded kraton polymer actuator and its characterization, RSC Adv. 5 (2015) 91564–91573. doi:10.1039/C5RA17776F.

[30] J.S. Y. Bar-Cohen, S. Leary, A. Yavrouian, K. Oguro, S. Tadokoro, J. Harrison, J. Smith, No Title, in: Proc. SPIE Smart. Struct. Mater. Symp., 2000: p. 140.

[31] M. Shahinpoor, Conceptual design, kinematics and dynamics of swimming robotic structures using ionic polymeric gel muscles, Smart Mater. Struct. 1 (1992) 91.

[32] K. Oguro, Y. Kawami, H. Takenaka, Bending of an ion-conducting polymer film-electrode composite by an electric stimulus at low voltage, J. Micromachine Soc. 5 (1992) 27–30.

[33] Inamuddin, A. Khan, R.K. Jain, M. Naushad, Development of sulfonated poly(vinyl alcohol)/polpyrrole based ionic polymer metal composite (IPMC)

actuator and its characterization, Smart Mater. Struct. 24 (2015) 95003. http://stacks.iop.org/0964-1726/24/i=9/a=095003.

[34] M. Shahinpoor, K.J. Kim, Ionic polymer-metal composites: I. Fundamentals, Smart Mater. Struct. 10 (2001) 819.

[35] S. Nemat-Nasser, J.Y. Li, Electromechanical response of ionic polymer-metal composites, J. Appl. Phys. 87 (2000) 3321. doi:10.1063/1.372343.

[36] M. Shahinpoor, K.J. Kim, D.J. Leo, Ionic polymer-metal composites as multifunctional materials, Polym. Compos. 24 (2003) 24–33. doi:10.1002/pc.10002.

[37] M. Shahinpoor, Ionic polymer–conductor composites as biomimetic sensors, robotic actuators and artificial muscles—a review, Electrochim. Acta. 48 (2003) 2343–2353. doi:http://dx.doi.org/10.1016/S0013-4686(03)00224-X.

[38] A.J. Grodzinsky, Electromechanics of deformable polyelectrolyte membranes., Cambridge, 1974.

[39] B.J. Akle, D.J. Leo, Single-Walled Carbon Nanotubes -- Ionic Polymer Electroactive Hybrid Transducers, J. Intell. Mater. Syst. Struct. 19 (2007) 905–915. doi:10.1177/1045389X07082441.

[40] I.W. Hunter, S. Lafontaine, A comparison of muscle with artificial actuators, in: Tech. Dig. IEEE Solid-State Sens. Actuator Work., IEEE, n.d.: pp. 178–185. doi:10.1109/SOLSEN.1992.228297.

[41] H.-I. Kim, D.-K. Kim, J.-H. Han, Study of flapping actuator modules using IPMC, in: Y. Bar-Cohen (Ed.), 2007: p. 65241A. doi:10.1117/12.715633.

[42] Inamuddin, A. Khan, R.K. Jain, M. Naushad, Study and preparation of highly water-stable polyacrylonitrile-kraton-graphene composite membrane for bending actuator toward robotic application, J. Intell. Mater. Syst. Struct. 27 (2016). doi:10.1177/1045389X15596627.

[43] D. Pugal, K. Jung, A. Aabloo, K.J. Kim, Ionic polymer-metal composite mechanoelectrical transduction: review and perspectives, Polym. Int. 59 (2010) 279–289. doi:10.1002/pi.2759.

[44] A. Punning, M. Kruusmaa, A. Aabloo, Surface resistance experiments with {IPMC} sensors and actuators, Sensors Actuators A Phys. 133 (2007) 200–209. doi:http://dx.doi.org/10.1016/j.sna.2006.03.010.

[45] S.J. Kim, S.-M. Kim, K.J. Kim, Y.H. Kim, An electrode model for ionic polymer–metal composites, Smart Mater. Struct. 16 (2007) 2286.

[46] M. Shahinpoor, K.J. Kim, The effect of surface-electrode resistance on the performance of ionic polymer-metal composite (IPMC) artificial muscles, Smart Mater. Struct. 9 (2000) 543.

[47] M. Shahinpoor, K.J. Kim, Novel ionic polymer–metal composites equipped with physically loaded particulate electrodes as biomimetic sensors, actuators and artificial muscles, Sensors Actuators A Phys. 96 (2002) 125–132. doi:http://dx.doi.org/10.1016/S0924-4247(01)00777-4.

[48] C.K. Chung, P.K. Fung, Y.Z. Hong, M.S. Ju, C.C.K. Lin, T.C. Wu, A novel fabrication of ionic polymer-metal composites (IPMC) actuator with silver nano-powders, Sensors Actuators B Chem. 117 (2006) 367–375. doi:http://dx.doi.org/10.1016/j.snb.2005.11.021.

[49] B.J. Akle, No Title Electrochemical response in ionic polymer transducers: an experimental and theoretical study, Virginia Polytechnic Institute and State University, 2005.

[50] K. Kanamura, H. Morikawa, T. Umegaki, Observation of Interface Between Pt Electrode and Nafion Membrane, J. Electrochem. Soc. 150 (2003) A193. doi:10.1149/1.1531970.

[51] M. Tsuda, W.A. Diño, H. Kasai, Behavior of hydrogen atom at Nafion–Pt interface, Solid State Commun. 134 (2005) 601–605. doi:http://dx.doi.org/10.1016/j.ssc.2005.03.006.

[52] A. Gruger, A. Régis, T. Schmatko, P. Colomban, Nanostructure of Nafion® membranes at different states of hydration: An {IR} and Raman study, Vib. Spectrosc. 26 (2001) 215–225. doi:http://dx.doi.org/10.1016/S0924-2031(01)00116-3.

[53] K.A. Mauritz, R.B. Moore, State of Understanding of Nafion, (2004).

[54] D. Seeliger, C. Hartnig, E. Spohr, Aqueous pore structure and proton dynamics in solvated Nafion membranes, Electrochim. Acta. 50 (2005) 4234–4240. doi:http://dx.doi.org/10.1016/j.electacta.2005.03.071.

[55] G. Polizos, Z. Lu, D.D. Macdonald, E. Manias, State of Water in Nafion 117 Proton Exchange Membranes Studied by Dielectric Relaxation Spectroscopy, MRS Proc. 972 (2006) 972-NaN-7. doi:10.1557/PROC-0972-AA08-07.

[56] N.P. Berezina, N.A. Kononenko, A.A.-R. Sytcheva, N. V Loza, S.A. Shkirskaya, N. Hegman, A. Pungor, Perfluorinated nanocomposite membranes modified by polyaniline: Electrotransport phenomena and morphology, Electrochim. Acta. 54 (2009) 2342–2352. doi:http://dx.doi.org/10.1016/j.electacta.2008.10.048.

[57] S.S. K. Oguro, N. Fujiwara, K. Asaka, K. Onishi, *Smart Struct. Mater Symp. Proc. SPIE* ***3669*** (1999) 64-71.

[58] S. Nemat-Nasser, Y. Wu, Comparative experimental study of ionic polymer–metal composites with different backbone ionomers and in various cation forms, J. Appl. Phys. 93 (2003) 5255. doi:10.1063/1.1563300.

[59] Z. Lu, M. Lanagan, E. Manias, D.D. Macdonald, Two-Port Transmission Line Technique for Dielectric Property Characterization of Polymer Electrolyte Membranes, J. Phys. Chem. B. 113 (2009) 13551–13559. doi:10.1021/jp9057115.

[60] M.W. Verbrugge, Analysis of Promising Perfluorosulfonic Acid Membranes for Fuel-Cell Electrolytes, J. Electrochem. Soc. 137 (1990) 3770. doi:10.1149/1.2086299.

[61] S. Ford, G. Macias, R. Lumia, Single active finger IPMC microgripper, Smart Mater. Struct. 24 (2015) 25015.

[62] P. Choi, N.H. Jalani, R. Datta, Thermodynamics and Proton Transport in Nafion, J. Electrochem. Soc. 152 (2005) E123. doi:10.1149/1.1859814.

[63] Inamuddin, A. Khan, M. Luqman, A. Dutta, Kraton based ionic polymer metal composite (IPMC) actuator, Sensors Actuators A Phys. 216 (2014) 295–300. doi:http://dx.doi.org/10.1016/j.sna.2014.04.015.

[64] K. Onishi, S. Sewa, K. Asaka, N. Fujiwara, K. Oguro, Morphology of electrodes and bending response of the polymer electrolyte actuator, Electrochim. Acta. 46 (2001) 737–743. doi:http://dx.doi.org/10.1016/S0013-4686(00)00656-3.

[65] L. Naji, J.A. Chudek, R.T. Baker, Time-Resolved Mapping of Water Diffusion Coefficients in a Working Soft Actuator Device, J. Phys. Chem. B. 112 (2008) 9761–9768. doi:10.1021/jp803792c.

[66] N. Jin, B. Wang, K. Bian, Q. Chen, K. Xiong, Performance of ionic polymer-metal composite (IPMC) with different surface roughening methods, Front. Mech. Eng. China. 4 (2009) 430–435. doi:10.1007/s11465-009-0053-6.

[67] R. Tiwari, No Title Modeling and characterization of IPMC energy harvesters, Lambert Academic, Saarbrucken, 2010.

[68] S.-M. Kim, K.J. Kim, Palladium buffer-layered high performance ionic polymer–metal composites, Smart Mater. Struct. 17 (2008) 35011.

[69] I.-S. Park, K.J. Kim, Multi-fields responsive ionic polymer–metal composite, Sensors Actuators A Phys. 135 (2007) 220–228. doi:http://dx.doi.org/10.1016/j.sna.2006.07.014.

[70] J.-W. Lee, Y.-T. Yoo, The Structure and Performance of Ionic Polymer-Metal Composite Actuators Prepared via Electroless Plating Process Using Various Alcohols, Macromol. Symp. 249–250 (2007) 56–60. doi:10.1002/masy.200750309.

[71] D.Y. Lee, K.J. Kim, S. Heo, M.H. Lee, B.Y. Kim, Application of an Equivalent Circuit Model for Ionic Polymer-Metal Composite (IPMC) Bending Actuator Loaded With Multiwalled Carbon Nanotube (M-CNT), Key Eng. Mater. 309–311 (2006) 593–596. doi:10.4028/www.scientific.net/KEM.309-311.593.

[72] H. Takenaka, E. Torikai, Y. Kawami, N. Wakabayashi, Solid polymer electrolyte water electrolysis, Int. J. Hydrogen Energy. 7 (1982) 397–403. doi:http://dx.doi.org/10.1016/0360-3199(82)90050-7.

[73] R.P. HAMLEN, C.E. KENT, S.N. SHAFER, Electrolytically Activated Contractile Polymer, Nature. 206 (1965) 1149–1150. doi:10.1038/2061149b0.

[74] T.K.K. N. Fujiwara, Y. Nishimura, K. Oguro, E. Torikai, No Title, 1997.

[75] T.K.K. N. Fujiwara, Y. Nishimura, K. Oguro, E. Torikai, No Title, 1998.

[76] V. Palmre, D. Brandell, U. Mäeorg, J. Torop, O. Volobujeva, A. Punning, U. Johanson, M. Kruusmaa, A. Aabloo, Nanoporous carbon-based electrodes for high strain ionomeric bending actuators, Smart Mater. Struct. 18 (2009) 95028.

[77] N. Fujiwara, K. Asaka, Y. Nishimura, K. Oguro, E. Torikai, Preparation of Gold−Solid Polymer Electrolyte Composites As Electric Stimuli-Responsive Materials, Chem. Mater. 12 (2000) 1750–1754. doi:10.1021/cm9907357.

[78] E. Frackowiak, F. Béguin, Electrochemical storage of energy in carbon nanotubes and nanostructured carbons, Carbon N. Y. 40 (2002) 1775–1787. doi:http://dx.doi.org/10.1016/S0008-6223(02)00045-3.

[79] I.S. Park, R. Tiwari, K.J. Kim, Sprayed Sensor Using IPMC PAINT, Adv. Sci. Technol. 61 (2008) 59–64. doi:10.4028/www.scientific.net/AST.61.59.

[80] J.N. Barisci, G.G. Wallace, D. Chattopadhyay, F. Papadimitrakopoulos, R.H. Baughman, Electrochemical Properties of Single-Wall Carbon Nanotube Electrodes, J. Electrochem. Soc. 150 (2003) E409. doi:10.1149/1.1593045.

[81] J. Brufau-Penella, M. Puig-Vidal, P. Giannone, S. Graziani, S. Strazzeri, Characterization of the harvesting capabilities of an ionic polymer metal composite device, Smart Mater. Struct. 17 (2008) 15009.

[82] D.Y. Lee, I.-S. Park, M.-H. Lee, K.J. Kim, S. Heo, Ionic polymer–metal composite bending actuator loaded with multi-walled carbon nanotubes, Sensors Actuators A Phys. 133 (2007) 117–127. doi:http://dx.doi.org/10.1016/j.sna.2006.04.005.

[83] I. Kang,Y.Y. Heung, J.H. Kim, J.W. Lee, R. Gollapudi, S. Subramaniam, S. Narasimhadevara, D. Hurd, G. R. Krikera, V. Shanov, M.J. Schulz, D. Shi, J. Boerio, S. Mall, M.R. Wren, Introduction to carbon nanotube and nanofiber smart materials, Compos. Part B Eng. 37 (2006) 382–394.

[84] Y.Y. V.K. Nguyen, H.T. Lim, J.W. Lee, Effect of reduction temperature on electrode formation and performance of ionic polymer metal composites, Proc. SPIE Smart Struct. Mater. Symp, 2006: p. 616813.

[85] M.H.L. S. Heo, K.J. Kim, D.Y. Lee, S. Vemuri, No Title, in: Proc. SPIE Smart Struct. Mater. Symp, 2005: p. 194.

[86] D.Y. Lee, M.-H. Lee, K.J. Kim, S. Heo, B.-Y. Kim, S.-J. Lee, Effect of multiwalled carbon nanotube (M-CNT) loading on M-CNT distribution behavior and the related electromechanical properties of the M-CNT dispersed ionomeric nanocomposites, Surf. Coatings Technol. 200 (2005) 1920–1925. doi:http://dx.doi.org/10.1016/j.surfcoat.2005.08.024.

[87] R.J. Lawrance, L.D. Wood, *U.S. Patent 4 272 353* 9, 1981.

[88] S,.Surampudi, S. R. Narayanan, E. Vamos, H. Frank and G. Halpert, Advances in Direct Methanol Fuel Cells, J. Power Sources, 47 (1994) 377-385..

[89] C. Yi-Lin, C. Tse-Chuan, Metals and alloys bonded on solid polymer electrolyte for electrochemical reduction of pure benzaldehyde without liquid supporting electrolyte, J. Electroanal. Chem. 360 (1993) 247–259. doi:http://dx.doi.org/10.1016/0022-0728(93)87017-P.

[90] D.W. Dewulf, A.J. Bard, The electrochemical reduction of CO2 to CH4 and C2H4 at Cu/Nafion electrodes (solid polymer electrolyte structures), Catal. Letters. 1 (1988) 73–79. doi:10.1007/BF00765357.

[91] U. Johanson, U. Mäeorg, V. Sammelselg, D. Brandell, A. Punning, M. Kruusmaa, A. Aabloo, Electrode reactions in Cu–Pt coated ionic polymer actuators, Sensors Actuators B Chem. 131 (2008) 340–346. doi:http://dx.doi.org/10.1016/j.snb.2007.11.044.

[92] B.-K. Fang, M.-S. Ju, C.-C.K. Lin, A new approach to develop ionic polymer–metal composites (IPMC) actuator: Fabrication and control for active catheter systems, Sensors Actuators A Phys. 137 (2007) 321–329. doi:http://dx.doi.org/10.1016/j.sna.2007.03.024.

[93] G. Di Pasquale, L. Fortuna, S. Graziani, M. La Rosa, D. Nicolosi, G. Sicurella, E. Umana, All-Organic Motion Sensors: Electromechanical Modeling, IEEE Trans. Instrum. Meas. 58 (2009) 3731–3738. doi:10.1109/TIM.2009.2019321.

[94] T.-G. Noh, Y. Tak, J.-D. Nam, H. Choi, Electrochemical characterization of polymer actuator with large interfacial area, Electrochim. Acta. 47 (2002) 2341–2346. doi:http://dx.doi.org/10.1016/S0013-4686(02)00089-0.

[95] J. Sunghee, S. Jeomsik, K. Gyuseok, L. Sukmin, M. Museong, Effects of the electrode interface on the electric properties of IPMC for artificial muscles, in: R. Magjarevic, J.H. Nagel (Eds.), World Congr. Med. Phys. Biomed. Eng. 2006 August 27 -- Sept. 1, 2006 COEX Seoul, Korea ``Imaging Futur. Med., Springer Berlin Heidelberg, Berlin, Heidelberg, 2007: pp. 2973–2976. doi:10.1007/978-3-540-36841-0_752.

[96] B.J. Akle, D.J. Leo, M.A. Hickner, J.E. McGrath, Correlation of capacitance and actuation in ionomeric polymer transducers, J. Mater. Sci. 40 (2005) 3715–3724. doi:10.1007/s10853-005-3312-x.

[97] K.J. Kim, M. Shahinpoor, A novel method of manufacturing three-dimensional ionic polymer–metal composites (IPMCs) biomimetic sensors, actuators and artificial muscles, Polymer (Guildf). 43 (2002) 797–802. doi:http://dx.doi.org/10.1016/S0032-3861(01)00648-6.

[98] K.J. Kim, M. Shahinpoor, Development of three-dimensional polymeric artificial muscles, in: Y. Bar-Cohen (Ed.), 2001: p. 223. doi:10.1117/12.432650.

[99] B. Akle, D. Leo, M. Bennett, K. Wiles, J. McGrath, Direct assembly process for fabrication of ionomeric polymer devices, United States Patent Application 20060266642, 2006.

[100] D. Griffiths, V.B. Sundaresan, B. Akle, P. Vlachos, D. Leo, Micro deposition method: a novel fabrication method for ionic polymer metallic composites, in: Y. Bar-Cohen (Ed.), 2008: p. 69270C. doi:10.1117/12.776462.

[101] K.J. Kim, M. Shahinpoor, <title>Effect of the surface-electrode resistance on the actuation of ionic polymer-metal composite (IPMC) artificial muscles</title>, in: Y. Bar-Cohen (Ed.), 1999: pp. 308–319. doi:10.1117/12.349703.

[102] M. Bennett, D. Leo, Morphological and electromechanical characterization of ionic liquid/Nafion polymer composites, in: Y. Bar-Cohen (Ed.), 2005: p. 506. doi:10.1117/12.599849.

[103] T.A. Zawodzinski, Water Uptake by and Transport Through Nafion® 117 Membranes, J. Electrochem. Soc. 140 (1993) 1041. doi:10.1149/1.2056194.

[104] V. Panwar, C. Lee, S.Y. Ko, J.-O. Park, S. Park, Dynamic mechanical, electrical, and actuation properties of ionic polymer metal composites using PVDF/PVP/PSSA blend membranes, Mater. Chem. Phys. 135 (2012) 928–937. doi:http://dx.doi.org/10.1016/j.matchemphys.2012.05.081.

[105] V. Panwar, K. Cha, J.-O. Park, S. Park, High actuation response of PVDF/PVP/PSSA based ionic polymer metal composites actuator, Sensors Actuators B Chem. 161 (2012) 460–470. doi:http://dx.doi.org/10.1016/j.snb.2011.10.062.

[106] A.K. Phillips, R.B. Moore, Ionic actuators based on novel sulfonated ethylene vinyl alcohol copolymer membranes, Polymer (Guildf). 46 (2005) 7788–7802. doi:http://dx.doi.org/10.1016/j.polymer.2005.02.123.

[107] D. Kim, K.J. Kim, J. Nam, V. Palmre, Electro-chemical operation of ionic polymer–metal composites, Sensors Actuators B Chem. 155 (2011) 106–113. doi:http://dx.doi.org/10.1016/j.snb.2010.11.032.

[108] T. Kwon, J.-W. Lee, H. Cho, D. Henkensmeier, Y. Kang, S.M. Hong, C.M. Koo, Ionic polymer actuator based on anion-conducting methylated ether-linked polybenzimidazole, Sensors Actuators B Chem. 214 (2015) 43–49. doi:http://dx.doi.org/10.1016/j.snb.2015.03.007.

[109] K.J. Kim, M. Shahinpoor, Ionic polymer–metal composites: II. Manufacturing techniques, Smart Mater. Struct. 12 (2003) 65.

[110] S.J. Lee, M.J. Han, S.J. Kim, J.Y. Jho, H.Y. Lee, Y.H. Kim, A new fabrication method for IPMC actuators and application to artificial fingers, Smart Mater. Struct. 15 (2006) 1217.

[111] B. Kim, B.M. Kim, J. Ryu, I.-H. Oh, S.-K. Lee, S.-E. Cha, J. Pak, Analysis of mechanical characteristics of the ionic polymer metal composite (IPMC) actuator using cast ion-exchange film, in: Y. Bar-Cohen (Ed.), 2003: p. 486. doi:10.1117/12.484296.

[112] R.-J. Chung, T.-S. Chin, L.-C. Chen, M.-F. Hsieh, Preparation of gradually componential metal electrode on solution-casted Nafion^{TM} membrane, Biomol. Eng. 24 (2007) 434–437. doi:http://dx.doi.org/10.1016/j.bioeng.2007.07.003.

[113] Y. Zhang, C. Ma, L. Dai, Electrode Preparation and Electro-deformation of Ionic Polymer-metal Composite (IPMC), in: IEEE Int. Conf. Nano/Micro Eng. Mol. Syst., 2007.

[114] T. Johnson, F. Amirouche, Multiphysics modeling of an IPMC microfluidic control device, Microsyst. Technol. 14 (2008) 871–879. doi:10.1007/s00542-008-0603-6.

[115] M. Siripong, S. Fredholm, Q.A. Nguyen, B. Shih, J. Itescu, J. Stolk, A Cost-Effective Fabrication Method for Ionic Polymer-Metal Composites, MRS Proc. 889 (2005) 889-W04-3. doi:10.1557/PROC-0889-W04-03.

[116] T. NAKAMURA, T. IHARA, T. HORIUCHI, T. MUKAI, K. ASAKA, Measurement and Modeling of Electro-Chemical Properties of Ion Polymer Metal Composite by Complex Impedance Analysis, SICE J. Control. Meas. Syst. Integr. 2 (2009) 373–378. doi:10.9746/jcmsi.2.373.

[117] H.D. Çilingir, M. Papila, ``Equivalent'' Electromechanical Coefficient for IPMC Actuator Design Based on Equivalent Bimorph Beam Theory, Exp. Mech. 50 (2010) 1157–1168. doi:10.1007/s11340-009-9311-0.

[118] Y. Bar-Cohen, X. Bao, S. Sherrit, S.-S. Lih, <title>Characterization of the electromechanical properties of ionomeric polymer-metal composite (IPMC)</title>, in: Y. Bar-Cohen (Ed.), 2002: pp. 286–293. doi:10.1117/12.475173.

[119] S. Zamani, S. Nemat-Nasser, Controlled actuation of Nafion-based ionic polymer-metal composites (IPMCs)with ethylene glycol as solvent, in: Y. Bar-Cohen (Ed.), 2004: p. 159. doi:10.1117/12.539075.

[120] Y. Bar-Cohen, T. Xue, M. Shahinpoor, J. O. Simpson, and J. Smith, "Flexible, low-mass robotic arm actuated by electroactive polymers (EAP)," Proceedings of the SPIE International Smart Materials and Structures Conference, SPIE Paper No. 3329-07, San Diego, CA, 1-6 March 1998.

[121] K.O. M. Konyo, S. Tadokoro, T. Takamori, in: Proc. IEEE Int. Conf. Robot. Autom., 2000: p. 3416.

[122] M. Shahinpoor, Continuum electromechanics of ionic polymeric gels as artificial muscles for robotic applications, Smart Mater. Struct. 3 (1994) 367.

[123] R. Lumia, M. Shahinpoor, <title>Microgripper design using electroactive polymers</title>, in: Y. Bar-Cohen (Ed.), 1999: pp. 322–329. doi:10.1117/12.349689.

[124] N. Bhat, W.-J. Kim, Precision force and position control of an ionic polymer metal composite, Proc. Inst. Mech. Eng. Part I J. Syst. Control Eng. 218 (2004) 421–432. doi:10.1177/095965180421800601.

[125] S. Holmberg, P. Holmlund, R. Nicolas, C.-E. Wilén, T. Kallio, G. Sundholm, F. Sundholm, Versatile Synthetic Route to Tailor-Made Proton Exchange Membranes for Fuel Cell Applications by Combination of Radiation Chemistry of Polymers with Nitroxide-Mediated Living Free Radical Graft Polymerization, Macromolecules. 37 (2004) 9909–9915. doi:10.1021/ma0353641.

[126] V.K. Nguyen, Y. Yoo, A novel design and fabrication of multilayered ionic polymer-metal composite actuators based on Nafion/layered silicate and Nafion/silica nanocomposites, Sensors Actuators B Chem. 123 (2007) 183–190. doi:http://dx.doi.org/10.1016/j.snb.2006.08.013.

[127] M. Shahinpoor, K.J. Kim, Ionic polymer–metal composites: III. Modeling and simulation as biomimetic sensors, actuators, transducers, and artificial muscles, Smart Mater. Struct. 13 (2004) 1362.

[128] S.J. Kim, M.S. Kim, S.R. Shin, I.Y. Kim, S.I. Kim, S.H. Lee, T.S. Lee, G.M. Spinks, Enhancement of the electromechanical behavior of IPMCs based on chitosan / polyaniline ion exchange membranes fabricated by freeze-drying, 14 (2005) 889–894. doi:10.1088/0964-1726/14/5/025.

[129] R.K. Jain, U.S. Patkar, S. Majumdar, Micro gripper for micromanipulation using IPMCs (ionic polymer metal composites), J. Sci. Ind. Res. (India). 68 (2009) 23–28.

[130] G. Filipcsei, J. Fehér, M. Zrínyi, Electric field sensitive neutral polymer gels, J. Mol. Struct. 554 (2000) 109–117. doi:http://dx.doi.org/10.1016/S0022-2860(00)00564-0.

[131] J. Fehér, G. Filipcsei, J. Szalma, M. Zrínyi, Bending deformation of neutral polymer gels induced by electric fields, Colloids Surfaces A Physicochem. Eng.

Asp. 183–185 (2001) 505–515. doi:http://dx.doi.org/10.1016/S0927-7757(01)00562-3.

[132] K. Asaka, N. Fujiwara, K. Oguro, K. Onishi, S. Sewa, State of water and ionic conductivity of solid polymer electrolyte membranes in relation to polymer actuators, J. Electroanal. Chem. 505 (2001) 24–32. doi:http://dx.doi.org/10.1016/S0022-0728(01)00445-4.

[133] K. Onishi, S. Sewa, K. Asaka, N. Fujiwara, K. Oguro, The effects of counter ions on characterization and performance of a solid polymer electrolyte actuator, Electrochim. Acta. 46 (2001) 1233–1241. doi:http://dx.doi.org/10.1016/S0013-4686(00)00695-2.

[134] T.J. Lu, A.G. Evans, Design of a high authority flexural actuator using an electrostrictive polymer, Sensors Actuators A Phys. 99 (2002) 290–296. doi:http://dx.doi.org/10.1016/S0924-4247(02)00005-5.

[135] K. Jung, J. Nam, H. Choi, Investigations on actuation characteristics of {IPMC} artificial muscle actuator, Sensors Actuators A Phys. 107 (2003) 183–192. doi:http://dx.doi.org/10.1016/S0924-4247(03)00346-7.

[136] J.D. Nam, H.R. Choi, Y.S. Tak, K.J. Kim, Novel electroactive, silicate nanocomposites prepared to be used as actuators and artificial muscles, Sensors Actuators A Phys. 105 (2003) 83–90. doi:http://dx.doi.org/10.1016/S0924-4247(03)00066-9.

[137] J.W. Paquette, K.J. Kim, D. Kim, Low temperature characteristics of ionic polymer–metal composite actuators, Sensors Actuators A Phys. 118 (2005) 135–143. doi:http://dx.doi.org/10.1016/j.sna.2004.08.006.

[138] Y. Yun, V. Shanov, Y. Tu, M.J. Schulz, S. Yarmolenko, S. Neralla, J. Sankar, S. Subramaniam, A Multi-Wall Carbon Nanotube Tower Electrochemical Actuator, Nano Lett. 6 (2006) 689–693. doi:10.1021/nl052435w.

[139] A. Punning, M. Kruusmaa, A. Aabloo, A self-sensing ion conducting polymer metal composite (IPMC) actuator, Sensors Actuators A Phys. 136 (2007) 656–664. doi:http://dx.doi.org/10.1016/j.sna.2006.12.008.

[140] V.K. Nguyen, J.W. Lee, Y. Yoo, Characteristics and performance of ionic polymer–metal composite actuators based on Nafion/layered silicate and Nafion/silica nanocomposites, Sensors Actuators B Chem. 120 (2007) 529–537. doi:http://dx.doi.org/10.1016/j.snb.2006.03.015.

[141] E. Shoji, D. Hirayama, Effects of Humidity on the Performance of Ionic Polymer–Metal Composite Actuators: Experimental Study of the Back-Relaxation of Actuators, J. Phys. Chem. B. 111 (2007) 11915–11920. doi:10.1021/jp074611q.

[142] K.Y. Choi, H.G. Lim, S.R. Yun, J. Kim, K.S. Kang, The Cause of Nanohole and Nanoparticle Formation on Au-Electrode after Actuation of Electro-Active Paper Actuator, J. Phys. Chem. C. 112 (2008) 16204–16208. doi:10.1021/jp802635x.

[143] J.K. Park, R.B. Moore, Influence of Ordered Morphology on the Anisotropic Actuation in Uniaxially Oriented Electroactive Polymer Systems, ACS Appl. Mater. Interfaces. 1 (2009) 697–702. doi:10.1021/am8002268.

[144] J. Li, K.G. Wilmsmeyer, J. Hou, L.A. Madsen, The role of water in transport of ionic liquids in polymeric artificial muscle actuators, Soft Matter. (2009). doi:10.1039/b904443d.

[145] S. Liu, R. Montazami, Y. Liu, V. Jain, M. Lin, X. Zhou, J.R. Heflin, Q.M. Zhang, Influence of the conductor network composites on the electromechanical performance of ionic polymer conductor network composite actuators, Sensors Actuators A Phys. 157 (2010) 267–275. doi:http://dx.doi.org/10.1016/j.sna.2009.11.022.

[146] D.-J. Guo, S.-J. Fu, W. Tan, Z.-D. Dai, A highly porous nafion membrane templated from polyoxometalates-based supramolecule composite for ion-exchange polymer-metal composite actuator, J. Mater. Chem. 20 (2010) 10159. doi:10.1039/c0jm01161d.

[147] J.K. Park, P.J. Jones, C. Sahagun, K.A. Page, D.S. Hussey, D.L. Jacobson, S.E. Morgan, R.B. Moore, Electrically stimulated gradients in water and counterion concentrations within electroactive polymer actuators, Soft Matter. 6 (2010) 1444. doi:10.1039/b922828d.

[148] Y. Lian, Y. Liu, T. Jiang, J. Shu, H. Lian, M. Cao, Enhanced Electromechanical Performance of Graphite Oxide-Nafion Nanocomposite Actuator, J. Phys. Chem. C. 114 (2010) 9659–9663. doi:10.1021/jp101337h.

[149] J. Santos, B. Lopes, P.J.C. Branco, Ionic polymer–metal composite material as a diaphragm for micropump devices, Sensors Actuators A Phys. 161 (2010) 225–233. doi:http://dx.doi.org/10.1016/j.sna.2010.04.032.

[150] J.-W. Lee, Y.-T. Yoo, Preparation and performance of {IPMC} actuators with electrospun Nafion®–MWNT composite electrodes, Sensors Actuators B Chem. 159 (2011) 103–111. doi:http://dx.doi.org/10.1016/j.snb.2011.06.057.

[151] B. Gaihre, G. Alici, G.M. Spinks, J.M. Cairney, Synthesis and performance evaluation of thin film PPy-PVDF multilayer electroactive polymer actuators, Sensors Actuators A Phys. 165 (2011) 321–328. doi:http://dx.doi.org/10.1016/j.sna.2010.10.009.

[152] Y. Liu, R. Zhao, M. Ghaffari, J. Lin, S. Liu, H. Cebeci, R.G. de Villoria, R. Montazami, D. Wang, B.L. Wardle, J.R. Heflin, Q.M. Zhang, Equivalent circuit modeling of ionomer and ionic polymer conductive network composite actuators containing ionic liquids, Sensors Actuators A Phys. 181 (2012) 70–76. doi:http://dx.doi.org/10.1016/j.sna.2012.05.002.

[153] Y. Liu, M. Ghaffari, R. Zhao, J.-H. Lin, M. Lin, Q.M. Zhang, Enhanced Electromechanical Response of Ionic Polymer Actuators by Improving Mechanical Coupling between Ions and Polymer Matrix, Macromolecules. 45 (2012) 5128–5133. doi:10.1021/ma300591a.

[154] V. Panwar, S.Y. Ko, J.-O. Park, S. Park, Enhanced and fast actuation of fullerenol/PVDF/PVP/PSSA based ionic polymer metal composite actuators, Sensors Actuators B Chem. 183 (2013) 504–517. doi:http://dx.doi.org/10.1016/j.snb.2013.04.037.

[155] S. Kim, S. Hong, Y.-Y. Choi, H. Song, K. No, Effect of nucleation time on bending response of ionic polymer–metal composite actuators, Electrochim. Acta. 108 (2013) 547–553. doi:http://dx.doi.org/10.1016/j.electacta.2013.06.092.

[156] N. Terasawa, I. Takeuchi, Electrochemical and electromechanical properties of high-performance polymer actuators containing vapor grown carbon nanofiber and metal oxide, Sensors Actuators B Chem. 176 (2013) 1065–1073. doi:http://dx.doi.org/10.1016/j.snb.2012.10.043.

[157] J.-W. Lee, Y.-T. Yoo, J.Y. Lee, Ionic Polymer–Metal Composite Actuators Based on Triple-Layered Polyelectrolytes Composed of Individually Functionalized Layers, ACS Appl. Mater. Interfaces. 6 (2014) 1266–1271. doi:10.1021/am405090d.

[158] M. Itik, Repetitive control of a trilayer conjugated polymer actuator, Sensors Actuators A Phys. 194 (2013) 149–159. doi:http://dx.doi.org/10.1016/j.sna.2013.01.052.

[159] J. Kim, J.-H. Jeon, H.-J. Kim, H. Lim, I.-K. Oh, Durable and Water-Floatable Ionic Polymer Actuator with Hydrophobic and Asymmetrically Laser-Scribed Reduced Graphene Oxide Paper Electrodes, ACS Nano. 8 (2014) 2986–2997. doi:10.1021/nn500283q.

[160] Y. Wang, H. Chen, Y. Wang, Z. Zhu, D. Li, Effect of Dehydration on the Mechanical and Physicochemical Properties of Gold- and Palladium -Ionomeric Polymer-Metal Composite (IPMC) Actuators, Electrochim. Acta. 129 (2014) 450–458. doi:http://dx.doi.org/10.1016/j.electacta.2014.02.114.

[161] P.M. Welch, A Tunable Dendritic Molecular Actuator, Nano Lett. 5 (2005) 1279–1283. doi:10.1021/nl050422c.

[162] J. Shi, Z.-X. Guo, B. Zhan, H. Luo, Y. Li, D. Zhu, Actuator Based on MWNT/PVA Hydrogels, J. Phys. Chem. B. 109 (2005) 14789–14791. doi:10.1021/jp052677k.

[163] K.Y. Cho, H.G. Lim, S.R. Yun, J. Kim, K.S. Kang, Electric Field Frequency and Strength Effects on Au-Electrode Damage for an Electroactive Paper Actuator Coated with Polypyrrole, J. Phys. Chem. C. 112 (2008) 7001–7004. doi:10.1021/jp067012c.

[164] L. Zhao, L. Tong, C. Li, Z. Gu, G. Shi, Polypyrrole actuators with inverse opal structures, J. Mater. Chem. 19 (2009) 1653. doi:10.1039/b819831d.

[165] B.K. Juluri, A.S. Kumar, Y. Liu, T. Ye, Y.-W. Yang, A.H. Flood, L. Fang, J.F. Stoddart, P.S. Weiss, T.J. Huang, A Mechanical Actuator Driven Electrochemically by Artificial Molecular Muscles, ACS Nano. 3 (2009) 291–300. doi:10.1021/nn8002373.

[166] C.-A. Dai, C.-J. Chang, A.-C. Kao, W.-B. Tsai, W.-S. Chen, W.-M. Liu, W.-P. Shih, C.-C. Ma, Polymer actuator based on PVA/PAMPS ionic membrane: Optimization of ionic transport properties, Sensors Actuators A Phys. 155 (2009) 152–162. doi:http://dx.doi.org/10.1016/j.sna.2009.08.002.

[167] A. Liu, L. Zhao, H. Bai, H. Zhao, X. Xing, G. Shi, Polypyrrole Actuator with a Bioadhesive Surface for Accumulating Bacteria from Physiological Media, ACS Appl. Mater. Interfaces. 1 (2009) 951–955. doi:10.1021/am9000387.

[168] D. Brandell, H. Kasemägi, A. Aabloo, Poly(ethylene oxide)–poly(butadiene) interpenetrated networks as electroactive polymers for actuators: A molecular dynamics study, Electrochim. Acta. 55 (2010) 1333–1337. doi:http://dx.doi.org/10.1016/j.electacta.2009.04.070.

[169] X.-L. Wang, I.-K. Oh, S. Lee, Electroactive artificial muscle based on crosslinked PVA/SPTES, Sensors Actuators B Chem. 150 (2010) 57–64. doi:http://dx.doi.org/10.1016/j.snb.2010.07.042.

[170] M. Rajagopalan, J.-H. Jeon, I.-K. Oh, Electric-stimuli-responsive bending actuator based on sulfonated polyetherimide, Sensors Actuators B Chem. 151 (2010) 198–204. doi:http://dx.doi.org/10.1016/j.snb.2010.09.021.

[171] V. Palmre, E. Lust, A. Jänes, M. Koel, A.-L. Peikolainen, J. Torop, U. Johanson, A. Aabloo, Electroactive polymer actuators with carbon aerogel electrodes, J. Mater. Chem. 21 (2011) 2577. doi:10.1039/c0jm01729a.

[172] L. Chen, C. Liu, K. Liu, C. Meng, C. Hu, J. Wang, S. Fan, High-Performance, Low-Voltage, and Easy-Operable Bending Actuator Based on Aligned Carbon Nanotube/Polymer Composites, ACS Nano. 5 (2011) 1588–1593. doi:10.1021/nn102251a.

[173] M. Luqman, J.-W. Lee, K.-K. Moon, Y.-T. Yoo, Sulfonated polystyrene-based ionic polymer–metal composite (IPMC) actuator, J. Ind. Eng. Chem. 17 (2011) 49–55. doi:http://dx.doi.org/10.1016/j.jiec.2010.10.008.

[174] J. Torop, T. Sugino, K. Asaka, A. Jänes, E. Lust, A. Aabloo, Nanoporous carbide-derived carbon based actuators modified with gold foil: Prospect for fast response and low voltage applications, Sensors Actuators B Chem. 161 (2012) 629–634. doi:http://dx.doi.org/10.1016/j.snb.2011.10.079.

[175] J.-W. Lee, S.M. Hong, J. Kim, C.M. Koo, Novel sulfonated styrenic pentablock copolymer/silicate nanocomposite membranes with controlled ion channels and their {IPMC} transducers, Sensors Actuators B Chem. 162 (2012) 369–376. doi:http://dx.doi.org/10.1016/j.snb.2011.12.105.

[176] J. Tang, B. Yang, J. Liu, Y. Wang, L. Huang, Z. Huang, Y. Wang, Q. Zhu, L.A. Belfiore, Electric-field-actuation of in situ composites that contain silver-coated carbon fibers in sodium sulfonate ionomers, RSC Adv. 2 (2012) 8813–8820. doi:10.1039/C2RA20766D.

[177] R. Gao, D. Wang, J.R. Heflin, T.E. Long, Imidazolium sulfonate-containing pentablock copolymer–ionic liquid membranes for electroactive actuators, J. Mater. Chem. 22 (2012) 13473. doi:10.1039/c2jm16117f.

[178] B. Gaihre, S. Ashraf, G.M. Spinks, P.C. Innis, G.G. Wallace, Comparative displacement study of bilayer actuators comprising of conducting polymers, fabricated from polypyrrole, poly(3,4-ethylenedioxythiophene) or poly(3,4-propylenedioxythiophene), Sensors Actuators A Phys. 193 (2013) 48–53. doi:http://dx.doi.org/10.1016/j.sna.2012.12.007.

[179] G. Alici, A. Punning, H.R. Shea, Enhancement of actuation ability of ionic-type conducting polymer actuators using metal ion implantation, Sensors Actuators B Chem. 157 (2011) 72–84. doi:http://dx.doi.org/10.1016/j.snb.2011.03.028.

[180] B.-K. Lee, S.J. Park, D.S. Kim, Fabrication of ionic polymer actuator with graphene nanocomposite electrodes and its characterization, Curr. Appl. Phys. 13 (2013) 1520–1524. doi:http://dx.doi.org/10.1016/j.cap.2013.05.011.

[181] Y. Tang, Z. Xue, X. Zhou, X. Xie, C.-Y. Tang, Novel sulfonated polysulfone ion exchange membranes for ionic polymer–metal composite actuators, Sensors Actuators B Chem. 202 (2014) 1164–1174. doi:http://dx.doi.org/10.1016/j.snb.2014.06.071.

[182] Z. Zeng, H. Jin, L. Zhang, H. Zhang, Z. Chen, F. Gao, Z. Zhang, Low-voltage and high-performance electrothermal actuator based on multi-walled carbon nanotube/polymer composites, Carbon N. Y. 84 (2015) 327–334. doi:http://dx.doi.org/10.1016/j.carbon.2014.12.012.

Keywords

About the Editors

Assistant Professor Inamuddin

Dr. Inamuddin is currently working as Assistant Professor in the Department of Applied Chemistry, Aligarh Muslim University (AMU), Aligarh, India. He obtained his Master of Science degree in Organic Chemistry from Chaudhary Charan Singh (CCS) University, Meerut, India, in 2002. He received his Master of Philosophy and Doctor of Philosophy degrees in Applied Chemistry from AMU in 2004 and 2007, respectively. He has extensive research experience in multidisciplinary fields of Analytical Chemistry, Materials Chemistry, and Electrochemistry and, more specifically, Renewable Energy and Environment. He has worked under different research projects as project fellow and senior research fellow funded by University Grants Commission (UGC), Government of India, and Council of Scientific and Industrial Research (CSIR), Government of India. He has received Fast Track Young Scientist Award from the Department of Science and Technology, India, to work in the area of bending actuators and artificial muscles. He is running one major research projects funded by Council of Science and Technology, Lucknow (Uttar Pradesh), India. He has completed three major research projects sanctioned by University Grant Commission, Department of Science and Technology, and Council of Scientific and Industrial Research, India. He has published 72 research articles in international journals of repute and eight book chapters in knowledge-based book editions published by renowned international publishers. He has published five edited books with Springer, United Kingdom, three by Nova Science Publishers, Inc. U.S.A., one by CRC Press Taylor & Francis Asia Pacific and two by Trans Tech Publications Ltd., Switzerland. He is the member of various editorial boards of journals. He has attended as well as chaired sessions in various international and nation conferences. He has worked as a Postdoctoral Fellow, leading a research team at the Creative Research Initiative Center for Bio-Artificial Muscle, Hanyang University, South Korea, in the field of renewable energy, especially biofuel cells. He has also worked as a Postdoctoral Fellow at the Center of Research Excellence in Renewable Energy, King Fahd University of Petroleum and Minerals, Saudi Arabia, in the field of polymer electrolyte membrane fuel cells and computational fluid dynamics of polymer electrolyte membrane fuel cells. He is a life member of the Journal of the Indian Chemical Society. His research interest includes ion exchange materials, sensor for heavy metal ions, biofuel cells, supercapacitors and bending actuators.

Professor Ali Mohammad

Prof. Ali Mohammad is presently working as UGC-Emeritus Fellow in the Department of Applied Chemistry, Faculty of Engineering and Technology, Aligarh Muslim University, Aligarh, India after serving as chairman of the department of Applied Chemistry for 6 years. His scientific interests include physico-analytical aspects of solid-state reactions, micellar thin layer chromatography, surfactants analysis, and green chromatography. He is the author or coauthor of 250 scientific publications including research articles, reviews, and book chapters. He has supervised 53 students for Ph.D./M.Phil. and M. Tech. degrees. He has also served as Editor of Scientific Journal, "Chemical and Environmental Research" published in India since 1992 to 2012 and as the Associate Editor for Analytical Chemistry section of the Journal of Indian Chemical Society. He has published three edited books with Springer, United Kingdom, He has been a member of the editorial board of Acta Chromatographica, Acta Universitatis Cibiniensis Seria F. Chemia, Air Pollution, and Annals of Agrarian Science. He has attended as well as chaired sessions in various international and nation conferences. He is the life member of several Indian Scientific and Chemical Societies. He has also served as Visiting Professor in King Saud University, Riyadh, Kingdom of Saudi Arabia. Dr. Mohammad obtained his M.Sc. (1972), M.Phil. (1975), Ph.D. (1978), and D.Sc. (1996) degrees from Aligarh Muslim University, Aligarh, India.

Professor Abdullah M. Asiri

Prof. Abdullah M. Asiri is the Head of the Chemistry Department at King Abdulaziz University since October 2009 and he is the founder and the Director of the Center of Excellence for Advanced Materials Research (CEAMR) since 2010 till date. He is the Professor of Organic Photochemistry. He graduated from King Abdulaziz University (KAU) with B.Sc. in Chemistry in 1990 and a Ph.D from University of Wales, College of Cardiff, U.K. in 1995. His research interest covers color chemistry, synthesis of novel photochromic and thermochromic systems, synthesis of novel coloring matters and dyeing of textiles, materials chemistry, nanochemistry and nanotechnology, polymers and plastics. Prof. Asiri is the principal supervisors of more than 20 M.Sc. and six Ph.D theses; He is the main author of ten books of different chemistry disciplines. Prof. Asiri is the Editor-in-Chief of King Abdulaziz University Journal of Science. A major achievement of Prof. Asiri is the discovery of tribochromic compounds, a class of compounds which change from slightly or colorless to deep colored when subjected to small pressure or when grind. This discovery was introduced to the scientific community as a new terminology published by IUPAC in 2000. This discovery was awarded a patent from European Patent office and from UK patent. Prof. Asiri involved in many committees at the KAU level and also on the national level, he took a major role in the advanced materials committee working for KACST to identify the National plan for science and technology in 2007. Prof. Asiri played a major role in advancing the chemistry education and research in KAU, he has been awarded the best Researchers from KAU for the past five years. He also awarded the Young Scientist award from the Saudi Chemical Society in 2009, and also the first prize for the distinction in science from the Saudi Chemical Society in 2012. He also received a recognition certificate from the American Chemical society (Gulf region Chapter) for the advancement of chemical science in the Kingdome. Also he received a Scopus certificate for the most Publishing Scientist in Saudi Arabia in chemistry in 2008. He is also a member of the Editorial Board of various journals of international repute. He is the Vice- President of Saudi Chemical Society (Western Province Branch). He holds four USA patents, more than 800 Publications in international journals, seven book chapters, and ten books.

www.ingramcontent.com/pod-product-compliance
Ingram Content Group UK Ltd.
Pitfield, Milton Keynes, MK11 3LW, UK
UKHW020659041026
15590UKWH00001B/354

9 781945 291081